Science and Fiction

Science and Fiction – A Springer Series

This collection of entertaining and thought-provoking books will appeal equally to science buffs, scientists and science-fiction fans. It was born out of the recognition that scientific discovery and the creation of plausible fictional scenarios are often two sides of the same coin. Each relies on an understanding of the way the world works, coupled with the imaginative ability to invent new or alternative explanations—and even other worlds. Authored by practicing scientists as well as writers of hard science fiction, these books explore and exploit the borderlands between accepted science and its fictional counterpart. Uncovering mutual influences, promoting fruitful interaction, narrating and analyzing fictional scenarios, together they serve as a reaction vessel for inspired new ideas in science, technology, and beyond.

Whether fiction, fact, or forever undecidable: the Springer Series "Science and Fiction" intends to go where no one has gone before!

Its largely non-technical books take several different approaches. Journey with their authors as they

- Indulge in science speculation – describing intriguing, plausible yet unproven ideas;
- Exploit science fiction for educational purposes and as a means of promoting critical thinking;
- Explore the interplay of science and science fiction – throughout the history of the genre and looking ahead;
- Delve into related topics including, but not limited to: science as a creative process, the limits of science, interplay of literature and knowledge;
- Tell fictional short stories built around well-defined scientific ideas, with a supplement summarizing the science underlying the plot.

Readers can look forward to a broad range of topics, as intriguing as they are important. Here just a few by way of illustration:

- Time travel, superluminal travel, wormholes, teleportation
- Extraterrestrial intelligence and alien civilizations
- Artificial intelligence, planetary brains, the universe as a computer, simulated worlds
- Non-anthropocentric viewpoints
- Synthetic biology, genetic engineering, developing nanotechnologies
- Eco/infrastructure/meteorite-impact disaster scenarios
- Future scenarios, transhumanism, posthumanism, intelligence explosion
- Virtual worlds, cyberspace dramas
- Consciousness and mind manipulation

More information about this series at http://www.springer.com/series/11657

Barry B. Luokkala

Exploring Science Through Science Fiction

Second Edition

 Springer

Barry B. Luokkala (iD)
Department of Physics
Carnegie Mellon University
Pittsburgh, PA, USA

ISSN 2197-1188   ISSN 2197-1196   (electronic)
Science and Fiction
ISBN 978-3-030-29392-5   ISBN 978-3-030-29393-2   (eBook)
https://doi.org/10.1007/978-3-030-29393-2

This Springer imprint is published by the registered company Springer Nature Switzerland AG
The registered company address is: Gewerbestrasse 11, 6330 Cham, Switzerland

To Janet—my wife, my love, my inspiration

Preface to the First Edition

Science and science fiction have been passions of mine since childhood. But the combination of the two—using science fiction as a vehicle for teaching science—is something that I only began to explore in relatively recent years. We live in an increasingly technological world, in which it is increasingly important for people to be scientifically informed. But traditional courses in science at the college level are usually geared toward students who are already on their way toward careers related to science, technology, engineering, or mathematics—the so-called STEM fields. Although there are some notable exceptions, most of the offerings in science for students in the fine arts, humanities, and social sciences continue to be watered-down versions of courses intended for science or engineering students. Students who take such courses often do so simply to fulfill a degree requirement, and with a sense of trepidation. They often complete such courses with no more interest in science than when they signed up. The present work has one primary goal: to make science accessible to a broad audience, including both nontechnical students, as well as technically oriented students, with a view toward increasing public awareness of science and interest in science.

Content and Scope

The content of the book is organized around seven major questions, which are frequently addressed in works of science fiction: What is the nature of space and time? What is the universe made of? Can a machine ever become conscious? Are we alone in the universe? What does it mean to be human? How do we solve our problems? What lies ahead? There is a deliberate progression in these seven major questions, beginning with the most objective (changing perspectives through history concerning the nature of space and time, the fundamental building blocks of matter, and the properties of materials), followed by topics which have both objective and subjective dimensions (Is consciousness computable? Is there intelligent life elsewhere in the universe? What, if anything, sets humans apart from other closely related species?), and ending with questions that may be more controversial and speculative (How do we distinguish between science and pseudoscience? How is science used or misused in attempts to solve the problems facing society? What will the future hold for our technological society?). In addition to the major questions, each of the chapters includes suggestions for further exploration, with more specific questions and references to scientific literature.

Much of the emphasis in this book reflects my training as a physicist. Yet I do not subscribe to the narrow view of science held by Ernest Rutherford, a famous physicist of the early twentieth century, who proclaimed, "All of science is physics. The rest is stamp-collecting." A few of my colleagues have an even more narrow view toward science than Rutherford and will tell you that you aren't even doing physics unless you are doing their particular brand of physics. When it comes to a discussion of what is science, I am willing to embrace considerably more diversity. Yet there are limits. A serious discussion of what is science and what is not science is also included in this book. Not all forms of human inquiry are rightly included under the umbrella of science.

The Approach

Science fiction is used throughout this book as a springboard for discussing both the fundamental principles of science and cutting-edge science research. Short scenes from science fiction movies and television episodes are critiqued in light of our current understanding of science. Class discussion focuses on discerning the level of plausibility of the science depicted in each scene. To this end, four general categories are useful. A handful of examples from some of the best movies turn out to be solidly based on good science—the things that we see on the screen are actually known to happen, essentially as depicted or as described in the dialog. An example that falls into this first category is the relativistic time dilation described in the opening scene of the original (1967) version of the *Planet of the Apes*. The second category includes things which are possible in principle, but beyond our current technology. That is to say, it hasn't happened yet, but there is nothing in the laws of science to forbid it from happening. An example is the sentient android, Commander Data, in the TV series *Star Trek: The Next Generation*. The third category is stuff that just can't happen as shown. Much of science fiction may be lots of fun to watch, but is simply impossible, and we will explore the reasons why. An example of this is the extraordinarily rapid rate at which the air leaks out of an interplanetary spaceship, after being punctured by a micrometeoroid, in the 2000 movie *Mission to Mars*. Using a reasonable set of assumptions, we will come up with an estimate of how long it would really take and conclude that the screenwriters simply wanted to heighten the sense of excitement and danger. The fourth category is one that is growing rapidly as our technology advances: things which were purely science fiction at the time that the movie or TV episode was produced, but are now part of reality, or are expected to become part of reality in the very near future. When the movie *GATTACA* was produced in 1997, the Human Genome Project had not yet been completed, and the concept of rapid DNA sequencing, as shown in the movie, was purely in the imagination of the writers. It took the enormous government-funded research project 13 years to complete the sequencing of the first human genome. Full-genome sequencing can now be done in about 1 day at a cost of about $1000.

Plausibility Checks and Quantitative Estimations

Some of the sci-fi illustrations are conducive to doing what we might call a quantitative plausibility check. That is to say, by watching the movie scene we can gather enough information to do a rough calculation, and then decide if what we see could actually happen or not. In the text, I call these *Estimation Problems*. The objective in each case is *not* to obtain a specific answer, which is either right or wrong (the sort of problems encountered in most introductory physics textbooks). Instead, the idea is to come up with an estimate, by making a reasonable set of assumptions to supplement the information you can gather directly from the movie scene. For example, how much power would be required for a handheld weapon, such as a *Star Trek* phaser, to vaporize a human body? The answer that you get will depend on what assumptions you make about how long the vaporization process takes (from watching a relevant scene) and about what the human body is made of. The nontechnical student should not be intimidated by the frequent use of equations in some of the chapters. Where calculations can be done, example problems are worked out in the text, and additional problems are proposed for you to work out on your own. Solutions for these *Estimation Problems* are included in an appendix.

Movie, TV, and YouTube References

Collectively, the chapters include over 180 references to specific scenes in 130 different movies and television episodes, spanning over 100 years of cinematic history. In general, but with a few notable exceptions, I refrain from passing judgment on these works from an artistic perspective. Some of the very worst sci-fi movies ever made actually include some rather useful illustrations and are likely to promote very fruitful discussions in class. Each chapter includes a list of references to movie and TV scenes, in the order that they are cited in the text. The entire collection of movies and TV episodes cited in the text is also included alphabetically in appendices at the end of the book.

Most of the visual material is available on DVD and is referenced by scene number, for ease in selecting the desired scene for a classroom presentation. A few of the scenes from the most recent TV series episodes had not yet been released on DVD at the time of publication, or are not yet part of my personal DVD collection. These are cited by season number, episode number, and air date.

How to Use This Book

The book was created as the primary text for a one-semester undergraduate course on science and science fiction. But anyone with an interest in science and science fiction will enjoy reading it. The level of mathematical sophistication is that of high school algebra, with a small amount of trigonometry. The amount of material included in the book is actually more than can be covered in a single semester. This allows some freedom for the prospective instructor to pick and choose, according to the desired emphasis of the course.

The more casual reader may enjoy looking at a particular science topic to see what sci-fi references are used, or may prefer to look up a particular movie or TV episode in the appendices, to see what science concepts are included. Although the list of sci-fi references is extensive, it's not possible to include every work that was ever produced in a book of this scope. Nevertheless, I hope that there will be something for everyone.

Acknowledgments

The story of the origin of this book is a case study in the orchestration of events. I owe a debt of gratitude to the many individuals and groups of individuals, who have played a role in this orchestration, either knowingly or unawares.

The first of these events happened before I was born, on the campus of what was then known as the Carnegie Institute of Technology. My father enlisted in the army during the Second World War and was stationed at Carnegie Tech, in the Army Specialized Training Program. My mother was working at her first job right out of high school in one of the administrative offices on campus. Neither of them ever imagined when they first met that they would eventually marry and have a son, who would grow up to become a member of the faculty in the physics department, on that same campus, now called Carnegie Mellon University. It's fitting that I should acknowledge my parents. Obviously, without them I would not be here. I am grateful to them for providing the kind of environment in which someone interested in the sciences could thrive and for allowing me the freedom to choose my own path.

The earliest direct influence for the book can be traced back to an event in my seventh grade history class. The teacher, Mr. Joseph Karlik, was also the science and math teacher for seventh and eighth grade at our very small elementary school. He gave us the most unusual assignment for homework one day, which I seem to recall having to persuade my parents to allow me to complete. We had to watch an episode of the new TV series *Star Trek*, called *The City on the Edge of Forever*, which would air that evening. The episode imagines the possibility of time travel into the past (to Earth in 1930) and the changing of one small event (preventing a traffic accident), which results in all of history unfolding differently (delaying the entry of the United States into the war and allowing Nazi Germany to develop the A-bomb first). The discussion in our next history class focused on how things today might be very different if a significant event of the past never occurred. Although my interest in science fiction had already begun several years earlier, with the cartoon series *The Jetsons*; my real love of science fiction began with this homework assignment. I was intrigued by the possibility of traveling through space and time, and the *Star Trek* character *Mr. Spock* became one of my role models. My other childhood role model was the famous marine biologist Jacques Cousteau. His underwater documentaries, produced by National Geographic, were something that I never missed on TV. Through most of high school, I dreamed of becoming a marine biologist and sailing around the world on the *Calypso* to seek out and study new life forms in the Earth's oceans. I went to college intending to major in biochemistry, which (I hoped) would lead to graduate school in marine biology. But a C in honors college chemistry prompted me to reconsider whether that was the best path for me to pursue. As I pondered which direction I should take, I recalled my high school physics class and Mr. David Speer, who was without a doubt the best teacher I ever had in high school. I credit him, and the very positive experience I had in his classroom and laboratory, with the decision to change my major from biochemistry to physics. Thus began a long and complex path, which eventually led me to the physics department at Carnegie Mellon University.

Although he sometimes questions my sanity for doing things in nontraditional ways, I owe a very big debt of gratitude to Steve Garoff, my PhD thesis advisor and one of my closest colleagues in the physics department. Steve and I began our association in his second semester as a faculty member at Carnegie Mellon, when he was assigned to teach our advanced undergraduate laboratory course. I had already been working in the department for some years as a lecturer, primarily for our introductory lab courses, but had also been assisting regularly with our advanced lab course. Toward the end of the semester, Steve was lamenting to another colleague that he had lots of startup funds, but only one graduate student. I told him that by coincidence, I had been kicking around the idea of finally growing up and pursuing a PhD, and his research was very much of interest to me. The words had hardly left my mouth when Steve invited me to join his research group. My future would certainly have unfolded very differently, if Steve had not provided the opportunity to work with him—a decision, which I'm sure neither of us has ever regretted.

The basic framework for the book began to develop after the convergence of three events in the second half of the 1990s. The first of these events happened late in 1995, thanks to Janet, my one and only wife of more than 30 years now. At the time, she was working at Pinocchio Bookstore for Children, one of the last of a dying breed of independently owned bookstores in Pittsburgh. In this position, she had easy access to information about all the latest books for grownups, as well as for children. Knowing my love of science fiction, in general, and of *Star Trek* in particular, she surprised me one day with a hardcover copy of *The Physics of Star Trek*, by

Lawrence Krauss, which had just been published that year. Delightful to read, this brilliant work holds a permanent place on my list of all-time favorite books.

The second event in this trilogy came in November of 1996, when Lawrence Krauss, author of *The Physics of Star Trek*, visited Carnegie Mellon University to give a lecture, complete with video clips from some of his favorite *Star Trek* episodes. As I sat in the audience, enthralled by his presentation, I started thinking that something like this would make a great course. Following the presentation, I managed to get Krauss to sign my first edition copy of his book, but wondered when I would ever have the time or opportunity to follow through on developing a course on the topic.

Two years later, in the fall of 1998, the opportunity presented itself. Susan Henry, then dean of the Mellon College of Science at Carnegie Mellon (1991–2000), called upon the faculty to create mini-courses, whose primary purpose would be to keep first-year students interested in science. I proposed a course which would use science fiction as a springboard for discussing cutting-edge science research. Although I am a physicist, and my idea for the course was inspired by Lawrence Krauss and *The Physics of Star Trek*, there is much more to science than just physics, and science fiction encompasses much more than just *Star Trek*. I planned a 6-week mini-course around five major topics: the nature of space and time, properties of solid-state materials, robotics and artificial intelligence, the search for extraterrestrial intelligence, and the future of our technological society. Students would watch clips from a broad range of science fiction movies, spanning a century of cinematic history, as well as selected sci-fi television episodes, and critique the science content of the clips in light of our current knowledge. Class discussions would focus on assigned readings from current science literature. The proposal was received with great enthusiasm, and I've been offering the course once or twice every year since the spring of 1999. Special thanks are due to Walter Pilant, whose vast personal collection of episodes of *Star Trek* (both the original series and *The Next Generation*) provided many hours of useful background research and preparation for the course. Thanks also to Jeffrey Hinkelman, my dear friend and manager of Carnegie Mellon's video collection, for his invaluable advice on the use of video materials in class, for correcting my mistaken understanding of gender stereotypes in sci-fi movies of the 1950s, and for acquiring many of the titles that I use in the course for the library's permanent collection.

Serious work on the actual material for the book started with a second convergence of events, which began 10 years after the creation of the mini-course. The global economic crisis of 2009 resulted in the termination of a summer enrichment program in the sciences for talented high school students, of which I had served as the director for the previous 8 years. Official word of the termination came in March of 2009, leaving me in need of a creative outlet for the summer. I made a proposal to Gregg Franklin, newly appointed head of the physics department (2009–2013), to expand my popular mini-course on science and science fiction to a full-semester course. Unlike the mini-course, which was restricted to first-year students majoring in the sciences, this course would be open to anyone on campus, regardless of their major or year. The primary goal of the course would be to make science accessible to students in the fine arts and humanities, while keeping enough rigorous content to hold the interest of the more technically oriented students in science or engineering. This proposal was met with enthusiasm, not only by Gregg, but also by Kunal Ghosh, assistant head for undergraduate affairs in the physics department. Kunal has always been one of my biggest cheerleaders and for years had been encouraging me to expand the course and offer it to a broader audience. The timing couldn't have been more perfect to do just that. The announcement of the new course offering was made on campus just a few weeks before the May 2009 release of the new *Star Trek* movie, directed by J.J. Abrams. The university was quick to pick up on the connections, the most important of which was Zachary Quinto, a graduate of Carnegie Mellon's School of Drama (class of 1999), who was very well cast as the new Mr. Spock. I was honored to be interviewed by the university for a YouTube video promoting the expanded version of the course, which was uploaded in June 2009. A story about the course was also featured on the university's homepage. I'd like to thank Jocelyn Duffy and Carrie Chisholm for their work in promoting the course to the university community and to the outside world.

For three summers in a row (2009–2011), I offered the full-semester version of the course to an increasingly diverse audience and gradually refined the content. Many thanks to the students who took this early version of the course, as it flew under the radar, so to speak, and as we worked out what requirements it would fulfill for the university's various degree programs.

In April of 2011, I received an email out of the blue from Jace Harker, a publishing editor with Springer-Verlag, who was coming on a short visit to Pittsburgh. He asked if he could meet with me to discuss my work in undergraduate education in physics and to solicit ideas for new textbooks. Coincidentally, at my wife's suggestion, I had already been toying with the idea of writing my own book for the science and science fiction course. Jace was very enthusiastic about the possibility of publishing my lecture notes. But at that time they existed only in my head and in the form of a six-page spreadsheet of all the film references that I use day by day in the course. At his request, I created a detailed chapter outline, which he circulated for peer review at Springer. By the end of September, he sent me a draft of a contract to write the book. Thanks to my good friend Steve Paschall, who kindly offered to review the contract from an attorney's perspective. He raised several insightful questions, all of which were addressed to our satisfaction. The final contract was officially signed in October 2011.

Countless hours over the next 14 months were devoted to putting the ideas from my head onto the printed page, and doing the research to locate the important science references that I felt were needed to make a proper textbook. I owe a huge debt of gratitude to my wife, Janet, for her patience in watching lots of good (and not-so-good) science fiction with me, and for her valuable suggestions of material to include in the book. I'd like to extend a word of thanks to the Powe family—Janet's sister Treva, brother-in-law, Joe, and our niece, Cati—with whom we spent our 2012 Christmas vacation in Texas, for their patience as I spent hours at their dining room table, typing away on the first draft of the manuscript. Thanks also to the Quick family—Donnalynne, Janet's good friend from college, and her son, Austin (our godson) and daughter, Eliana—with whom we spent part of our summer vacation. They, too, had to put up with me, as I worked on the second draft at their home in California. The manuscript was delivered in nearly final form at the end of December 2012, with only three-and-a-half hours before the contract deadline. Many thanks to Mr. Ho Ying Fan at Springer for reading the manuscript and making helpful suggestions for revision, as well as for shepherding the manuscript through the editing and publication process.

Finally, I'd like to thank three groups of individuals, whose prayers and encouragement have sustained me through the entire journey: my Tuesday lunchtime discussion group, including David Anderson, Robin Capcara, John Dolan, Bob Griffiths, David Handron, John Ito, Christoph Mertz, and Gary Patterson; our Wednesday evening house group, consisting of Alan and Linda Komm, David, Debbie, and Karis Kornfield, Meredith Dobson, Jerry Martin, and Casey McDonald; and our Sunday morning fellowship group, which regularly includes Karen and Randy Woods, Carla and Jeff Sivek, Pam and Lou D'Abruzzo, Art Burt and Virginia Phillips.

Pittsburgh, PA   Barry B. Luokkala
May 2013

Preface to the Second Edition

The invitation to write the first edition of this book came as a surprise. I had never undertaken a writing project of this magnitude before. (The finished product ended up being longer than my physics PhD thesis.) But the request to write a second edition exceeded my expectations. While the reviews of the first edition were generally quite favorable, there were a few criticisms, which I have endeavored to take into consideration in preparing this revised and updated edition. I'm also grateful to many students and several colleagues, who carefully read the first edition and made helpful suggestions for changes. My sincerest thanks to Sam Harrison and colleagues at Springer for overseeing the editing and production of the second edition.

The overall structure is essentially unchanged. The material continues to be organized around the seven major questions of the first edition, spanning topics from the nature of space and time to the future of our technological society. The content within several chapters has been rearranged to enhance the clarity and flow. Some of the original artwork has been redrawn and 17 new illustrations have been added. Several new subtopics have been included and existing topics expanded. Some of these were inspired by sci-fi movies released since the publication of the first edition (*Interstellar*, *Chappie*, *Arrival*, to name just a few) or by works that had already been released prior to the first edition, but which I had not yet seen (The *Fringe* TV series, *Despicable Me*, *Robot and Frank*). Others were prompted by new discoveries or significant advances in science and technology (e.g., gravitational waves, machine learning, quantum computing, and controversial uses of gene editing). As was the case in the first edition, a few of the new references are to movies that do not fall into the category of science fiction (*Hidden Figures*, *Harry Potter and the Chamber of Secrets*), but nevertheless serve as good illustrations of the science concepts that have been added. And some of the new references are to movies that are definitely not new (*The Andromeda Strain*, *The Day the Earth Caught Fire*), but the subject matter remains very relevant to important issues facing us today. In all, more than 40 new references to specific scenes in more than 30 new (or newly rediscovered) movies and TV series episodes have been included.

To borrow a line from the 1995 version of *Sabrina* (not a science fiction movie but one of my favorites), "More isn't always better. Sometimes it's just more." In the process of writing the second edition, I constantly had to be reminded of this bit of wisdom. Every new sci-fi movie or scientific discovery that came out since the first edition had to pass the test. If I chose to include it, would it be better or would it just be more? As a result, there were a number of possibilities for new material which, for various reasons, didn't make the final cut. I owe a special debt of gratitude to my wife, Janet, for helping me to think this through and to exercise some restraint. I'm also grateful to her for keeping an eye out for interesting science news items and passing them on to me. Many of them have been included as new Exploration Topics at the end of several chapters. My hope is that this revised and expanded edition will be genuinely better than the first and not just more.

Pittsburgh, PA   Barry B. Luokkala
July 2019

Contents

About the Author

Barry B. Luokkala is a teaching professor and director of undergraduate laboratories in the department of physics at Carnegie Mellon University. He received his BS and MS degrees in Physics at the University of Pittsburgh, where he did experimental research in physics and chemistry of the ionosphere. He received his PhD in Experimental Condensed Matter Physics at Carnegie Mellon University. He also serves as program director for the Pennsylvania Governor's School for the Sciences and has been a science consultant for the Sloan Foundation Screenplay Competition in Carnegie Mellon's School of Drama.

Chapter 1
Introduction: Discerning the Real, the Possible, and the Impossible

Believe none of what you hear and half of what you see.
—*Benjamin Franklin*

Why, sometimes I've believed as many as six impossible things before breakfast.
—*The Queen of Hearts*
Alice in Wonderland

A major goal of the present work is to increase public awareness and appreciation of science, but the approach is somewhat unorthodox. We will use science fiction as a vehicle for exploring actual science and as a springboard for discussing some of the exciting topics that are currently being researched. Our examples will be drawn almost exclusively from film and television. Occasional references to a few of the classic works of science fiction literature are also included.

As we consider each of these sci-fi examples, we will attempt to discern whether what we see on the screen falls into one of the following categories:

- *realistic*—something that is it solidly grounded in real science, regardless of whether or not it has actually happened.
- *possible in principle*—something that has never happened before, but the laws of science do not necessarily rule it out. However improbable it may seem, it might simply be beyond our current technology.
- *impossible*—something that is total fantasy and impossible by any science we know.

When it comes to science fiction, the advice attributed to Benjamin Franklin, "Believe none of what you hear and half of what you see" may not be strict enough. We may end up believing even less than half of what we see. But there are some rare exceptions, in which the science is particularly well done, and you may be surprised at some of the things that are actually possible (at least in principle). We will also encounter a number of examples which fall into a fourth category:

© Springer Nature Switzerland AG 2019
B. B. Luokkala, *Exploring Science Through Science Fiction*, Science and Fiction, https://doi.org/10.1007/978-3-030-29393-2_1

- *science fiction predicts the future*—many things currently exist that were purely in the imagination of the writers when the movie or television episode was produced. Thanks to breakthroughs in science and technology, what was once science fiction may now be real.

We begin with a few examples, which will set the tone for the rest of the book, while at the same time conveying a sense of the history of science fiction as a genre.

1.1   The First Sci-Fi Movie

The earliest motion pictures, produced in the late 1800s, were typically only a few minutes in length and collectively covered a broad range of topics, from the mundane to the exotic. The first motion picture of significant length (roughly 20 min) also happens to be the first science fiction movie ever made and is well worth examining in detail. Produced in 1902, *Le Voyage dans la Lune* (A Trip to the Moon) was directed by George Méliès, who began his career as a stage magician [1, 2]. The 2011 movie *Hugo* is, in part, a somewhat fictionalized account of the life of George Méliès (played by Ben Kingsley) [3]. Because of its place in cinematic history (the first sci-fi movie ever made and the first film of any kind of significant length) and its subject matter (a trip to the moon, more than 60 years before such a thing was ever attempted in reality), *Le Voyage dans la Lune* provides an ideal starting point from which to launch our exploration of science through science fiction.

First, let's cover some of the key points of the story, as presented in the movie. The version described here is the one that is included in the excellent DVD collection, *Landmarks of Early Film*, which includes not only the silent motion picture, but also the accompanying narrated script and musical score [4]. In *Le Voyage dans la Lune,* Méliès weaves together elements from two sci-fi novels: the already well-known *De la Terre a la Lune* (Jules Verne, 1865) and the very recently published *First Men in the Moon* (H.G. Wells, 1901). Méliès also plays a leading role in the movie as the president of a council of "astronomers." In the opening scene, the president proposes a trip to the Moon. The means of locomotion, a capsule launched from a giant gun, is borrowed directly from Verne's novel, in which the gun is described as 900 ft long, with an inner diameter of 9 ft [5]. The president's proposal is received enthusiastically, except for a lone dissenter, who is ultimately persuaded by intimidation (the president throws his books and papers at him). However, as we explore the science in more detail, it will become clear that the rest of the council should have paid more attention to the dissenter.

The scenes which follow depict the construction of the space capsule and the casting of the giant gun. One event, in particular, might spark considerable discussion on matters of science, technology, industrial safety, and public policy. In a clear violation of modern occupational safety standards, the soon-to-be space travelers are shown walking through the construction site, without any personal protective devices (hardhats, safety glasses, lab coats, etc.). One of them is accidentally pushed into an open tub of nitric acid. Méliès surely included this for its slap-stick entertainment value. But imagine the biological, medical, and legal consequences of such an incident. It's not difficult to understand why, in today's society, it is increasingly rare for factories to offer guided tours of their facilities.

When the construction is completed, there is much pomp and circumstance, including a parade, the waving of French flags, and the playing of *La Marseillaise*. The capsule is loaded into the breach of the giant gun (Fig. 1.1), the fuse is lit, and instantly a puff of smoke appears out of the muzzle. The Moon comes into view, and soon the details of the face of the man in the Moon (the face of Méliès) become clear. The landing is shown, at first comically, as the capsule pierces the eye of the moon, and then somewhat more seriously, as the capsule glides gently onto the surface of the Moon.

The astronomers exit the space capsule to find a breathable atmosphere, gravity comparable to that on Earth, and snowfall. Numerous celestial oddities appear, including the rising of the Earth over the lunar horizon. As they explore a subterranean cavern, the astronomers find giant mushrooms and discover that an umbrella planted in the ground will take root and transform into a giant mushroom. The astronomers encounter an aggressive (or possibly just curious and hyperenthusiastic) race of beings, called the Selenites, or inhabitants of the Moon (a concept and terminology borrowed from Wells). They defend themselves against the Selenites (or is it an unwarranted imperialist attack on the indigenous population?) by striking them with their umbrellas. In so doing they discover that these are exceedingly fragile beings, which instantly disintegrate into a puff of smoke. The astronomers are eventually outnumbered, captured, and brought before the Selenite king. They manage to escape, vaporizing more Selenites in the process, and return to their space capsule, only to realize that they have no means of propulsion to get back to Earth. No worries. One of the astronomers (the president, himself?) tugs on a rope attached to the nose of the capsule, pulling it off the edge of the Moon, and it simply falls back to Earth. They splash down in the ocean and are recovered by a steamship, which tows them back to safety.

Fig. 1.1 Imagining a trip to the moon. A custom-built artillery shell (**a**) is loaded into the breach of a giant gun (**b**). The shell is designed to accommodate a handful of human passengers and is equipped with all the comforts of home. The gun is 900 ft long. The shell must reach escape velocity before leaving the muzzle of the gun. Will the travelers survive the launch?

1.2 Exploring the Science in *Le Voyage dans la Lune*

As we explore the science in this movie, we should keep in mind that Méliès was not primarily concerned with getting the science right. Rather, as a professional magician, he was more interested in exploring the kinds of illusions he could create with this new medium of motion pictures. Thus, Méliès was a pioneer of motion picture special effects. Nevertheless, it is fair game for us to critique the science content of the movie, and to discover how much of it, if any, is plausible.

Let's begin with the launch mechanism for the space capsule. Unlike actual spacecraft, which have been built on Earth since the mid-twentieth century, the space capsule in the movie carries no fuel and is not self-propelled. It is fired from a giant gun. Is this a plausible mechanism for achieving human spaceflight? Simply put, could the passengers in the space capsule survive a launch of this sort? Extensive research has been done on the biological effects of large accelerations—what happens to the human body when you experience a large increase in speed over a short period of time (as in a rocket launch), or when you are traveling at high speed and suddenly change direction (as in a fighter jet). Throughout this book, you will be invited to come up with estimates of various things, based on information presented in a movie or TV episode scene. But the information that you are able to gather by watching the scene may not be enough. You may need to make some additional assumptions, in order to calculate the result. The launch mechanism in *Le Voyage dans La Lune* provides a good illustration of the kind of information you can gather by watching the movie scene and the kind of additional assumptions you will need to make, in order to do a calculation. In particular, is there enough information in the movie to make an estimate of how much acceleration the passengers in the space capsule will experience during the launch? If not, what additional assumptions do we need to make, in order to do the calculation? Finally, we can compare the result of our estimate to known limits on the amount of acceleration that the human body can tolerate and decide whether or not the giant gun approach to spaceflight is plausible.

1.2.1 Motion with Uniform Acceleration

When an object experiences uniform (constant) acceleration, a, the position of the object, x, and the velocity of the object, v, at any time t, are described by the following equations:

$$x = x_0 + v_0 t + \tfrac{1}{2} at^2 \tag{1.1a}$$

$$v = v_0 + at. \tag{1.2a}$$

The constant x_o is the initial position, at time $t = 0$, and v_o is the initial velocity. We are free to choose the starting time, $t = 0$, to be any time that is convenient. The simplest choice is to define $t = 0$ to be the time at which the space capsule is at rest in the breach of the giant gun. This means that the initial velocity, v_o is zero. We are also free to choose our coordinates to make the initial position convenient. The simplest choice is $x_o = 0$. With these choices, the two equations (1.1a) and (1.2a) are simplified considerably, giving us:

$$x = \tfrac{1}{2}at^2 \tag{1.1b}$$

$$v = at. \tag{1.2b}$$

We want to come up with an estimate of the acceleration, a, of the space capsule. How much information do we know, and what additional information do we need in order to answer the question, Will the astronomers survive the launch?

We will consider two fairly straight-forward ways of estimating the acceleration of the space capsule, both of which involve using *information presented in the movie*, plus an additional set of *reasonable assumptions*, which are not explicitly presented in the movie. The first important observation to make is that the space capsule in *Le Voyage dans la Lune* has no internal propulsion system. It's just a giant artillery shell fired from a giant gun. So one reasonable assumption to make is that in order to leave Earth and travel to the Moon, the space capsule must achieve *escape velocity*: the minimum velocity needed to go into a stable orbit around the Earth. It's also important to realize that the space capsule must achieve escape velocity *before* it leaves the muzzle of the gun. Once the capsule leaves the gun, the expanding gas from the explosion of the gunpowder is no longer of any use to increase the speed of the capsule. In actuality, a little more than escape velocity is needed, if the capsule is to overcome the effects of air resistance. Once the capsule leaves the muzzle of the gun, not only is there no more propelling force from the expanding gas of the explosion, but also is there a resistive force that will slow it down as it travels through the Earth's atmosphere. But since all we want is an estimate of the acceleration, we can ignore air resistance. Escape velocity will be good enough for our purpose.

Example 1.1: Estimating the Acceleration of the Space Capsule (Simple Approach)

The simplest approach to estimating the acceleration of the space capsule in *Le Voyage dans la Lune* is to take a guess for the time, t, that the capsule spends inside the gun. Based on what we see in the movie, it takes about 1 s from the moment the gun is fired until the capsule leaves the gun. We know that the final velocity of the capsule must be equal to escape velocity (approximately 11.2 km/s). So we can solve Eq. (1.2b) for the acceleration, a, and substitute our values for the time, t, and velocity, v.

$$a = v/t$$
$$= (11.2\,\mathrm{km/s})/(1\mathrm{s}) \tag{1.3}$$
$$= 11,200\,\mathrm{m/s^2}$$

Example 1.2: Acceleration of the Space Capsule Using Data from Jules Verne's Novel

An alternative approach to estimating the acceleration involves making another reasonable assumption, which is not explicitly presented in the movie. Recall that the movie is based, in part, on a novel by Jules Verne, in which the length of the giant gun is said to be 900 ft. Instead of taking a guess for the time, t, at which the capsule leaves the muzzle of the gun after it is fired, we could use the known final velocity (escape velocity) and the distance traveled to reach escape velocity (the length of the gun). We can combine the two equations (1.1b) and (1.2b) to eliminate the time, t. If we solve Eq. (1.2b) for t, and substitute into Eq. (1.1b), we get

$$a = v^2/2x. \tag{1.4}$$

We can now calculate the acceleration using escape velocity for v and the length of the gun for x. But in order to do the calculation, we need to put all quantities in a consistent set of units (e.g., velocity in m/s, and distance in m). We convert the length of the gun from feet to meters using the approximate conversion factor of 0.305 m/ft: (900 ft)(0.305 m/ft) = 274.5 m. Finally, using Eq. (1.4), we calculate the acceleration:

$$a = (11.2\,\mathrm{km/s})^2/2(274.5\mathrm{m})$$
$$= 228,488\,\mathrm{m/s^2}$$

Note that the results of Examples 1.1 and 1.2 do not agree with each other. If you are puzzled by this apparent discrepancy, keep in mind that we made *different assumptions* in each case. In the first example, we simply took a guess for the time, t, based on what we saw in the movie. In the second example, we used information that was not actually presented in the movie, but which came from the novel on which the movie was based. The result that you get when you do any calculation will depend on the assumptions that you make. When you are asked to do calculations later in this book, be sure to state your assumptions clearly.

Example 1.3: Comparing Space Gun Acceleration to the Acceleration Due to Gravity

Having estimated the acceleration experienced by the space travelers in *Le Voyage dans la Lune* by two different methods, we are now in a position to ask whether or not they will survive the launch. Let's first compare the estimated acceleration to the average acceleration due to gravity on Earth: $g = 9.8$ m/s². If we divide the acceleration from Example 1.1 by 9.8 m/s^2, we find that the travelers will experience an acceleration of more than 1100 times the acceleration due to gravity. Similarly, the result from Example 1.2 turns out to be over 23,000 times the acceleration due to gravity. Is this safe? How does this compare to real-life space launches from Earth and to the maximum acceleration that the human body can tolerate without serious damage? The answers to these questions are left as a topic for exploration.

Exploration Topic 1.1: The Biological Effects of Large Acceleration (Is It Safe to Launch Humans into Space from a Giant Gun?)

(a) Consult a reliable source of information, such as NASA's web site, to find out how much acceleration is experienced by real-life astronauts, when they are launched into space. The acceleration is typically expressed as a multiple of *g*, the acceleration due to gravity on Earth, and is sometimes referred to as the number of "*G*"s.

(b) The 1979 movie *Moonraker* includes a scene in which British secret agent James Bond (played by Roger Moore) is exposed to near-lethal acceleration in a flight training centrifuge [6]. According to the scene, most people will pass out if they experience an acceleration of seven times the acceleration due to gravity (without the benefit of a special pressurized flight suit to maintain blood flow to the brain). Twenty times the acceleration due to gravity is lethal. Do some research to verify (or refute) this information. How many "*G*"s can a human tolerate without passing out? What is the maximum number of "*G*"s that can be tolerated without serious or permanent injury? Does it make a difference whether the acceleration is along the head-to-foot direction through the body or the front-to-back direction? Why or why not? How many "*G*"s are fatal to humans?

(c) How does the acceleration experienced by the astronomers in the giant space gun (the results of Examples 1.1, 1.2 and 1.3) compare to a typical NASA space launch? Is the giant space gun a plausible approach to human space flight?

The results of our calculations suggest that the council of astronomers should have listened to the lone dissenter and would have done well to explore other options for their trip to the Moon. But remember that the director, Méliès, was concerned primarily about entertainment (creating illusions), and not about getting the science right. Despite the completely implausible (lethal!) launch mechanism, the astronomers in the movie actually do survive the launch and land on the Moon.

1.2.2 Imagining Human Exploration of the Moon

As we have just seen, the mechanism of human space flight imagined in *Le Voyage dans la Lune* is completely implausible—the astronomers would have died before they even left the Earth. Although humans did, in fact, land on the Moon more than 60 years later (Apollo 11, July 20, 1969), the way that it was done was nothing at all like the 1902 movie. But can we find any similarity between the events imagined in the movie and the actual events of the Apollo missions? We now turn our attention to the many things that the astronomers experienced when they arrived on the Moon.

Except for the presence of mountains and craters, very little was known about conditions on the Moon when this movie was produced. Would the surface be solid enough for the astronomers to walk on it, or would it be covered with a thick layer of dust? Would there be an atmosphere (and therefore, weather patterns)? If there is an atmosphere, would it be breathable? Would there be any kind of life forms, or even intelligent life? With very few scientific constraints, Méliès was free to imagine what the astronomers would find, and to create his own fantasy world. Two things that Méliès portrays are worth discussing in some detail.

It was well-known, even in 1902, that the same side of the Moon always faces the Earth. The Moon rotates on its own axis with exactly the same period as its orbit around the Earth. One of the first things that the astronomers see when they land on the Moon is the Earth rising over the lunar horizon. Is this possible? Why or why not? Compare this to what the Apollo astronauts saw from the surface of the Moon (recorded in the iconic photo of the Earth against the black sky, which has been labeled the *blue marble*).

Newton's Universal Law of Gravitation was also well-known in 1902. Yet when the astronomers escape from the Selenites, and return to their space capsule, their way of getting back to Earth was simply to fall off the edge of a cliff. Does this make sense, given what we know (and what was known at the time) about the way gravity works? We will discuss this in more detail in the next chapter.

Finally, a bit of prescience on the part of Jules Verne and George Méliès: the splashdown in the ocean and recovery by ship. Although it was apparently unplanned in Verne's novel (a ship just happened to be nearby when the capsule fell into the Pacific) and it's not clear from the brief treatment in the movie whether it was planned or accidental, this is almost exactly the way that NASA planned the recovery of all of their space capsule astronauts, from the Mercury, Gemini, and Apollo missions. Verne's imagination was 100 years ahead of its time!

1.3   The First Literary Work of Science Fiction

Our exploration of science will be aided almost exclusively by examples from science fiction film and television series. But science fiction as a genre is considerably older than either of these relatively recent entertainment media. Television is a product of the early-to-mid twentieth century, and motion picture technology is only a little over 100 years old, dating back to the late nineteenth century. Some historians of science fiction trace the origins of the literary genre back only slightly before the beginning of motion pictures, to Jules Verne, whose early works include *Journey to the Center of the Earth* (1864) and *From the Earth to the Moon* (1865). Others may go back almost another half-century, to Mary Shelley's *Frankenstein* (1818), which we will discuss in more detail in a later chapter. But there is a work of speculative fiction with a genuinely scientific foundation, which was written by a practicing scientist in the early part of the seventeenth century. Johannes Kepler, whose laws of planetary motion revolutionized our concept of the solar system, wrote a story with the simple title *Somnium* (Dream). Published posthumously in 1635 by his nephew, Ludwig Kepler, *Somnium* recounts the elder Kepler's dream about reading a book, which he had found in a market. The book tells the story of a youth from Iceland, who, by a curious chain of events, spends 5 years in Denmark as an assistant to the famous astronomer, Tycho Brahe. Upon returning to his native Iceland, the narrator and his mother are transported to another planet in the solar system, called Levania, and thus are able to observe the motion of the other bodies in the solar system from a different frame of reference. Although the trip itself is accomplished by magic arts, the account includes considerable technical details concerning the precautions that must be taken to ensure the safety of the travelers, and how the solar system appears from this new perspective.

Like Earth, Levania also has a moon, but this moon can only be seen from half of the surface of Levania. This suggests that the period of the moon's orbit around Levania must be equal to the period of rotation of Levania on its own axis, so that the moon remains forever on the same side of Levania. (The reverse is true of the Earth and its Moon.) Unlike Earth, which experiences 365 solar days per year, Levania only experiences 12 solar days per year. It is not exactly clear whether this means that 1 day on Levania is equivalent to a month on Earth, or if Levania's year is only 12 Earth days long [7].

Kepler's Dream addresses a very interesting scientific question for the early seventeenth century: what would it be like to observe the motion of the planets and the stars from a different point of view, other than the Earth? The irony of the work is that it was published in Latin, which suggests that it was probably intended to be taken seriously. But it is a story about a dream about reading a book, which the dreamer found in a marketplace, making it fairly clear that the author is not suggesting that it is true.

1.4   Reference Frames, Revisited

From a scientific perspective, there is an interesting connection between the first literary work of science fiction and the first science fiction film. Kepler's *Somnium* is about moving reference frames, written by someone who made his mark in the history of science by accurately describing the motion of the planets around the Sun. *Le Voyage dans la Lune* includes a scientifically inaccurate scene, in which the Earth is observed from the surface of the Moon, and appears to rise over the horizon. As we've already seen, this doesn't happen because the Moon rotates on its axis with exactly the same period as its orbit around the Earth. So from a fixed point on the lunar surface, the Earth always appears in the same place in the sky. But the continents on Earth appear to move in and out of view as the Earth rotates on its axis.

At the opposite end of the scientific accuracy spectrum is *2001: A Space Odyssey*. Directed by Stanley Kubrick and released in 1968—just 1 year before the first Apollo Moon landing—the movie is remarkable for getting the science right, as well as for its artistic beauty. An early scene shows a number of small satellites in orbit around Earth and a large rotating space station. (As we will see in the next chapter, the rotation of the space station provides simulated gravity for passengers around the rim.) A Pan American space shuttle, en route to the space station, moves into the field of view. Inside the cabin of the space shuttle, which has no artificial gravity, we see a pen floating freely. The flight attendant, wearing hook-and-loop "Grip Shoes", walks along the aisle, plucks the pen from the air, and returns it to the pocket of the lone sleeping passenger. The camera then cuts again to the view from space, and we see for the first time the shuttle approaching the rotating space station, with the Earth in the distance. The problem at hand is more complicated than anything any real-life astronaut had to accomplish up to that point in history: how to dock a space shuttle with an orbiting space station, which is not only moving, but rotating. The camera cuts to the shuttle cockpit, and we see things from the point of view of the shuttle pilot. The space station appears to be rotating and moving slowly across the field of view, as seen through the cockpit window. The camera then focuses on the instrument console, where a computer-generated rectangle rotates on the screen, with respect to fixed cross-hairs. Presumably, the rotating rectangle represents the rectangular-shaped docking bay on the axis of the rotating space station. Next the camera cuts to a perspective from inside the docking bay, and we see the space shuttle moving across a rotating field of stars in the background. The shuttle gradually matches its orientation to the orientation of the docking bay. The camera cuts again to the point of view of an external observer, watching the whole process, and we see both the shuttle and the space station in synchronized rotation. (Fig. 1.2) Back to the cockpit of the shuttle, and we see the space station again, but this time the docking bay no longer appears to be rotating. We're seeing the rotating space station from a frame of reference, which is in synchronous rotation, making it appear stationary. The only thing that now remains is for the shuttle to enter the docking bay. The entire scene is played out to the music of Johann Strauss' Blue Danube waltz, conveying the sense of a dance in space [8].

On May 25, 1961, President John F. Kennedy gave his famous speech in which he proposed a project to land an astronaut on the Moon before the end of the decade. Seven years later, in 1968—the same year that *2001: A Space Odyssey* was released in theaters—Apollo 7's mission included practicing the docking maneuvers that would be used in the actual Apollo 11 lunar landing mission the following year. The separation and rejoining of the modules of the Apollo spacecraft involve the same concepts as the shuttle docking scene in *2001: A Space Odyssey*. Although neither of the Apollo modules would be intentionally rotating, it is still essential to keep the same orientation of both modules throughout the docking maneuver.

Fig. 1.2 Reference frames: an Earth-orbiting space station, in the shape of a giant wheel, rotates to create artificial gravity around the rim. A space shuttle (lower left) approaches and must match its own rotation to that of the space station, in order to land in the docking bay, on the axis of the space station

1.5 Roadmap to the Rest of the Book

The material of this book is organized around *Seven Major Questions*—seven recurring themes in science fiction, which will serve as springboards for exploring science concepts and current research. Each chapter includes a set of exploration topics, with references for further reading. In Chap. 2, we take up the first of the seven major questions: *What is the nature of space and time?* We will explore the physics of space travel and time travel within the framework of classical Newtonian physics, as well as Einstein's special and general theories of relativity. Chapter 3—*What is the universe made of?*—is an exploration of matter, energy, and the fundamental interactions, or *forces* of physics. Beginning with the Standard Model of particle physics and working up the scale through the atomic nucleus, atoms, gases, liquids, and solids, we will examine some interactions of energy and matter. In Chap. 4, we take up the question, *Can a machine become self-aware?* We will explore some of the branches of the cognitive sciences, a highly interdisciplinary field, which includes specialists in computer science, robotics, artificial intelligence, neuroscience, and cognitive psychology, all focused on understanding how humans think and learn. Chapter 5 examines the science behind the search for extraterrestrial intelligence, as we take up the question *Are we alone in the universe?* In Chap. 6 we will transgress the boundaries of science and philosophy, as we explore the question *What does it mean to be human?* The focus will be primarily on biological sciences and biomedical technology, but a complete answer to the question may take us beyond the domain of science. Chapter 7 addresses the question *How can we solve our problems?* We will explore some of the many ways in which science and technology are brought to bear on the problems facing the world. We will also consider some complex problems, which are of a fundamentally human nature and are not likely to be solved by science and technology, alone. Finally, with the help of some science fiction visions of things to come, Chap. 8 raises the question *What lies ahead?* We will take a look back at things that once were purely science fiction, but are now part of everyday life, and then look ahead to the future of our technological society.

References

1. Frayling, C.: Mad, Bad and Dangerous? The scientist and the cinema, p. 48. Reaktion Books, London (2005)
2. Duncan, D.W.: Package essay for *Landmarks of Early Film*. Image Entertainment, Inc. (1994)
3. *Hugo* (Martin Scorsese, Paramount 2011). Fictionalized account of the life of motion picture pioneer Georges Méliès
4. *Le Voyage dans la Lune* (A Trip to the Moon) (Georges Méliès, 1902) in *Landmarks of Early Film*, Image Entertainment, Inc. 1994 [DVD chapter 25]
5. Verne, J.: From the Earth to the Moon and a Journey Around It (English translation), p. 76. Charles Scribner's Sons, New York (1886)
6. *Moonraker* (Lewis Gilbert, MGM 1979). James Bond investigates the hijacking of an American space shuttle. Lethal acceleration in flight training centrifuge [DVD scene 5]
7. Lear, J.: Kepler's Dream, with the full text and notes of Somnium, Sive Astrononomia Lunaris, Johannis Kepleri, p. 87. University of California Press, Berkeley and Los Angeles (1965)
8. *2001: A Space Odyssey* (Stanley Kubrick, MGM 1968). Reference frames [DVD scenes 6 and 10]

Chapter 2
What Is the Nature of Space and Time? (The Physics of Space Travel and Time Travel)

> *People assume that time is a strict progression of cause to effect. But actually, from a nonlinear, non-subjective viewpoint, it's more like a big ball of wibbly-wobbly, timey-wimey … stuff.*
> —The 10th Doctor
> Doctor Who, "Blink" [1]

The most successful science fiction television series in the history of the medium is undoubtedly *Doctor Who*. The lead character, whom one refers to as "the Doctor", is a Time Lord, who travels through space and time in a sentient device called the TARDIS (Time And Relative Dimension In Space). From the outside, the TARDIS looks like a 1950s British police box (Fig. 2.1). Painted blue and slightly larger than the classic red telephone booths, which can still be found in England, the police public call boxes could be used by anyone at no charge to phone the police in case of emergency. But from the inside, the TARDIS is more like the size of a small house. Evidently, the door to the TARDIS connects the exterior of a relatively small object (the police box) to the interior of a large object (the space/time machine). Hence, the phrase that is frequently used to describe the TARDIS: "It's bigger on the inside." A device such as the TARDIS is possible only in the realm of the imagination. But from a scientific perspective, what is the nature of space and time? Is time travel possible? The answers to these questions have changed considerably over the last few centuries.

2.1 Changing Perspectives Through History

In this chapter will consider three historical perspectives on the nature of space and time, and of the force of gravity. We begin our exploration by turning to another highly successful science fiction television series. A recurring theme in *Star Trek: The Next Generation* is the quest of the android, Commander Data, to become more like his human shipmates. We will take a closer look at Data and his quest in Chap. 3

© Springer Nature Switzerland AG 2019
B. B. Luokkala, *Exploring Science Through Science Fiction*, Science and Fiction, https://doi.org/10.1007/978-3-030-29393-2_2

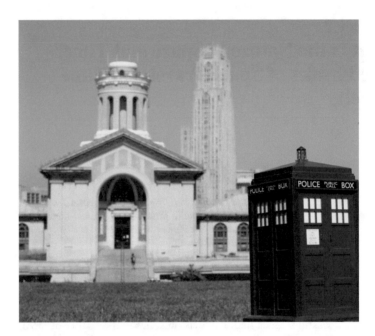

Fig. 2.1 A mysterious blue police box, from mid-twentieth century England, appears briefly on the campus of a major American research university, in the early twenty-first century. (Photo by the author)

and again in Chap. 6. Here we will use Data's creativity and scientific curiosity as a lead-in to an exploration of the nature of space and time.

In the opening scene of the episode *Descent, Part I*, Data creates a *holodeck* simulation (a virtual reality environment, which we will discuss more in Chap. 7) to enable him to play poker with three of the most famous people in the history of physics: Sir Isaac Newton (1643–1727), Albert Einstein (1879–1955), and Stephen Hawking (1942–2018) [2]. Data's primary concern was to learn more about these three specific human personalities, and by extension, to understand more about what it means to be human. But here we are interested in comparing the three views of the nature of space and time represented by these three figures from the history of physics, as well as their different ways of understanding the force of gravity [3]. A brief summary of the three views is presented in Table 2.1. We will discuss each of them in some detail in the sections which follow.

2.2   Classical Physics: Newton's Laws

The foundation of classical physics was laid with the publication of the *Philosophiae Naturalis Principia Mathematica* (Mathematical Principles of Natural Philosophy). Newton's first book of the *Principia* (1687) included statements of his three Axioms

Table 2.1 The nature of space and time, according to Newton, Einstein, and Hawking

Classical Physics represented by Sir Isaac Newton (1642–1727)	• *Space* and *time* are separate and independent quantities • The speed of anything is relative to one's frame of reference • The force of gravity acts instantaneously between any two objects; varies directly as the product of their masses; and inversely as the square of the distances between the objects
Modern Physics represented by Albert Einstein (1879–1955)	• Space and time constitute a continuous, four-dimensional fabric: *spacetime* • The speed of light is absolute (same in all reference frames) • Nothing can travel faster than the speed of light in a vacuum • Moving clocks run slow, relative to clocks at rest (*time dilation*) • The phenomenon which we experience as gravity is simply a property of *distorted spacetime*, near a large mass • Clocks in a strong gravitational field run slow, relative to clocks far from a source of gravity
"Future Physics" represented by Stephen Hawking (1942–2018)	• The concept of spacetime as a continuous four-dimensional fabric may not be adequate • *Black holes* (enormous amounts of matter compressed into an infinitesimal space) cause such extreme distortions of spacetime that *quantum mechanics* must be used, instead of classical physics, to describe the resulting curvature • Ongoing quest for a consistent theory of *quantum gravity*! • Is there such a thing as a *graviton*? • Does the *multiverse* hypothesis have merit?

or Laws of Motion, which, according to Motte's translation [4], may be summarized as follows:

1. Every body continues in its state of rest, or of uniform motion in a right line, unless it is compelled to change that state by forces impressed upon it.
2. The change of motion is proportional to the motive force impressed and is made in the direction of the right line in which that force is impressed.
3. To every action there is always opposed an equal reaction, or the mutual actions of two bodies upon each other are always equal, and directed to contrary parts.

The second law, as originally stated, included no explicit mention of the mass of the body. By *change of motion*, Newton was referring to the change in momentum, which he understood to be the product of the mass times the velocity. If mass is constant, then we simply have the change in velocity (or acceleration). The second law may then be written in equation form as force equals mass times acceleration:

$$F = ma. \tag{2.1}$$

It's important to make a distinction between *vector* quantities, such as force and acceleration, which have both *magnitude* and *direction*, and a *scalar* quantity, such as mass, which has only magnitude. Thus it would be possible for multiple forces to act on an object from various directions, in such a way as to make the net force equal to zero. An object's state of rest or motion will change, according to Newton's first law, only if there is a net (nonzero) force acting on it. Similarly, the acceleration of an object, according to Newton's second law, will be nonzero only if there is a net force acting on the object from the outside. Forces which are purely internal to a system cannot change the state of rest or motion of the system. Newton's Laws of Motion affect us every moment of every day of our lives. For some illustrations of these concepts, let's consider the following science fiction movie scenes.

2.2.1   Illustration of Newton's First Law of Motion (Changing an Object's State of Motion Requires an **External** Force)

The final movie in the *X-Men* trilogy, *X Men III: The Last Stand*, revolves around the discovery of a "cure" for mutant superpowers. A young boy, held captive on Alcatraz Island, holds the key to the cure. The Brotherhood of mutants are not about to allow themselves to be rounded up and deprived of their powers. In opposition to the oppressive public policy, the leader of the Brotherhood, Magneto (played by Ian McKellen) plans an assault on Alcatraz Island. Since Alcatraz is accessible only by water or by air, most of the mutant Brotherhood are in need of an alternate form of transportation. Magneto uses his creativity—and his superpowers—to relocate the Golden Gate Bridge, which carries U.S. highway 101 between San Francisco and Sausalito, roughly three miles to the west of the island. But how can he possibly do this within the framework of classical Newtonian physics? [5].

According to the story, Magneto has the ability to "manipulate magnetic fields and metal." We will explore some of the properties of solid state materials, including magnetic materials, in more detail in Chap. 3. For now, however, let us temporarily suspend disbelief in superpowers and stipulate that Magneto actually does have this extraordinary ability. Can he use these powers to relocate the Golden Gate Bridge, without violating Newton's Laws of Motion?

As the scene opens, we observe traffic on the bridge turned into chaos. Cars and trucks are pushed out of the way (but without any physical contact), as Magneto walks onto the bridge—hands raised, palms facing forward—leading the first wave of the Brotherhood. Having brought traffic to a standstill, Magneto then turns around, stretches one arm toward the near end of the bridge, and the other toward the far end, and (again without physical contact) uses his power to rip the end of the bridge away from the road connecting it to the shore. Finally, Magneto uses his power to rip the entire bridge off of its supporting piers and moves it three miles to the east, to enable the rest of the Brotherhood to walk across the bay to Alcatraz Island.

Discussion Topic 2.1: Magneto and Newton's First Law

Think carefully about each of the three scenarios from *X-Men III*, described above. Magneto uses his superpowers to do the following: (a) push cars out of the way while walking on the bridge, (b) rip the end of the bridge away from the land, while standing on the bridge, and (c) move the entire bridge three miles to the east while standing on the bridge. Discuss how these feats might or might not be possible within the framework of Newton's First Law of Motion. In each case, identify the *system* (object at rest or in motion) and the source of the *external* force acting on the system to change its state of rest or motion.

2.2.2   Illustration of Newton's Second Law of Motion (F = ma) and the Law of Gravity

Next we consider a scene, which does not involve superpowers, but rather a mountain-climbing accident, combined with a bit of futuristic technology. Toward the beginning of *Star Trek V: The Final Frontier*, James T. Kirk (played by William Shatner, who also directed the movie) is free-climbing El Capitan, a popular landmark in Yosemite National Park, while on shore leave from his usual post as Captain of the U.S.S. Enterprise [6]. He is distracted from his ascent by his First Officer, Mr. Spock (Leonard Nimoy), who suddenly comes alongside Kirk, hovering on *jet boots*. Spock fails to understand the logic of mountain climbing, and questions whether Kirk appreciates the *gravity* of his situation. Kirk assures him that *gravity* is foremost on his mind (obvious pun intended). Unfortunately, the philosophical discussion comes to an abrupt end, when Kirk slips off the rock face and falls to his apparent doom. Spock, with the aid of his *jet boots*, dives to the rescue. Quickly overtaking Kirk, and grabbing him by the ankle, Spock brings him safely to a stop at the last possible moment, just inches above the rocks at the base of the mountain.

Kirk falling off the face of the mountain provides a dramatic illustration of another of Newton's famous laws: the Universal Law of Gravitation. The law states that the magnitude of the force of gravity acting between two objects depends on the product of their masses and inversely on the square of the distance between them, or

$$F = G \frac{m_1 m_2}{r^2}. \tag{2.2}$$

The constant, G, in Eq. (2.2) is the universal gravitational constant. When Newton published this law in the *Principia*, along with his three laws of motion, he was accused of plagiarism by his contemporary, Robert Hooke (1635–1703). Hooke is perhaps better known for his law concerning the force exerted by springs, which we will examine in the next chapter, when we explore materials science.

For a small object (such as Kirk) falling over a short distance near the surface of the Earth, Eq. (2.2) can be expressed as

$$F = mg, \tag{2.3}$$

where g is the acceleration due to gravity at the particular location. Note that Eq. (2.3) is just a special case of Newton's more general Second Law of Motion

(Eq. 2.1), with the acceleration, a, replaced by the local acceleration due to gravity, g. The precise value of g depends on both latitude and altitude, with an average value of around the globe of approximately 9.8 m/s².

Example 2.1: Kirk, Spock, and Jet Boots

Using information in the movie scene, plus a few reasonable assumptions, we can come up with an estimate of the force which Spock must exert on Kirk's ankle, to stop his fall.

We will ultimately use Eq. (2.1), $F = ma$, to calculate the force. The first ingredient we need is an estimate of Kirk's mass. Let's take $m = 80$ kg as a reasonable estimate. Next we need to estimate the acceleration, a. Recall from Chap. 1 that acceleration is the change in velocity over the change in time, $\Delta v / \Delta t$. The force that Spock exerts on Kirk's leg brings him quickly to a stop. So his final velocity is zero. All we need are estimates of Kirk's initial velocity, at the instant that Spock grabs his ankle, and the amount of time, Δt, required to stop him from that instant. The time interval is easy to estimate by watching the movie scene, but how do we estimate Kirk's initial velocity? As an object falls under the influence of gravity, it's speed increases. But as it falls through the air, the force of air resistance opposes the force of gravity. Air resistance is a force, which increases with speed. So if an object falls long enough, eventually the force of air resistance will equal the force of gravity. If the net force is zero, there is no longer any acceleration, and the object will fall at a constant speed, which we call *terminal velocity*. Let's make the reasonable assumption that Kirk has fallen long enough to reach *terminal velocity* before Spock catches up with him. Terminal velocity actually depends on the shape and size of the object and the density of the air, but a reasonable value is about 56 m/s for an adult human. If we assume that it takes about 1 s for Spock to stop Kirk completely, we are now in a position to calculate the force.

$$F = ma$$
$$= m\left(\Delta v / \Delta t\right)$$
$$= 80\text{kg}\left(56\text{m} / \text{s}\right) / \left(1\text{s}\right)$$
$$= 4480\text{kgm} / \text{s}^2$$

The result is expressed in units of kg m/s², which is the unit of force named for Sir Isaac Newton (N). But most of us are used to thinking about force in units of pounds. So if we want to appreciate what effect this force is likely to have on Kirk's leg, let's convert from Newtons to pounds, using the approximate conversion factor, 1 N = 0.225 lb. This gives a force of about 1007 lb!

Estimation 2.1: Kirk, Spock, and Jet Boots, Revisited

We have just estimated that Spock would have to exert a force of a thousand pounds on Kirk's ankle in order to stop him before he hits the rocks at the bottom of El

Capitan. But what about forces exerted on Spock? Make any additional assumptions you need to estimate the force that the jet boots must exert on Spock's ankles, in order to stop both Spock and Kirk. What are the implications for Spock?

2.2.3   Illustrations of Newton's Third Law of Motion (Action-Reaction)

A simple but dramatic illustration of Newton's Third Law of motion (action-reaction) is rocket-propelled space flight. Hot gas is pushed out of the back of the rocket, which pushes the rocket (and the remaining fuel) in the opposite direction. The same effect is experienced in the recoil of a gun, when a bullet is fired, as the cartoon character Woody Woodpecker demonstrated in the 1950 movie *Destination Moon*. The animated short was incorporated into the movie as a teaching tool, to convince skeptics of the plausibility of space flight to the moon, more than 10 years before President John F. Kennedy proposed the real-life Apollo mission [7].

Discussion Topic 2.3: The "Iron Man" Scene from *The Martian*

Can air at atmospheric pressure be used efficiently as a means to change the state of motion for an object in space? Consider the "Iron Man" scene in the 2015 movie *The Martian*. Biologist Mark Watney (played by Matt Damon) was left stranded on Mars and presumed to be dead. When the rest of the crew, on their way back to Earth, discover that he is still alive, they change course and head back to Mars in an attempt to rescue him. Watney manages to launch himself into orbit using a Mars lander. But two problems remain to be solved. In order to make the intercept, the (very large) spaceship *Hermes* needs to slow down and Watney needs to increase his altitude. Slowing down the *Hermes* involves blowing all the air out of a compartment at the bow of the ship. Watney punctures the sleeve of his space suit and flies like Iron Man toward the *Hermes* [8]. Would there be enough air in the (room-size) compartment of the Hermes to slow it down enough to make a difference? Would Watney have enough control over the force and direction of the air leak and enough air supply to survive the flight?

Newton's Third Law applies to pulling, as well as pushing, as illustrated in the opening scene of the original *Star Wars* movie. After Princess Leia's small ship is disabled by Darth Vader's huge star destroyer, some sort of tractor beam is used to pull the disabled ship inside. Regardless of the nature of the force exerted by the large ship on the smaller one, the small ship will exert an equal and opposite force on the large ship (action-reaction). If there are no other forces acting from outside the system (large ship + small ship), the center of mass of the system will not move. They both pull toward each other. But since the large ship has much greater mass, what we see in the movie is the small ship being pulled inside the nearly stationary large ship [9].

2.2.4   Applications of Newton's Laws

As an application of Newton's Laws of motion, we consider the giant city-destroying spaceships from *Independence Day*. Early scenes in the movie show the approach toward Earth of an object, at first thought to be an asteroid, with a diameter of 550 km, and a mass of approximately one-quarter the mass of the Moon. But the object begins to slow down, which suggests that it is not of natural origin. Soon a number of smaller objects break off from the primary object and descend to the surface of the Earth. These are roughly disk-shaped, with a diameter of 15 miles. They proceed to position themselves over the centers of several major cities, with as yet unknown intent. Their arrival prompts a broad range of responses, from fear and panic, to simple curiosity, and even delight at the prospect of being taken away by alien visitors. But at the conclusion of an ominous countdown, each one fires a powerful beam of energy, destroying the city below [10].

The attack ships are shown hovering over the surface of the Earth, at a constant low altitude, and either moving at a slow, constant speed, or stopped above some significant U.S. landmark, such as the Empire State Building in New York, or the White House in Washington, DC. An illustration is shown in Fig. 2.2. Newton's First and Second Laws of motion tell us that the net force on each attack ship must be zero, if the acceleration is zero. But the Earth's gravity is pulling the ship down. So in order to stay aloft (that is, no change in its state of motion or rest, and no acceleration), each ship must also be experiencing an upward force, which is equal in magnitude to the force of gravity, but opposite in direction. Given that such a force must be acting on the ship in order to keep it aloft, what are the implications for the objects on the surface of the Earth, directly below the ship? To understand this, we must invoke Newton's Third Law: action and reaction. The Earth's gravity acts on the ship, pulling it downward toward the surface of the Earth. By Newton's Third Law, the ship is also exerting a gravitational force on the Earth, pulling it upward. This is the action-reaction pair for the force of gravity between the two objects. If there is some other mysterious force acting upward on the ship, which prevents the ship from being pulled down to the surface of the Earth, Newton's Third Law tells us that there must also be a reaction force of some kind. What could that reaction force be? Several possibilities come to mind.

One possible mechanism for keeping the attack ships aloft is to have some object pulling on the attack ships from above—the mother ship perhaps? In this case, the reaction force would be the attack ship pulling downward on that object. But the mother ship can only be in one place in orbit around the Earth at a time. So it can't be exerting an upward force on all of the attack ships at various places around the world all at once. Another possibility is that the invaders have discovered a form of antigravity and are somehow able to cancel the gravitational interaction between the Earth and the attack ships. But to the best of our current understanding, the force of gravity can only be an attractive force and never repulsive. So within the realm of known physics, the only remaining option is for the attack ships to be exerting pressure downward on the Earth, in much the same way that a hovercraft uses a cushion

Fig. 2.2 Can an alien spacecraft, 15 miles wide, hover over a city without crushing everything underneath it?

ol air to float over the surface of the Earth. If the attack ships are pushing down on the surface of the Earth to counter the force of gravity, then anything in between the ship and the surface of the Earth will feel the force. The magnitude of this force must be equal to the weight of the ship (which we can estimate from data given in the movie). The resulting pressure can be calculated by dividing the force by the surface area under the ship.

Example 2.2: Giant City-Destroying Spaceships

Let's use the data presented in the movie Independence Day to estimate the pressure underneath the giant city-destroying attack ships. One possible complication is the fact that we are given dimensions of both the mother ship (550 km in diameter) and the attack ships (15 miles wide), but we are only given the mass of the mother ship (1/4 the mass of the Moon). One reasonable assumption we can make is that the attack ships are approximately disc-shaped. (Recall that the volume of disc is $4\pi r^2 h$, where r is the radius and h is the height of the disc.) Another reasonable assumption we could make is that the density (mass per unit volume) of the mother ship and the attack ships might be comparable to each other. Although this is not quite true, let's further assume that the mother ship is roughly spherical in shape. (Recall that the volume of a sphere is $4/3\pi r^3$.) We can then take the following approach to solve the problem:

Assumption: Density of mother ship = density of attack ship = mass/volume = constant.

$$\frac{m_{\text{attack\_ship}}}{4\pi r^2_{\text{attack\_ship}} h} = \frac{m_{\text{mother\_ship}}}{\frac{4}{3}\pi r^3_{\text{mother\_ship}}}$$

Rearranging and simplifying, we obtain

$$m_{attack\_ship} = \frac{\left(m_{mother\_ship}\right)\left(4\pi r_{attack\_ship}^2 h\right)}{\frac{4}{3}\pi r_{mother\_ship}^3}$$

$$m_{attack\_ship} = \frac{3h\left(m_{mother\_ship}\right)\left(r_{attack\_ship}^2\right)}{r_{mother\_ship}^3}.$$

The movie scene shows the attack ships hovering over the surface of the Earth. In other words, no acceleration in the vertical direction. According to Newton's Second Law, if there is no acceleration, there can be no net force. But the force of Earth's gravity is pulling the ship downward, so the ship must somehow be pushing against the Earth with an equal and opposite force to keep itself hovering. The force required to keep one of these ships aloft (balanced against the force of gravity) is $F = mg$. The resulting pressure under one of these ships is the force divided by the area of the disc, or

Pressure = force/area

$$P = \frac{mg}{A} = \frac{3h\left(m_{mother\_ship}\right)}{r_{mother\_ship}^3}\frac{\left(r_{attack\_ship}^2\right)g}{\pi r_{attack\_ship}^2},$$

which simplifies to

$$P = \frac{3hg\left(m_{mother\_ship}\right)}{\pi r_{mother\_ship}^3}.$$

In the end, the radius of the attack ship cancels out. All we need to do is to take a guess as to the height, h, of one of the attack ships. Then it remains simply to put all of the quantities given in the movie into the same system of units and look up a value for the mass of the moon. The completion of this calculation is left as an exercise.

Estimation 2.2: Pressure Underneath the *Independence Day* Attack Ships

Complete the calculation, which we set up in Example 2.2. Be sure to convert all quantities into a consistent set of units (mass in kg, linear dimensions in m, acceleration due to gravity in m/s^2). Using data gathered from the movie, together with a reasonable set of additional assumptions, the resulting pressure will come out to a ridiculously large number. Compare this pressure to the pressure at the bottom of the Mariana Trench—the deepest part of the ocean. Do the attack ships really need to use a death ray to destroy the major cities of Earth?

Example 2.3: Armored Hovercraft and Landmines in *Die Another Day*

In stark contrast to the scenario in *Independence Day*, a considerably more realistic illustration of the pressure underneath a hovering object is found in the 2002 movie *Die Another Day*. British secret agent James Bond (Pierce Brosnan), on a mission to prevent war with a potentially hostile nation, discovers an illegal arms deal financed by African conflict diamonds. A chase ensues involving armored hovercraft, which are able to float over a minefield [11]. How realistic is this scenario?

According to data posted by the Federation of American Scientists (FAS), the landmine known as the M15 AT is "13–1/8 in. in diameter, 4–7/8 in. high, and weighs 30 lb with 22 lb of high explosives." This type of mine is a "track-breaker" (presumably designed to destroy the treads of an armored tank) and is "activated by 350 lb of pressure" [12]. It's important to note that a *pound* is not a unit of pressure. It is a unit of weight or force. If a force of 350 lb is applied to the pressure plate, the mine will explode. Would one of the armored hovercraft in *Die Another Day* be able to float over such an explosive device? To answer this question, we will need to estimate three things: first, the surface area of the mine's pressure plate; second, the trigger pressure of the mine; and third, the pressure under the armored hovercraft.

Landmine pressure plate area, A: From a photo available on the FAS website, the diameter of the pressure plate is approximately 2/3 the overall diameter of the M15 AT mine, or about 9 in. Thus, the area of the pressure plate is given by

$$A = 4\pi r^2$$
$$= 4\pi \left(4.5\text{in.}\right)^2$$
$$= 64\text{in.}^2$$

Landmine trigger pressure: Given that a force, F, of 350 lb is sufficient to trigger the mine, the trigger pressure is given by

$$P_{\text{trigger}} = F / A$$
$$= \left(350\text{lb}\right) / \left(64\text{in.}^2\right)$$
$$= 5.5\text{lb} / \text{in.}^2$$

Pressure under the armored hovercraft: In order to determine whether it is plausible for one of the hovercraft to float over the landmine without triggering it, we need to make some *reasonable assumptions* about the hovercraft. An illustration is shown in Fig. 2.3. Let's suppose that the weight of one of these armored vehicles is about 5000 lb, which is about the same as the "curb weight" of a typical heavy-duty pickup truck. Let's also assume that the dimensions of the armored vehicle are comparable to the pickup truck, or 200 in. long by 80 in. wide. If the weight is distributed evenly over the surface area of the hovercraft, the pressure is given by

$$P_{\text{hovercraft}} = \text{Weight} / \text{Area}$$
$$= (5000\text{lb}) / (200\text{in.} \times 80\text{in.})$$
$$= 0.3\text{lb} / \text{in.}^2$$

Since the trigger pressure of the landmine is many times greater than the estimated pressure under the hovercraft, the scene in *Die Another Day* is definitely realistic. Assuming that the properties of the mines in the movie are comparable to the M15 AT track-breakers, the armored hovercraft could easily float over the mines without exploding them.

You might be wondering about the apparent inconsistency. Why is it that the hovercraft in *Die Another Day* can safely glide over the minefield, without triggering the mines, but the attack ship from *Independence Day* would crush everything underneath it. Check all of the assumptions that went into the *Independence Day* calculation carefully and see if you can find the flaw. The answer will be revealed in the next chapter.

Estimation 2.3: Newton's Law of Gravitation Applied to *Le Voyage dans la Lune*

In the previous chapter, we analyzed the mechanism of human spaceflight imagined by Jules Verne and illustrated in the 1902 movie *Le Voyage dans la Lune*. We concluded that launching a space capsule out of a giant canon would be fatal to the astronauts. We also mentioned that the return to Earth by falling off the edge of the

Fig. 2.3 Can an armored hovercraft glide over a minefield without triggering the mines?

Moon (Fig. 2.4) was also suspect, given an understanding of Newton's Law of Gravitation. Use Eq. (2.2) to compare the force of the Moon's gravity acting on the space capsule to the force of Earth's gravity acting on the space capsule. Does this explain why the Apollo astronauts needed rocket propulsion in order to leave the Moon and return to Earth?

2.2.5 The Validity of Newton's Laws of Motion

Even though Newton formulated his laws of motion over three centuries ago, they continue to serve us perfectly well for most everyday purposes. Only under a few extreme conditions, do they need to be modified or replaced by more sophisticated concepts. These include the extremely small, the extremely massive, and the extremely fast. Classical physics begins to break down at the nanoscale (dimensions of order 10^{-9} m). The behavior of subatomic particles cannot be explained without the concepts of quantum mechanics, a subject which is beyond the scope of this book. But even objects consisting of a few hundred or a few thousand atoms behave differently, compared to the same material in bulk quantities. We will touch on the physics of *nanoparticles* briefly in the next chapter. But first we turn our attention to the other two regimes in which Newton's laws need to be refined or replaced: at speeds approaching the speed of light and for masses on a planetary scale or larger.

Fig. 2.4 Can a space capsule return to Earth by falling off the edge of the Moon?

2.3   Modern Physics: Einstein and Relativity

Newton's Laws of Motion and his Universal Law of Gravitation transformed our understanding of the way in which the universe works. Without them, spaceflight to the Moon and the planets simply would not be possible. Although they apply perfectly well in the realm of most of everyday experience and always will, they are not the last word. In particular, Newton's concept of space and time as separate and independent quantities is only an approximation to reality. As we will see shortly, an important piece of modern technology, which was unimaginable in Newton's time, but has become commonplace in the twenty-first century, simply would not work without another refinement to our understanding of space and time. We will also discover that this same refinement opens another possibility, which Newton never could have anticipated: time travel!

The second of Commander Data's holodeck-simulated trio of famous physicists is Albert Einstein. Einstein's special theory of relativity, published in 1905, included the concepts of *time dilation* (moving clocks run slow relative to clocks at rest), *length contraction* (to an observer at rest, moving objects appear shorter along the direction of relative motion), and the equivalence of matter and energy (the famous equation $E = mc^2$). His general theory of relativity, published in 1916, described space and time not as separate and independent quantities, but as a single four-dimensional quantity, *spacetime*. In the vicinity of a large mass, this *spacetime* is distorted: space is stretched out and clocks run slower the deeper they are into the distortion (closer to the mass). According to Einstein, the phenomenon which we experience as gravity is simply a property of this distorted *spacetime*. But the concept of existence in four dimensions, and the possibility of time travel predates the publication of Einstein's special theory of relativity by 10 years, and shows up in a work of science fiction. H.G. Wells published his debut novel, the "Time Machine," in 1895, the first chapter of which includes the following:

> "You must follow me carefully. I shall have to controvert one or two ideas that are almost universally accepted. The geometry, for instance, they taught you at school is founded on a misconception."

> "Is that rather a large thing to expect us to begin upon?" said Filby, an argumentative person with red hair.

> "I do not mean to ask you to accept anything without reasonable ground for it. You will soon admit as much as I need from you. You know of course that a mathematical line, a line of thickness *nil*, has no real existence. They taught you that? Neither has a mathematical plane. These things are mere abstractions."

> "That is all right," said the Psychologist.

> "Nor, having only length, breadth, and thickness, can a cube have a real existence."

> "There I object," said Filby. "Of course a solid body may exist. All real things –"

> "So most people think. But wait a moment. Can an *instantaneous* cube exist?"

"Don't follow you," said Filby.

"Can a cube that does not last for any time at all, have a real existence?"

Filby became pensive. "Clearly," the Time Traveler proceeded, "any real body must have extension in *four* directions: it must have Length, Breadth, Thickness, and Duration. But through a natural infirmity of the flesh, which I will explain to you in a moment, we incline to overlook this fact. There are really four dimensions, three which we call the three planes of Space, and a fourth, Time. There is, however, a tendency to draw an unreal distinction between the former three dimensions and the latter because it happens that our consciousness moves intermittently in one direction along the latter from the beginning to the end of our lives." [13]

We now turn our attention to several science fiction TV and movie scenes, which will help us to develop an appreciation of these concepts, and to see how they are relevant to some technology which has become part of everyday life.

In the opening scene of "The City on the Edge of Forever," an episode from the first season of *Star Trek* (the original series), the U.S.S. Enterprise is buffeted by "ripples in time"—waves of distortion of *spacetime*, emanating from a previously unexplored planet. While responding to an emergency on the bridge, the ship's chief medical officer, Dr. McCoy (DeForest Kelley), accidentally receives an overdose of a substance which, in small quantities can save a person's life, but in large quantities induces extreme paranoia. The doctor, followed by several of the crew, teleports down to the surface of the planet, where they discover the source of the *spacetime* distortions: a sentient time portal, which calls itself the "Guardian of Forever." The doctor leaps through the portal, into the past, and all of history is changed. The rest of the episode is devoted to determining what McCoy did to change history and to set things right [14].

Moving through space is an everyday occurrence. Not necessarily interstellar space, as in *Star Trek*, but the everyday space within the walls of your house or the streets of your neighborhood. It's usually not as dramatic as Magneto moving the Golden Gate Bridge, or Kirk falling off the face of a mountain, but it's something that most of us experience literally every day of our lives. What about traveling through time? "The City on the Edge of Forever" deals with traveling back into the past and changing the course of history. There is no conceivable way of doing this within the framework of Newtonian physics, in which space and time are completely separate quantities and time flows always in one direction. But according to Albert Einstein, space and time are not independent quantities. Rather, they constitute a four-dimensional fabric of *spacetime*. If this is really so, what are some of the implications of this interconnectedness? Is it possible to travel far into the future or back into the past? What sort of device could the Guardian of Forever be and could something like it actually exist within the framework of Einstein's four-dimensional spacetime? We will explore these and other questions in the following sections.

2.3.1 Special Relativity and Time Dilation

Einstein revolutionized our concept of space and time. The big idea that led to this breakthrough was the realization that the laws of physics, specifically Maxwell's equations of electricity and magnetism, can only be consistent in all frames of reference if the speed of light is the same in all frames of reference. According to Newton, space and time were absolute quantities. From this *classical* perspective, your perception of the speed of anything would depend on your relative state of motion with respect to absolute space. But according to Einstein, the speed of light is the one absolute quantity. From this *modern* perspective, your perception of space and time are altered by motion. Another episode from the original series of *Star Trek* will serve to illustrate the concept. In "Wink of an Eye," the starship Enterprise is taken over by a race of aliens, called the Scalosians. These beings move so quickly that the only direct evidence of their presence is an annoying buzzing sound, like that of a flying insect. A few drops of strange liquid in Captain Kirk's coffee cause him to be "accelerated" to their speed. As a result, the Captain perceives the rest of his crew as nearly stationary, in extremely slow motion. The queen of the Scalosians explains that Kirk is to become their king. Kirk doesn't like the idea and prepares to stun the queen with his *phaser* weapon. But in their fast-moving frame of reference, the phaser beam propagates slow enough for the alien queen to step out of the way, unharmed. This would make sense if Newton's view of absolute space were correct. The phaser is a directed energy weapon. But what kind of energy does a phaser emit? If it is any form of electromagnetic energy, the phaser beam must propagate through space at the speed of light. Now if the perceived speed of anything is relative to your own speed, as it is in Newton's view, then you could step out of the way of an incoming electromagnetic energy beam. But if Einstein's view is correct, and the speed of light is the same in all frames of reference, then what we have just described would be impossible. Fortunately for the Scalosian queen, *Star Trek*'s phaser weapon emits a particle beam and not electromagnetic energy. And according to Einstein, nothing that has mass can travel at the speed of light. So as long as the Scalosian queen can move faster than the particle beam emitted by the phaser (which must propagate at a speed less than the speed of light), she could (in principle) step out of the way of the beam [15]. But the important question remains: which view of the nature of space and time—Newton's or Einstein's—is the better description of reality? We have more to explore.

One of the best film illustrations of Einstein's time dilation effect is found in the opening scene of the original version of *Planet of the Apes*. The pilot of a space ship (played by Charlton Heston) is recording his final log entry before placing himself into suspended animation, along with the rest of the crew, for the extended voyage. According to the chronometer, 6 months have passed since the launch from Cape Canaveral. Six months, as experienced onboard the ship, that is. The chronometer also displays the date back on Earth, which is nearly 700 years in the future, relative

Fig. 2.5 Illustration of
relative motion. The
coordinate axes labeled x'
and y' are moving at speed
v, relative to the coordinate
axes labeled x and y (the
rest frame)

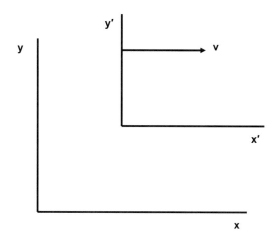

to the date on the ship [16]. From the information provided, it is possible to calculate the speed of the ship, according to Einstein's equation for time dilation:

$$t = \frac{t'}{\sqrt{1 - v^2 / c^2}}, \tag{2.4}$$

where t is the time as experienced in a frame of reference at rest (on Earth), t' is the
time in the moving frame (onboard the spaceship), v is the relative speed, and c is
the speed of light. The situation is illustrated in Fig. 2.5.

Example 2.3: Relativistic Time Dilation in *Planet of the Apes*

Here we will use the information provided in the opening scene of *Planet of the
Apes* to calculate the speed of the spaceship, according to Einstein's concept of time
dilation (Eq. 2.4). We will do this in two ways and compare the results. First, let's
calculate the average speed of the ship over the entire trip, up to the point of the
opening monolog. The ship's chronometer displays three important pieces of information: the launch date (01-14-1972), the current time onboard the ship (07-14-1972),
and the current time on Earth (03-23-2673). In Eq. (2.4), the proper time, t, is the
elapsed time in the rest frame of reference (on Earth). Subtracting the launch date
from the current date on Earth gives approximately 702 years (including an extra
day for every leap year). Meanwhile, in the moving frame of reference, the elapsed
time, t', is the difference between the current time onboard the ship and the launch
date, which is 6 months. We can solve Eq. (2.4) for the average speed of the ship, v,
first by rearranging:

$$\sqrt{1 - v^2 / c^2} = \frac{t'}{t}.$$

Then squaring both sides, we get

$$1 - v^2 / c^2 = \frac{t'^2}{t^2}.$$

Rearranging terms again gives

$$\frac{v^2}{c^2} = 1 - \frac{t'^2}{t^2}.$$

Solving for v, we obtain

$$v = c\sqrt{1 - t'^2 / t^2}.$$

Finally, substituting values for the elapsed times gives the average speed of the ship:

$$v = c\sqrt{1 - (0.5\text{year})^2 / (702\,\text{years})^2}$$
$$v = 0.999999746c$$

So the average speed of the ship up to the point of the monolog is just slightly less than the speed of light. It is important to emphasize once again that according to Einstein, nothing that has mass can actually travel at the speed of light. (Only massless particles, such as photons, can travel at the speed of light.)

The camera shot of the ship's chronometer, along with Charlton Heston's scientific monolog, would have been plenty to satisfy just about any fan of serious science fiction. But the director (Franklin J. Schaffner) and screenwriters (Michael Wilson and Rod Serling) went even further. As Charlton Heston gets ever more philosophical about human life on Earth, and his own place in the universe, we see more shots of the chronometer, showing the days passing on Earth. The rate at which time passes on Earth is roughly three days for every 90 s of ship time. Converting days into seconds gives 3 day = 3 × 24 h × 3600 s/h = 259,200 s of Earth time for every 90 s of ship time. We can use this information to calculate the current speed of the ship at the time of the monolog:

$$v = c\sqrt{1 - (90\text{s})^2 / (259,200\text{s})^2}$$
$$v = 0.99999994c$$

which is somewhat faster than the average speed of the ship, as calculated from the 702 years Earth time and 6 months ship time. This is direct evidence that the ship has not been traveling at a constant speed for its entire trip but has undergone some acceleration. It's rare for a science fiction film to treat the science at this level of detail! Unfortunately, the difference between the current speed of the ship and the average speed turns out to be rather small. This suggests that the acceleration must have occurred early in the trip, and over a very short time—probably short enough to be lethal to the passengers, unless some unexplained technology exists to protect them from the effects of enormous acceleration.

Estimation 2.4: Relativistic Time Dilation Measured in a Passenger Jet

Einstein's time dilation effect has been confirmed experimentally, by using two identical atomic clocks, initially set to the same time, one of which is flown on a trip around the world by jet, while the other remains at rest in the airport. When the moving clock returns to the airport, it is found to be slow, relative to the clock that remained at rest, by exactly the amount predicted by Eq. (2.4), given the average speed of the jet.

Use a typical cruising speed of a conventional passenger jet (not supersonic), and the amount of time for a nonstop flight around the world, to calculate the difference in time $(t - t')$ that would result when the flight is completed, and the flying clock is compared to the clock at rest.

Estimation 2.5: Relativity and Fusion-Powered DeLorean

The 1985 movie *Back to the Future* invites another simple calculation using Einstein's time dilation equation. A fusion-powered DeLorean, with the help of something called a *flux capacitor*, is used as a time machine. One simply sets the chronometer to the desired date (past or future) and accelerates up to the magic speed of 88 miles per hour. In the first demonstration of the time machine, the inventor, Dr. Emmett Brown, uses two stopwatches: one travels in the DeLorean, around the neck of Dr. Brown's dog (named Einstein), and the other remains at rest in the parking lot. The stopwatches are synchronized at the start of the experiment. When the experiment is completed, the stopwatches differ by 1 min. The difference is in the right direction—the one that traveled had slowed down. But the magnitude of the difference can't be explained just in terms of relative speed of the two reference frames [17].

Use Eq. (2.4) to calculate how fast the DeLorean had to be traveling in order for the elapsed time in the moving frame of reference to be only 1 min, while the elapsed time in the rest frame was 2 min. Alternatively, how long would the trip have had to take (in the rest frame) if the actual speed of the DeLorean had been only 88 miles per hour?

Since we know that moving clocks run slow relative to clocks at rest, we can imagine a scenario in which two identical twins might not be the same age. Suppose one of the twins goes off into space at a substantial fraction of the speed of light, while the other twin remains on Earth. Time pass more slowly for the twin on the high-speed flight, while time passes at the "normal" rate for the twin who remained on Earth. If the twin on the space flight reverses course and returns to Earth, she will discover that her twin, who remained on Earth, is significantly older. How much older will depend on their relative speed. This scenario is known as the *twin paradox*.

Estimation 2.6: Relativistic Time Dilation and the Twin Paradox

A year-long experiment conducted aboard the International Space Station (ISS) provides us with a real-life application of the twin paradox. From March 27, 2015, through March 1, 2016, U.S., astronaut Scott Kelly spent nearly a year aboard the International Space Station, while his identical twin brother, Mark Kelly, remained on Earth. Let's make the simplifying assumption that the twins, Scott and Mark, were exactly the same age when Scott took off for the ISS. Visit NASA's official website for the ISS and look up its speed, relative to the surface of the Earth. Use this speed and the time that Scott Kelly spent aboard the ISS to calculate how much older his twin brother, Mark, would be when Scott returned to Earth.

Discussion Topic 2.2

Based on these calculations, what do you conclude about the possibility (at least in principle) of traveling into the future? If you believe it might be possible, how would it be done? Does the concept of time dilation, as expressed by Eq. (2.4), allow for the possibility of traveling into the past? Why not?

2.3.2   General Relativity and Distortion of Spacetime

Einstein's General Theory of Relativity describes the distortion of *spacetime* in the presence of large masses, such as stars or planets. Space is stretched out and time runs more slowly, compared to regions of space that are far away from large masses. The gravitational time dilation effect is expressed in equation form as follows:

$$t_r = t_\infty \sqrt{1 - \frac{2Gm}{rc^2}},$$

(2.5)

where G is Newton's universal gravitational constant, m is the mass of the object creating the gravitational field, r is the distance from the center of the massive object, and c is the speed of light. The subscripts, r and ∞, indicate the time as measured at a distance r from the gravitational source, and the time measured far from any source of gravitational distortion, respectively.

We have already seen that the effect of time dilation due to relative speed (Eq. 2.4) can actually be measured, thanks to the high precision of atomic clocks. Shortly, we will consider an example of present-day technology which simply would not work without taking into account both time dilation due to relative speed and time dilation due to gravity. But first we pause to consider the question, what exactly is gravity?

In the classical (Newtonian) worldview, gravity is a force which acts instantaneously, over an arbitrary distance, between any two objects with mass. But according to Einstein, the effect which we experience as gravity is simply the result of the distortion of the four-dimensional fabric of space in the vicinity of a large mass.

Furthermore, if you are inside a closed box (one from within which you are unable to see the outside world), you would not be able distinguish between the effects of distorted space (gravity) and uniform acceleration. This effect, which is known as Einstein's *principle of equivalence*, can be experienced and measured using the simple equipment shown in Fig. E.1, in the Appendix of Artifacts.

In order to experience Einstein's principle of equivalence first-hand, all you have to do is take a ride on an elevator and measure the apparent force on an object during the brief moments of acceleration, when the elevator begins to move and when it comes to a stop. If you did not know that the elevator was accelerating, the change in the measured force could just as easily be interpreted as a momentary change in the distortion of space, resulting in an increase or decrease in the apparent strength of the force of gravity acting on the object.

Table 2.2 shows the results of a field project conducted on a university campus, to measure the apparent change in force on an object in an elevator, during the brief moments of acceleration. An object, whose weight at rest is approximately 0.5 N, was suspended from a spring scale, with a full-scale reading of 1.00 N, and a resolution of 0.01 N. The scale reading was recorded during acceleration of the elevator under two different conditions: when the elevator is starting from rest and moving upward, and when the elevator is coming to a stop while moving downward. Measurements were taken on nine different elevators around campus. The results are expressed as a multiple of g, the acceleration due to gravity.

The results in Table 2.2 indicate that the acceleration experienced in a typical passenger elevator is approximately one-tenth of the acceleration due to gravity. There appears to be a slightly larger acceleration when stopping, going downward, compared to starting and going upward, but the difference is not statistically significant. The field project could be turned into a contest to find which elevator in which department on campus gives the smoothest ride or the roughest ride. But the main point of the project is to illustrate the equivalency of acceleration and the gravitational distortion of spacetime.

Einstein's principle of equivalence can be put to practical use to create artificial gravity, as illustrated by two scenes in the movie *2001: A Space Odyssey*. As mentioned in the previous chapter, early in the movie we see a space station, shaped like a large wheel, in orbit around Earth. The wheel rotates to simulate the force of gravity for passengers around the inside of the rim. Later in the movie, the effect is illustrated on a smaller scale. It is well known that muscles will atrophy under conditions of microgravity. In order to reduce this effect, an interplanetary spaceship might be built with a section which rotates at a constant speed. An astronaut could

Table 2.2 Average acceleration experienced in a passenger elevator when starting from rest and moving upward, and when stopping, going downward

	Starting, going UP	Stopping, going DOWN
Mean	0.090	0.110
Standard deviation	0.039	0.044

The results are expressed as a multiple of *g*, the local acceleration due to gravity exercise by jogging around the inside surface of this rotating reference frame. The force required to keep objects (and astronauts) in uniform circular motion on the inside of the rotating walls would be indistinguishable from the force of gravity [18].

Discussion Topic 2.3

The 2013 movie *Elysium* is a future dystopia in which the human race is divided into two social classes. The under class struggle to survive on the surface of the Earth, which has become inhospitable, while members of the elite have abandoned Earth and live on Elysium—a space station of gigantic proportion in orbit around the Earth. Elysium is a ring world, with a population comparable to a small city. It rotates to create artificial gravity and is so large that it needs no enclosure to keep the air inside the rim [19]. Discuss the engineering problems associated with constructing such a large space station. What would have to be done to get the air inside the rotating rim and to keep the air from escaping into space without any kind of enclosure? Do you think such a construction project is plausible? Why or why not?

The effects of time dilation at high speed and distortion of spacetime near large masses are not normally thought of as having an impact on everyday life. Neither are the *James Bond* movies typically included in the category of science fiction. Nevertheless, we can use a couple of scenes from another of the Pierce Brosnan *Bond* films as a lead-in to a discussion of a very practical, everyday device, which makes use of both special relativity and general relativity. In *Tomorrow Never Dies*, a demented media mogul attempts to start a war between China and the United Kingdom by sinking a British naval vessel in the South China Sea. The crew of the ship, which had drifted off course into Chinese territorial waters, believed themselves to be in international waters, based information from their GPS (Global Positioning Satellite) system. After repeated warnings from the Chinese, the ship was sunk, not by the Chinese MIGs, which were overflying the ship at the time, but by a self-propelled sea drill, launched from a nearby boat, equipped with stealth technology, also owned by the media mogul. Later in the movie, James Bond recovers a stolen GPS encoder, which was used to broadcast a fake GPS signal from one of the media satellites [20].

The GPS system depends crucially on accurate timekeeping. As bizarre it may seem, the gravitational distortion of spacetime is a real effect, which must be taken into account, along with the time dilation effect of Special Relativity, in order to make the GPS system work properly. Clocks on the satellites in orbit are moving faster than clocks on the surface of the Earth, and thus will run more slowly by about one part in ten billion, due to the time dilation effect of Special Relativity (Eq. 2.4). But the clocks in orbit are also farther away from the spacetime-stretching gravitational field of the Earth. Clocks that are closer to a large mass will run more slowly, according to Eq. (2.5). The time-dilating effect of General Relativity makes the clocks that are on the surface of the Earth run slower by about five parts in ten billion, compared to clocks in orbit (farther from the center of the Earth). These two time-dilating effects (gravitation and relative speed) are in the opposite direction and are of different magnitudes. Both must be taken into account, in order to make accurate positioning calculations.

Estimation 2.7: Relativity, GPS, and the Importance of Time-Keeping

The GPS satellites are in geosynchronous orbit at a certain altitude above the surface of the Earth. Thus, they are moving at a greater linear speed (larger radius of orbit) compared to the speed of a fixed point on the surface of the Earth. Go to your favorite sources of reliable information and find the radius of the Earth and the altitude at which the GPS satellites orbit the Earth. Express both of these distances in units of meters. Use these values to calculate the linear speed of a point on the surface of the Earth and a GPS satellite. Use Eq. (2.4) to calculate the relativistic time dilation effects for the satellite, compared to a GPS receiver on Earth. Use Eq. (2.5) to calculate the gravitational time dilation effects for the GPS satellite and a GPS receiver on Earth.

Discussion Topic 2.4

The question remains, is the scenario in *Tomorrow Never Dies* plausible? Could a fake GPS signal be broadcast by a media satellite, orbiting near a real GPS satellite, in such a way as to confuse a naval vessel, and send it off course? Discuss what else might have to happen, in addition to broadcasting the fake signal, in order for this to work.

2.3.3   Black Holes

One way of describing a black hole is to say that it is a *singularity of density*: a finite amount of mass in an infinitesimally small volume of space. Such an object is one of many solutions to Einstein's (general relativity) *field equation*, a discussion of which we will defer until the next chapter. But a black hole is not just a theoretical concept. They are known to exist and were originally referred to as "gravitationally completely collapsed objects." A black hole can be formed when a star of sufficient mass reaches the end of its life and collapses under its own gravity. The resulting concentration of mass distorts the fabric of spacetime to such an extent that anything which approaches close enough to a black hole will never escape. The closest distance of safe approach is known as the *event horizon*. Beyond the event horizon, not even light can escape. Our own solar system will never suffer this fate, since the minimum mass required for a star to collapse into a black hole is roughly three or four times the mass of our Sun.

The term *black hole* was first introduced into the scientific literature by physicist John Wheeler, in the same year that a similar-sounding term was introduced into popular culture by the original series of *Star Trek*. In the episode "Tomorrow is Yesterday," (original air date: 26 January 1967) the Enterprise is caught in the gravitational field of what Captain Kirk refers to as a *black star* [21]. In the process of breaking away, the ship and crew are sent back into the past (the situation to which Kirk later referred in the motion picture *Star Trek IV: The Voyage Home*, when he said that they had done it before). It is not clear in this case whether the writers of *Star Trek* have influenced the language of science, or if they were somehow anticipating a technical term that was yet to be published. Although Wheeler is often credited with having coined the term black hole, he insists that he did not invent it, but simply popularized it. In his memoirs, published in 1998, he gives the exact date and location of the talk in which he first used the term: 29 December 1967, at a Sigma Xi-Phi Beta Kappa lecture at the New York Hilton hotel. But he credits an unknown member of the audience at a previous talk, which he gave in the fall of 1967, at NASA's Goddard Institute for Space Studies in New York, for proposing the term, as a simpler way of referring to a "gravitationally completely collapsed object" [22].

Black holes of tens or hundreds of solar masses exist throughout the galaxy, and supermassive black holes may exist at the centers of galaxies throughout the universe. In fact, history was made on April 10, 2019, when the Event Horizon Telescope collaboration released the first direct image of a supermassive black hole. Located at the center of galaxy Messier 87, the black hole is 6.5 billion times more massive than our Sun [23]. But black holes can also be created on a much smaller scale in high energy particle collisions. This effect led to considerable public outcry about the safety of particle accelerators, and the wisdom of conducting research in high energy physics.

The Large Hadron Collider (LHC) at CERN, near Geneva, Switzerland, is the largest particle accelerator in the world. Straddling the border between France and Switzerland, the main accelerator ring of the LHC is 16.7 miles in circumference and is designed primarily to collide high energy protons. We will consider the LHC and its research objectives in more detail in the next chapter. Before the LHC was scheduled to be turned on for the first time in September, 2008, there was widespread concern over the possibility that a black hole would be created. People feared that a black hole would suck in all the matter around it and destroy the world. Among the more tongue-in-cheek responses was the painting of a landmark fence on the campus of Carnegie Mellon University, by some of our students, in commemoration of the scheduled start-up of the LHC, on 10 September 2008 (Fig. 2.6). The words on the fence include a famous song lyric "It's the end of the world, as we know it, and I feel fine."

The public fear and misperception of the LHC may have been fueled, in part, by a sci-fi movie, released just a few years earlier. The 2005 movie, *The Black Hole*, depicts the devastation that results when a black hole is created at a fictitious particle research facility near St. Louis, MO [24]. The proposed solution to the problem is more typical of sci-fi movies of the 1950s, than the early twenty-first century: full nuclear strike!

Fig. 2.6 Carnegie Mellon University's *Senior Fence*, painted by students, in commemoration of the start-up of the LHC (10 September 2008), with the words "It's the end of the world, as we know it, and I feel fine." (Photo by the author)

Unfortunately for the people of St. Louis, and for the rest of the world, this is a completely counterproductive proposal. If you throw more matter and energy, of whatever form, into a black hole, you are not likely to blow it up. You'll just make it bigger.

So the question remains, is it possible for a high energy particle accelerator to create a black hole? And if so, what is the best thing to do about it? The answer to the first question is yes! Microscopic black holes are, indeed, created in high energy collisions, of the sort that happen all the time in the Earth's upper atmosphere. High energy protons from supernova explosions race through the galaxy. When one of these particles collides with matter in the Earth's ionosphere, the result can be a microscopic black hole—a singularity of density. But thanks to the research of Stephen Hawking, we understand that such microscopic black holes do not last long enough to accrete enough additional mass to cause the kind of devastation depicted in the movie *The Black Hole*. This effect, first described in a 1974 paper by Hawking, now bears his name: Hawking radiation [25]. So the answer to the second question is just leave them alone, and they will evaporate.

After a number of technical difficulties associated with its initial start-up, the LHC operated successfully for 10 years, gathering enormous amounts of data, which will help to answer some fundamental questions about the structure of the universe. We will take up some of these questions later in this book. The good news is that during its 10 years of operation, no black holes were ever observed eating France or Switzerland. For those who are interested in more details about the connection between black holes and particle accelerators, the press office at CERN have addressed the issue and the concerns very thoroughly on their web site [26]. The LHC shut down in December, 2018, for 2 years of planned upgrades and is expected to resume operation in 2021.

The concept of an event horizon can be formalized with reference to Eq. (2.5) (gravitational time dilation). The quantity $2Gm/c^2$ in Eq. (2.5) has dimensions of length and is known as the Schwarzschild radius:

$$r_s = \frac{2Gm}{c^2}. \tag{2.6}$$

It is possible for an object to have such a large mass that its Schwarzschild radius is larger than its physical radius. Anything that approaches the object closer than the Schwarzschild radius (or event horizon) can never escape. Thus, a black hole need not be infinitely dense. It just has to have a sufficiently large mass to create an event horizon at a radius greater than its physical radius.

Estimation 2.8: What Would It Take for the Moon to Become a Black Hole?

The animated comedy *Despicable Me* tells the story of Gru, a creative genius turned to a life of crime after many failed attempts to receive validation for his remarkable talent. His latest scheme is to commit the greatest crime in history: to steal the Moon. Gru uses a "shrink ray" to reduce the Moon to the size of a grapefruit [27]. Consult your favorite source of reliable information to find the mass of the Moon. Use

Fig. 2.7 If you shrink the Moon to the size of a grapefruit, are you in danger of creating a black hole?

Eq. (2.6) to calculate the Schwarzschild radius for the Moon. If Gru shrinks the Moon down to the size of a grapefruit, is he in danger of creating a black hole? (Fig. 2.7).

The 2009 movie *Star Trek* treats black holes in a way that has absolutely no basis in reality, but will provide a prelude for our discussion of *wormholes* later in this chapter. Director J.J. Abrams creates an alternate universe scenario, touched off by a catastrophic accident. A star within the territory of the *Romulan Empire* is about to become a super-nova and (so we are told) threatens to destroy the galaxy. The proposed solution is to create a black hole inside the star and stop the supernova from expanding. Unfortunately, the plan is executed too late to prevent the destruction of *Romulus*, the home planet of the *Romulans*. But what does occur changes the course of history. Or to put it more precisely, it transports the audience into an alternate universe, in which the flow of history is completely different from the one that generations of *Star Trek* fans knew [28]. Two ships, one piloted by Spock, the *Vulcan* ambassador (played by Leonard Nimoy) who was attempting to stop the supernova, and another piloted by Nero, a *Romulan* miner (played by Eric Bana), get caught in the gravitational field of the black hole and are pulled in. The Romulan ship and then the Vulcan ship go through the black hole and end up in the past. Their subsequent actions change the course of history.

As brilliant as this movie is, from an entertainment perspective, this scenario could never occur. Based on our understanding of the properties of black holes, both ships would be destroyed by the extreme distortion of the fabric of spacetime. Spock and Nero would die and history would continue to unfold undisturbed.

In contrast to *Star Trek*'s completely incorrect treatment of a black hole, the 2014 movie *Interstellar* deals with the subject in a way that is both more scientifically accurate and more speculative. Set in the mid-twenty-first century, the movie depicts Earth as becoming uninhabitable. A blight is killing plant life and once fertile farmland is slowly turning to dustbowls. Humans must find another planet on which to live or face extinction. None of the planets in our solar system will do, so the only solution for the survival of the human race is interstellar exploration. This presents many technical difficulties, not the least of which is time. How do you travel to distant star systems and back again before Earth becomes completely uninhabitable and everyone left behind dies? We'll take this up later in our discussion of wormholes. But for now, we focus on the supermassive black hole, which has been given the name "Gargantua". The protagonists are about to explore a planet in orbit around Gargantua, but must take into account the effect of gravitational time dilation. The distortion of spacetime is so severe that 1 h on the planet will correspond to 7 years on Earth [29]. Given this information, could we use Eq. (2.5) to calculate anything about this black hole? Unfortunately, the movie itself does not provide enough detail. Neither the mass of Gargantua, nor the radius of orbit of the planet, is provided in the movie. But for those who are interested, a source of information is available which cites a calculation by physicist Kip Thorne and estimates the mass of Gargantua to be "100 million solar masses" [30]. Thus, with this extra information, one could calculate the Schwarzschild radius of Gargantua and the radius of orbit of the planet that the protagonists are about to explore. The details are left as an exercise.

It is a fact of general relativity that the distortion of spacetime near a large mass will make clocks run more slowly. But could traveling sufficiently close to a massive black hole ever make clocks run backward, thus enabling time travel into the past? The answer, unfortunately for sci-fi fans, is no. In the same way that traveling faster than the speed of light does not result in negative time, but rather imaginary time (Eq. 2.4), so traveling closer to a black hole than the event horizon would also not result in negative time, but imaginary time (Eq. 2.5). If this is the case, what hope is there for someone who wants to travel back in time without violating the laws of physics? We will take up this question toward the end of the chapter.

2.3.4   Gravitational Waves

Einstein's General Theory of Relativity, published in 1916, predicted that the motion of extremely massive objects could produce wavelike disturbances in the fabric of spacetime. In contrast to Newton's concept of gravity as a force which acts instantaneously over an arbitrary distance, Einstein predicted that gravitational disturbances should propagate at a finite speed, namely the speed of light. However, he believed that the effect of gravitational waves would be too small to measure. The opening scene of *Star Trek VI: The Undiscovered Country* provides an illustration of something like Einstein's prediction of gravitational waves. The catastrophic explosion of the Klingon moon, Praxis, creates a shockwave through space, which rocks the Federation starship Excelsior [31]. We have already considered the distortion of spacetime due to the presence of a large mass, which results in the experience of the force of gravity. But suppose something catastrophic happens to a large mass, such as the explosion of Praxis, or many orders of magnitude larger, the explosion of a supernova? Could this cause a wave of distorted spacetime, which might propagate outward and be felt (or measured) at a large distance away? 100 years after Einstein's prediction, the effect which he believed to be too small to measure was confirmed experimentally and announced to the public. On September 14, 2015, the twin research facilities in the United States, known as LIGO (Laser Interferometer Gravitational-Wave Observatory), made the first direct experimental observations of gravitational waves caused by the merger of two black holes. Their sister facility in Italy, Virgo, was offline for upgrades at the time. The results were published in February 2016, confirming Einstein's prediction of 100 years earlier [32]. The work resulted in the awarding of the Nobel Prize in Physics for 2017 to Rainer Weiss, Barry Barish, and Kip Thorne "for decisive contributions to the LIGO detector and the observation of gravitational waves."

The LIGO facilities, one of which is located in Hanford, Washington, and the other in Livingston, Louisiana, measured minute displacements of their laser-reflecting mirrors. The mirrors are located at opposite ends of 4 km-long arms and carefully aligned to create a pattern of interference of the laser light. A schematic illustration is shown in Fig. 2.8. The distortion of space due to the passage of a

Fig. 2.8 Simplified schematic diagram of the LIGO gravitational wave interferometer

gravitational wave would affect each of the two arms of the laser interferometer differently, resulting in a shift of the bright and dark fringes of the interference pattern. An oscillating signal in the detector at one of the two facilities might be dismissed as a glitch in the instrumentation. But identical signals were detected at both the Hanford and Livingston facilities, thus confirming that the effect was real.

The basic configuration of the LIGO detector is similar to a Michelson interferometer, the apparatus designed and used by Michelson and Morley in 1887 to demonstrate that light does not require a material medium in which to propagate. A light source (in this case a Neodymium YAG laser) is directed onto a beam splitter, which divides the light intensity equally into two beams at right angles to each other. Mirrors at the far ends of each of the two "arms" reflect the light back to the beam splitter, which recombines them into a detector, creating an interference pattern.

The extraordinary sensitivity needed to detect gravitational waves is due, in part, to the placement of an additional mirror in each of the arms of the interferometer. The pair of mirrors in each arm form a resonance cavity, also known as a Fabry-Perot interferometer, which builds up the intensity of the laser light. The sensitivity of the instrument to small shifts in wavelength, or to small changes in the physical space between the mirrors, is related to the separation between the mirrors. As a spectroscopic tool, the Fabry-Perot interferometer can easily measure small changes in wavelength given by

$$\delta\lambda = \frac{\lambda^2}{2t}$$

where λ is the wavelength of the light illuminating the instrument, $\delta\lambda$ is the small change in wavelength, and t is the separation between the mirrors. A tabletop Fabry-Perot interferometer has a typical mirror separation of 1 cm and is sensitive to shifts in wavelength on the order of 10^{-11} m, making it a useful instrument for high-resolution optical spectroscopy. For some order-of-magnitude comparisons, the wavelength of visible light is in the neighborhood of 5×10^{-7} m and the separation between atoms in a solid is about 10^{-10} m. The LIGO interferometer is not a spectroscopic tool (the wavelength of the Nd:YAG laser is a constant at 1064 nm). Instead, it is used to measure small changes in relative lengths of the two arms, due to the effect of gravitational waves on the mirrors (test masses). The 4 km separation between the pair of mirrors in each arm gives it a sensitivity on the order of 10^{-16} m, which is smaller than the radius of a proton!

Additional gravitational wave events from other sources have since been observed and posted to the official LIGO website [33]. You can join the ongoing research effort by downloading software from Einstein@home and donating idle time on your home computer to analysis of the data [34].

2.3.5   Warp Drive: Faster Than Light, But Not Faster Than Light

One of Einstein's most important contributions to our understanding of the nature of space and time is that nothing can travel faster than the speed of light in a vacuum. At least not with respect to one's local surroundings. But are there any loopholes in this law of Nature? Is it possible, for example, by some trick of General Relativity, to distort the fabric of spacetime, and create a traveling wave, which one could ride, like a surfer riding a water wave? As we have just seen, Einstein predicted gravitational waves in 1916 and their existence was confirmed experimentally 100 years later in 2016. But according to Einstein, naturally occurring gravitational waves propagate at the speed of light. Could such a wave be made by artificial means to travel through the galaxy faster than light, while the wave rider would not have to travel very fast at all, with respect to the wave, itself? The writers of *Star Trek*, supported by at least one respectable physicist, would say yes! Warp drive, a most clever and essential piece of *Star Trek* technology, is what enables travel from one star system to another in a short time, without the complications of time dilation. Without warp drive, it would not be possible for the starship Enterprise to engage in its mission of exploration. But how does it work and is it even remotely plausible? Lawrence Krauss devotes an entire chapter in *The Physics of Star Trek* to the details [35]. A brief summary of the basic concept will suffice here.

The engines of a starship create a warp field—a wavelike distortion of space, in which the space behind the ship is made to stretch out, and the space in front of the ship is compressed. The starship rides along on this wave of distorted space, like a surfer on a wave of water. How to create such a wave of distorted space involves a clever combination of special and general relativity. We have already seen that the GPS system simply would not work properly without taking into account the effects of both special relativity and general relativity. The time dilation effect experienced by the satellites in orbit, as well as the distortion of the fabric of spacetime by the Earth's gravity, must be included in the calculations, in order to achieve accurate positioning. But warp drive actually makes use of both special relativity and general relativity in order to function. One of the ideas of Special Relativity is the equivalency of matter and energy, according to Einstein's equation

$$E = mc^2. \tag{2.7}$$

A basic idea of General Relativity is that space is significantly distorted by the presence of large amounts of matter and energy. In principle, any desired distortion of space can be achieved by the appropriate configuration of matter and energy. Perhaps even more remarkable is that the concept of warp drive is not limited to the minds of the writers of science fiction. At least one serious physicist has explored the possibility in mathematical detail and has even published in a respected scientific journal. In a five-page letter to the editor of the journal *Classical and Quantum*

Gravity, Miguel Alcubierre has shown that "within the framework of general relativity and without the introduction of wormholes, it is possible to modify a spacetime in a way that allows a spaceship to travel with an arbitrarily large speed" [36]. Thus, *Star Trek's* warp drive is a bit of science fiction which seems to be possible, in principle, but simply beyond our technological capability.

A particularly imaginative episode of *Star Trek: The Next Generation* raises questions about the effect of modern technology on the environment. *Force of Nature* takes the concept of spacetime as a four-dimensional fabric very literally and asks whether or not the distorting effects of warp drive might ultimately be bad for the fabric of spacetime. Might it be possible, by overuse of warp engines, to create a rift in the fabric of spacetime, much like overstressing a piece of physical material? If so, what might be the consequences? Like any environmental issue, the problem is highly complex and interdisciplinary. Scientific investigation, sometimes delayed by bureaucracy, is necessary to determine the cause. The solution is a delicate balance of politics, economics, and environmental stewardship [37].

2.4 "Future Physics": Stephen Hawking, Wormholes, and Quantum Gravity

The third of Commander Data's trio of famous physicists is Stephen Hawking, who at one time held the Lucasian Chair of Mathematics at Cambridge University, a position previously held by Sir Isaac Newton [38]. Unlike the first two of Data's holodeck-simulated physicists, both of whom were long deceased at the time, the role of Stephen Hawking was played by the actual Stephen Hawking, who just passed away in 2018 but was still very much alive in 1993, when the "Descent" episode of *STTNG* was filmed. Hawking devoted much time to the study of solutions to Einstein's equations of general relativity, including black holes and wormholes. His major contribution to our understanding of space and time has to do with the fact that a black hole is enormously massive and creates huge distortions of the fabric of spacetime over exceedingly small distances. Einstein's general relativity works well when dealing with very large massive objects but fails completely at the small (subatomic) scale. Quantum mechanics, on the other hand, has been very successful at describing physics at the subatomic level but (so far) is completely incompatible with Einstein's description of gravity. The need for a quantum theory of gravity arises because the sharp distortion of spacetime near the center of a black hole occurs over subatomic distances. As we will discuss in the next chapter, a consistent theory of quantum gravity, whose predictions can easily be tested experimentally, does not yet exist and is the subject of ongoing research.

2.4.1   Reports of Faster-Than-Light Particles and Theories of Quantum Gravity

As we have already seen, one of Einstein's major contributions to our understanding of space and time is the constancy of the speed of light in all reference frames and the notion that nothing can travel faster than light in a vacuum. In September of 2011, the CERN research center, near Geneva, Switzerland, announced that neutrinos produced at the Large Hadron Collider seemed to have traveled to a detector in Gran Sasso, near Rome, Italy, faster than light. Details were provided in a report for Reuters by Robert Evans [39]. If this had proved to be true, it would have meant another major revolution in thinking about space and time. Proposed explanations for the anomalous results included support for a theory of quantum gravity [40]. Even the writers of TV's *The Big Bang Theory* were quick to pick up on the potentially earthshaking news. Only 6 weeks after the public announcement, in the opening scene of the 11/3/2011 episode "*The Isolation Permutation*," Dr. Sheldon Cooper proposed the following topic for discussion: "Faster-than-light particles from CERN: paradigm-shifting discovery, or another Swiss export, as full of holes as their cheese?" [41]. It wasn't until several months later that the definitive explanation was found. In the end, the effect was attributed to a faulty connection between a GPS receiver and a computer [42].

2.4.2   Wormholes

In our previous discussion of the 2009 *Star Trek* movie, we concluded that Ambassador Spock and Nero could not have passed through a black hole and lived to tell about it, nor could they have passed into a different parallel universe, by way of a black hole. They would die. But is there some other way in which the events of the movie might have taken place? The younger version of Spock (Zachary Quinto), who is the Captain of the Enterprise in the alternate timeline, hypothesizes that if someone had the technology to create a black hole, then (theoretically) they might manipulate it to create a wormhole: a shortcut through spacetime [43]. The trouble with this hypothetical explanation is that the so-called "red matter," which was used to create the black hole, was not being "manipulated." It was apparently being used as directed to create the black hole. Nevertheless, the idea of creating a shortcut through the fabric of spacetime brings us into another dimension of theoretical physics, which presents a whole new set of possibilities, as well as its own set of complications.

When the astronomer Carl Sagan was writing his novel, "Contact," about the search for extraterrestrial intelligence (a topic which we will take up in Chap. 5), he needed a way for his main character to travel enormous distances through the galaxy in a short time. He initially made the same mistake that was made in the new *Star*

Trek movie: he used a black hole as the means of transportation. But he had the foresight to consult with Kip Thorne, a physicist who specializes in general relativity, for some advice. In Sagan's own words, Thorne replied to his inquiry with "… pages of tightly reasoned equations, which was more than I had expected." He recommended that Sagan use a wormhole, instead of a black hole [44].

A wormhole, simply put, is a shortcut through space. Within the framework of Einstein's general relativity, it is theoretically possible to distort space in such a way that you could travel from Pittsburgh to London as quickly as you could step through a doorway. But there are numerous technical difficulties associated with wormholes. Unlike black holes, which are known to exist and occur naturally throughout the universe, wormholes are purely hypothetical and (so far) have not been confirmed to exist experimentally. As far as using a wormhole as a means of space travel or time travel, there are even more daunting obstacles to be overcome. If they do exist, they occur naturally only at the quantum scale. A most unfortunate property of wormholes is that they are extremely unstable. They flash in and out of existence in a fraction of a second. There is no known way of keeping a wormhole open long and wide enough for it to be useful as a means of travel. For these reasons, at least for the time being, we need to rely on the imaginations of the science fiction writers.

In *Contact*, a coded message from an alien civilization includes the plans to construct a giant machine, which will open a wormhole to another part of the galaxy. How this machine does what it does is never explained in the movie, but the enormous scale of the device (larger than any of our current space launch facilities) serves to underscore the fact that stable wormholes do not exist in nature. The operation of the machine may involve the creation of some kind of exotic matter, with negative energy, in order to keep the wormhole from collapsing [45]. A more recent movie, which also acknowledges the difficulty associated with creating a stable wormhole, is *Interstellar*. We have already discussed the supermassive black hole, named Gargantua, whose gravitational time dilation effect presents a challenge. But how did the astronauts travel to Gargantua in a reasonable amount of time? A wormhole conveniently appeared near Saturn, taking the astronauts to another galaxy in a matter of minutes [46]. Since stable wormholes don't occur naturally, the characters in the movie hypothesize that a mysterious "they" must have created it using technology from the future.

At the opposite extreme of the plausibility scale are the wormholes created in the movie *Jumper* [47]. According to the story, "jumpers" are humans who have somehow evolved the natural ability to spontaneously create their own wormholes and "jump" to any place on Earth. Equally speculative is the portable wormhole-opener, used by the Paladins, in their efforts to track down and eliminate the jumpers. If the imagination allows for the giant wormhole-creating machine in *Contact*, could this same technology then be miniaturized down to the scale of a suitcase? The question of miniaturization will come up again in the next two chapters, as we consider such technologies as lasers and portable data storage devices.

Discussion Topic 2.4

The *Time Travel* episode of the PBS *NOVA* series includes the suggestion that a stable wormhole (if it could exist) could also be turned into a time machine, simply by taking one mouth of the wormhole on a round trip, at relativistic speed. As we have already discovered, time would slow down for the moving mouth of the wormhole, according to Eq. (2.4). But once the moving mouth of the wormhole returns to the point of departure, will the wormhole really be able to take a person into a different time, or has the moving mouth simply aged more slowly than the mouth of the wormhole that remained at rest (like Charlton Heston in *Planet of the Apes*)? Is this proposal as simple as it seems?

If spacetime really is a single four-dimensional fabric, there is no reason, in principle, why a wormhole might not connect different times, as well as different places. The concept of a wormhole time machine is taken up in the movie *Timeline*, in which an interdisciplinary team of researchers are able to travel back in time to 1357, in the midst of the 100-Years War. This wormhole to another time, according to the story, was created accidentally in the process of developing a teleportation device, intended to send materials almost instantaneously to other places around the world [48].

2.5   Loopholes for Time Travel into the Past?

As we have already seen in our discussion of special relativity, Einstein's time dilation equation allows for the possibility of time travel into the future. All you have to do is make a round trip at relativistic speed, and you will find yourself far in the future without having aged very much, as Charlton Heston did in *Planet of the Apes*. But this does not allow for time travel into the past. Are there any other possibilities, within the framework of physics as we know it, which might enable someone to travel backward in time?

In *Star Trek IV: The Voyage Home*, science and technology are used to solve a fundamentally human problem. A giant probe of enormous power and unknown origin approaches Earth, disabling starships as it goes. It takes up orbit around the planet and begins to vaporize the oceans. Analysis of the transmission from the probe leads to the conclusion that it is attempting to make contact with humpback whales. But there is a serious problem: it is the twenty-third century, and humpback whales had been hunted to extinction at some time in the twenty-first century. The only solution to the problem: travel back in time to Earth in the twentieth century, find some humpback whales, bring them forward in time to the twenty-third century to appease the probe and save the Earth from certain destruction [49]. The movie is tremendously entertaining and does an admirable job of portraying interdisciplinary collaboration to solve a complex problem. As we shall discover in a later chapter, some of the science and technology is actually plausible. Unfortunately, the time travel scenario, upon which the entire plan depends for success, is not.

It involves what is described as a slingshot around the Sun, in order to pick up enough speed to go into "time warp." The slingshot effect is a real technique used by NASA to change the direction of space probes, in order to explore the solar system more efficiently, but not to send them back into the past. As we have already seen, taking a round trip at high speed is a plausible way of traveling into the future. But speed alone does not enable travel into the past. The absolute limit on speed is the speed of light. In principle, according to Eq. (2.4), time can be made to run arbitrarily slowly, by going arbitrarily close to the speed of light. But the time interval, as perceived in the moving frame of reference, can never be negative. A different mechanism must be invoked, other than the time dilation of Special Relativity, in order to travel back into the past. Can General Relativity be of any help here? We know that the mass of the Sun stretches the fabric of spacetime, and that clocks close to the surface of the Sun will run more slowly, compared to clocks far from the Sun. But just as the time dilation effect of Special Relativity can never enable a clock to run backwards, so the stretching of time by a large mass, according to General Relativity, goes only in one direction. Is there any plausible explanation for time travel, as portrayed in *Star Trek IV*? One scenario, proposed by Kurt Gödel, involves traveling through a universe in which the fabric of spacetime is spinning. According to the best observational data, the universe in which we actually live is expanding, but not spinning. So this proposal is purely hypothetical. But suppose that the fabric of spacetime in the vicinity of a huge spinning mass, such as a star, could be dragged along with the rotation of the mass. What then? If the effect occurs at all, it would likely require much more mass than the mass of our Sun to make it noticeable.

Einstein's equations of general relativity allow for a curious situation known as a closed timelike curve. Like the concept of spinning space, this scenario was also first explored by Kurt Gödel. In principle, an object could follow a trajectory through spacetime and return to its starting point, not just in terms of its spatial coordinates, but in time, as well. An object trapped on a closed timelike curve would repeat the same set of events over and over again. This possibility formed the basis for the "Cause and Effect" episode of *Star Trek: The Next Generation*. The scriptwriters call the phenomenon a temporal causality loop, but the idea is the same. Members of the crew of the Enterprise begin to get feelings of déjà vu. While playing poker, they seem to know which cards are going to be dealt next. The ship's doctor knows who is about to walk into sickbay, and so forth. Eventually they put the pieces together and figure out that the Enterprise seems to be stuck in a loop in time and destined to repeat the same catastrophic event over and over again [50].

A temporal causality loop, or properly, a closed timelike curve, is something, which, for the moment, at least, remains in the realm of the imagination of science fiction writers and theoretical physicists. But if it were possible to create such a loop through spacetime, could one then travel back in time and change the events of the past? In *Cause and Effect*, the crew of the Enterprise eventually figure out how to send a message to themselves, which they could pick up in the next cycle of the loop, and avoid the repeating catastrophe.

If it is possible to change one small event in the past, what other ripples in history would be created as a result? These ideas are explored in such movies as *The Butterfly Effect* and *Donnie Darko* [51, 52]. But according to physicist Igor Novikov, time travel into the past, if it is possible, must be constrained to follow what he calls the principle of consistent causality. As recounted in the PBS NOVA episode, *Time Travel*, Novikov was inspired to explore the possibilities of time travel by colleague Kip Thorne, and his willingness to help Carl Sagan with the scientific foundation of his novel "Contact." Novikov suggests that time travel can only work without violating causality. That is to say, no event of the past can be changed in any way that would affect events in the future [44]. This opens the possibility of yet another *Star Trek* time travel scenario. In the "Trials and Tribbleations" episode of *Star Trek: Deep Space 9*, members of the crew have traveled roughly 75 years into the past and are witnessing the events of an episode of the original series of *Star Trek* ("The Trouble with Tribbles"). While taking every precaution not to do anything that might alter the flow of history, one of the main characters finds himself contemplating the possibility that he might be his own great-grandfather [53]. Igor Novikov may be on to something significant about the nature of space and time. If his consistent causality principle is correct, then time travelers into the past would be as helpless to alter the flow of history as we would be helpless to defy the law of gravity. This would rule out the scenario in *Primer*, in which two engineers accidentally invent a time machine, and proceed to use their discovery for personal profit. Day by day they invest in the stock market, based on the foreknowledge they have gained, thanks to their ability to travel in time [54].

Another nod to Igor Novikov's ideas about time travel appears in *X-Men: Days of Future Past*. Mutants are being hunted down and killed by "Sentinel" robots, who seem to adapt to the mutants' superpowers and even acquire their powers and use them against the mutants. The solution to the problem does not exactly involve time travel, but rather sending the consciousness of one of the mutants back into his body in the past. The suggestion is made that "there is a theory in quantum physics that time is immutable." Nevertheless, the mutants find a way around this supposed limitation and prevent the creation of the Sentinels [55]. Of course, just as we did in our discussion of Newton's Laws of Motion, we need to suspend disbelief in mutant superpowers to accept this. In a similar vein, but not involving mutant superpowers, is the 2000 movie *Frequency*. As in *Days of Future Past*, the movie does not involve actual time travel, but the sending of information into the past. A police officer (Jim Caviezel) discovers that he is able to communicate with his father (Dennis Quaid) by means of his father's ham radio, despite the fact that his father died 30 years earlier. Changing one event in the past sets off a chain of murders that need to be solved and, if possible, prevented [56].

Novikov's principle of consistent causality would not, however, rule out the scenario in "No Time Like the Past," an episode of the TV series *The Twilight Zone*. Frustrated with present circumstances, a scientist invents a time machine and makes multiple attempts to go back and change events of the past. One of the most notable is an attempt to assassinate Adolf Hitler, which is prevented by his rifle misfiring and other inconveniences [57].

2.6   The Multiverse Hypothesis

In the spirit of science fiction, we will take this opportunity to depart from the realm of science, into the realm of pure speculation. The miraculous passage of Spock and Nero through the black hole, which we have already discussed from a scientific perspective, resulted in what has been referred to as an *alternate universe*—a different playing out of events from the one that we are used to. For example, in the universe of the original series of *Star Trek*, James T. Kirk (William Shatner) was the captain of the U.S.S. Enterprise. In the alternate universe of the 2009 *Star Trek* movie, Spock (Zachary Quinto) is the captain, and Kirk (Chris Pine) is marooned on a desolate planet for mutiny. The question might arise, did Ambassador Spock (Leonard Nimoy) and Nero go back in time in their own universe and change the flow of history, or did they pass into a completely different *parallel universe*? Related to the quest for a consistent theory of quantum gravity are theories of the so-called *multiverse*, that is to say, multiple parallel universes, in which reality plays out differently in each one. But there is no possibility of communication between any two of them—they are *causally disconnected* from one another. Nothing that happens in our universe can have any effect on similar events occurring in another universe, and vice versa. So this does not present a plausible explanation for the events in the movie. By the same token, since there is no possibility of information exchange between these hypothetical parallel universes, there is also no way to know whether or not they exist. The multiverse hypothesis can never be tested experimentally.

This does not (nor should it!) prevent the writers of science fiction from having fun with the concept. A brilliant illustration of the most extreme version of the multiverse hypothesis is the "Parallels" episode of *Star Trek: The Next Generation* [58]. In the opening scene, Lt. Worf (Michael Dorn) is returning to the U.S.S. Enterprise from a *Klingon bat'leth* tournament, having won champion standing. But his sense of pride in his victory is tempered with a sense of dread, as he anticipates his fellow crewmembers might be planning a surprise birthday party for him. As the episode unfolds Worf begins to suspect that things are not as they ought to be. Details of his surroundings change. Relationships with crewmembers are different. Even interplanetary political situations are not the way he knew them. He finds himself shifting from one quantum reality to another. Toward the end of the episode, multiple quantum realities begin to converge at the same place. As Commander Data observes, "Everything that can happen does happen."

Data is no doubt making a reference to the many-worlds interpretation of quantum physics, due to Hugh Everett, III. According to this interpretation, every choice that we are faced with causes the universe to branch into parallel worlds in which each possible choice is played out somewhere by some parallel version of ourselves [59].

A more restrained version of the multiverse hypothesis is explored in the TV series *Fringe*, in which there seem to be two parallel universes. A brilliant scientist Walter Bishop (John Noble) develops a way to observe events in the alternate universe. In order to convince the government to fund his research, he copies some of the advanced technology that he observes in the alternate universe and shares it with the military, while assuring them that it is "theoretically impossible" to crossover into the alternate universe. When his young son dies of an incurable illness, he observes his alternate self (whom he refers to as "Walternate") developing a cure. He extends his viewing technology to build a device, which opens a wormhole into the alternate universe. The resulting breach between the parallel universes triggers a series of bizarre events on both sides [60].

In his book *The Hidden Reality*, physicist and acclaimed author Brian Greene distinguishes nine variations on the multiverse theme, which differ markedly both in terms of the overall makeup of the landscape, and in terms of the nature of the individual member universes [61]. And not all variants of the multiverse hypothesis should be dismissed as mere flights of fancy. Our observable universe is finite. That is to say, we are only able to see out as far as light could have traveled to us, since the beginning of the universe. There is no reason (at the moment) to believe that the universe does not extend beyond our visual horizon, and that there might be other regions of this hypothetically bigger universe. But again, since there is no way for us to receive information from these hypothetical other regions, we have no way of knowing what they might be like. The conditions might be similar to our known universe, or very different [62].

2.7   Exploration Topics

Exp-2.1: Time Travel

What property of the universe was predicted by H.G. Wells in his 1895 novel "The Time Machine" 10 years before Albert Einstein published his special theory of relativity?

What experimental evidence confirms the effect of relativistic time dilation (clocks slow down at high speed) and suggests the possibility of time travel into the future?

Why was Einstein opposed to the concept of *closed timelike curves* and the possibility of time travel into the past?

We do not yet know if closed timelike curves actually exist. But if they do, what are some of the implications?

Reference:
* Tim Folger, "A Brief History of Time Travel" Scientific American, September 2015

Exp-2.2: Gravitational Waves and the Cosmic Microwave Background Radiation

Why are gravitational waves so difficult to detect (so much so that Albert Einstein doubted whether they would ever be observed)?

What is the Cosmic Microwave Background radiation (CMB)?

What specific properties of the CMB might provide evidence of gravitational waves?

Why are astrophysicists not all in agreement about the interpretation of the BICEP2 observations of the CMB? What are the alternative explanations?

What astronomical phenomena are most likely to produce directly observable gravitational waves?

Reference:
- Lawrence M. Krauss, "A Beacon from the Big Bang" Scientific American, October 2014

Exp-2.3: Gravitational Waves: Prediction and Direct Observation

When was the phenomenon of gravitational waves first predicted and who made the prediction?

Why are gravitational waves so difficult to detect?

When and where did the first confirmed observation of gravitational waves take place?

How do the LIGO and VIRGO facilities work?

What is the biggest technical difficulty that had to be overcome in order to make ground-based gravitational wave observatories viable?

How were the researchers who made the first observation of gravitational waves sure that they had observed a real effect and not just a glitch in their instrumentation?

Why did the researchers choose to keep their discovery secret for several months before announcing the discovery to the public?

Describe some of the proposed alternatives to ground-based gravitational wave observatories. What advantages and disadvantages do the alternatives have, compared to LIGO or VIRGO?

References:
- Ross D. Anderson, "An Ear to the Big Bang" Scientific American, October 2013
- Robert Garisto, "How gravitational waves went from a whisper to a shout" Physics Today, August 2016

Exp-2.4: Black Holes, Black Stars, and Naked Singularities

Why has it been so difficult to observe a black hole directly?

What new techniques are being developed to enable direct observation of a black hole?

What are the differences between a black hole and a black star?

How would an observer be able to distinguish between a black hole and a black star?

In what ways are naked singularities similar to black holes?

What are the observable differences between black holes and naked singularities?

If naked singularities exist in space, what could be learned by observing them?

References:

- Remo Ruffini and John A. Wheeler, "Introducing Black Holes" Physics Today, January 1971
- Avery E. Broderick and Abraham Loeb, "Portrait of a Black Hole" Scientific American, December 2009
- Carlo Barcelona, Stefano Liberati, Sebastiano Sonego and Matt Visser, "Black Stars not Holes" Scientific American, October 2009
- Pankaj S. Joshi, "Naked Singularities" Scientific American, February 2009

Exp-2.5: Wormholes and Time Travel

Watch the PBS *NOVA* episode, "Time Travel", which proposes that a wormhole might be used as a time machine. Based on your understanding of special relativity and time dilation, write an argument either for or against this proposed mechanism for time travel.

Exp-2.6: Improving the GPS

What important physics principles are essential for an effective GPS?
    Why might there be a need to improve the current system?

References:

- Thomas A. Herring, "The Global Positioning System" Scientific American, February 1996
- Neil Ashby, "Relativity and the Global Positioning System" Scientific American, May 2002
- Per Enge, "Retooling the Global Positioning System" Scientific American, May 2004.

Exp-2.7: The Multiverse Hypothesis

Briefly describe the difference between a *Level 1* multiverse and a *Level 2* multiverse.
    What is the fundamental problem with all multiverse theories, which prevents any of their claims from ever being directly tested or verified?

Reference:

- George F. R. Ellis, "Does the Multiverse Really Exist?" Scientific American, August 2011

References

1. *Doctor Who*, "Blink" (Steven Moffatt, BBC 2007). 10th Doctor's description of time ("wibbly-wobbly, timey-wimey…stuff") [DVD third series, episode 10, scene 8]

Changing Perspectives Through History

2. *Star Trek: The Next Generation* – "Descent, Part I" (Alexander Singer, Paramount 1993). Data's poker game with Newton, Einstein, and Hawking [DVD season 6, disc 7, opening scene]
3. Although using Data's poker game as a springboard for discussing the nature of space and time should be obvious to any physicist, who is also a fan of science fiction, Lawrence Krauss deserves the credit for being the first (to my knowledge) to make such a connection in print. I am indebted to him for his brilliant and entertaining book *The Physics of Star Trek* (Basic Books, 1995), which served, in part, as the inspiration for the course, upon which the present work is based

Classical Physics: Newton's Laws

4. Motte, A.: Newton's Principia, English translation revised, p. 13. University of California Press, Berkeley, CA (1964)
5. *X-Men III: The Last Stand* (Brett Ratner, 20th Century Fox 2006). Magneto's assault on Alcatraz [DVD scene 18]
6. *Star Trek V: The Final Frontier* (William Shatner, Paramount 1989). Kirk falls from El Capitan; rescued by Spock with jet boots [DVD scene 2]
7. *Destination Moon* (Irving Pitchel, George Pal Production 1950). Woody Woodpecker illustrates rocket propulsion [DVD scene 3]
8. *The Martian* (Ridley Scott, 20th Century Fox 2015). An astronaut uses his scientific knowledge to survive being stranded on Mars. The rescue involves some illustrations of Newton's Third Law of Motion. [DVD scenes 29 and 30]
9. *Star Wars, Episode IV: A New Hope* (George Lucas, Lucasfilm/20th Century Fox 1977). Tractor beam: internal force [DVD opening scene]
10. *Independence Day* (Roland Emmerich, 20th Century Fox 1996). Giant spaceships descend upon earth and proceed to destroy the major cities with death rays. But are death rays really needed? Data presented in the film, on mass and dimensions of the ships, enable calculation of the pressure underneath the ship [DVD scenes 1 (July 2), 2 (data on mother ship: diameter 550 km, mass ¼ mass of moon), 4 (mother ship collides w/ satellite), 8 (deployment of attack ships, roughly disk-shaped, 15 miles wide), 24 (death ray used on several major U.S. cities), 27 (conventional air strike), 40 (nuclear attack)]
11. *Die Another Day* (Lee Tamahori, MGM 2002). Armored hovercraft float over mine field [DVD scenes 3 and 4]
12. Federation of American Scientists – M15 AT landmine data. https://fas.org/man/dod-101/sys/land/atm.htm

Modern Physics: Einstein and Relativity

13. Wells, H.G.: "The Time Machine" reprinted in Seven Science Fiction Novels of H.G. Wells, pp. 3–4. Dover Publications, New York
14. *Star Trek* (The Original Series) – "City on the Edge of Forever" (Written by Harlan Ellison, Directed by Joseph Pevney, Paramount 1967). Time portal creates ripples in spacetime [DVD, vol. 14, ep. #28, opening scene and scene 2]

15. *Star Trek* (The Original Series) – "Wink of an Eye" (Written by Lee Cronin, Directed by Judd Taylor, Paramount 1968). Alien being moves fast enough to step out of the way of an incoming phaser beam [DVD, vol. 34, ep. #68, scene 3]
16. *Planet of the Apes* (Franklin J. Schaffner, Twentieth Century Fox 1967). Time dilation [DVD opening scene]
17. *Back to the Future* (Robert Zemeckis, Universal 1985). Time travel [DVD scene 5]
18. *2001: A Space Odyssey* (Stanley Kubrick, MGM 1968). Einstein's Principle of Equivalence: artificial gravity and accelerated reference frame [DVD scenes 6 and 14]
19. *Elysium* (Neill Blomkamp, Tristar 2013). Einstein's Principle of Equivalence on a larger scale: rotating space station large enough to hold an atmosphere with no obvious enclosure [DVD scene 12]
20. *Tomorrow Never Dies* (Roger Spottiswoode, MGU/UA 1997). GPS navigation requires both Special Relativity and General [DVD scenes 5 and 19]
21. *Star Trek* (The Original Series) – "Tomorrow is Yesterday" (Written by D.C. Fontana, Directed by Michael O'Herlihy, Paramount 1967). The Enterprise and her crew are sent back into the past, after breaking away from the gravitational field of a *black star* [DVD, Time Travel Fan Collective, disc 1, opening scene]
22. Wheeler, J.A., Ford, K.: Geons, Black Holes & Quantum Foam, pp. 296–297. W.W. Norton and Company, New York (1998)
23. Event Horizon Telescope produces first ever direct image of a black hole. https://eventhorizon-telescope.org/
24. *The Black Hole* (Tibor Takacs, Equity Pictures (made for television) 2005). A particle accelerator creates a black hole, which begins to eat St. Louis [DVD scenes 9 and 10]
25. Hawking, S.W.: Black hole explosions? Nature. **248**, 30–31 (1974)
26. Public safety information on the Large Hadron Collider. http://press.web.cern.ch/public/en/LHC/Safety-en.html
27. *Despicable Me* (Chris Renaud, Pierre Coffin, Universal 2010). Gru uses a "shrink ray" to reduce the Moon to the size of a grapefruit. Is this small enough to make a black hole? [DVD scene 17]
28. *Star Trek* (J.J. Abrams, Paramount 2009). Black hole, time travel [Ambassador Spock's account: DVD toward the end of scene 9]
29. *Interstellar* (Christopher Nolan, Warner Brothers and Paramount, 2014). Gravitational time dilation near a supermassive black hole [DVD scene 9]
30. Details about Gargantua, the supermassive black hole in Interstellar. https://interstellarfilm.fandom.com/wiki/Gargantua
31. *Star Trek VI: The Undiscovered Country* (Nicholas Meyer, Paramount 1991). Space and time, gravitational waves [DVD scene 2 following title credits]
32. Abbot, B.P., et al.: Observation of gravitational waves from a binary black hole merger. Phys. Rev. Lett. **116**, 061102 (2016)
33. LIGO. http://www.ligo.caltech.edu/
34. Einstein@home. https://einsteinathome.org/
35. Krauss, L.M.: The Physics of Star Trek, pp. 53–61. Basic Books, New York (1995)
36. Alcubierre, M.: The warp drive: hyper-fast travel within general relativity. Class. Quantum Grav. **11**, L73–L77 (1994)
37. *Star Trek: The Next Generation* – "Force of Nature" (Written by Naren Shankar, Directed by Robert Lederman, Paramount 1993). Warp drive has detrimental effects on the fabric of space-time [DVD season 7 disc 3, scene 5, or full episode]

"Future Physics": Stephen Hawking, Wormholes, and Quantum Gravity

38. Website for Stephen Hawking. http://www.hawking.org.uk/
39. Evans, R.: Faster than light particles may be physics revolution. Reuters, 23 Sep 2011. http://www.reuters.com/article/2011/09/23/science-light-idUSL5E7KN2NE20110923
40. Tamburini, F., Lavender, M.: Apparent Lorentz violation with superluminal Majorana neutrinos at OPERA arXiv:1109.5445v7 [hep-ph], 2 Nov 2011. http://arxiv.org/abs/1109.5445
41. *The Big Bang Theory* – "The Isolation Permutation" (Mark Cendrowski, Warner Brothers, 2011). Faster-than-light particles from CERN? Season 5, episode 8, air date 11/3/2011
42. Cartlidge, E.: Error undoes faster-than-light neutrino results. Science Insider, 22 Feb 2012. http://news.sciencemag.org/scienceinsider/2012/02/breaking-news-error-undoes-faster.html
43. *Star Trek* (J.J. Abrams, Paramount 2009). Black hole/wormhole, time travel [young Spock's hypothetical explanation: DVD toward end of scene 8]
44. *NOVA*: "Time Travel" (Judith Bunting, BBC/WGBH 1999). Including interviews with Carl Sagan and Kip Thorne, the episode proposes using a wormhole as a time machine [VHS tape, use the first half hour through billiard balls causality]
45. *Contact* (Robert Zemeckis, Warner Brothers 1997). Wormhole [DVD scenes 32 and 33]
46. *Interstellar* (Christopher Nolan, Warner Brothers and Paramount, 2014). Wormhole to another galaxy created by beings from the future [DVD scene 4]
47. *Jumper* (Doug Liman, 20th Century Fox 2008). Humans with the unusual ability to create wormholes, and jump spontaneously from one place on Earth to another, are hunted down by others who hate them because of their ability. The hunters use portable devices to keep the wormhole open, enabling them to follow the jumpers [DVD scenes 19 and 20]
48. *Timeline* (Richard Donner, Paramount 2003). Wormhole to the past [DVD scene 4]

Loopholes for Time Travel into the Past?

49. *Star Trek IV: The Voyage Home* (Leonard Nimoy, Paramount 1986). Time travel by "slingshot" around the Sun at warp speed: is this possible, even in principle, in the kind of universe in which we live? [DVD scenes 5 and 6]
50. *Star Trek: The Next Generation* – "Cause and Effect" (Jonathan Frakes, Paramount 1992). Temporal causality loop (closed, timelike curve) [DVD season 5 disc 5, opening scene, then skip to scenes 4 and 5]
51. *The Butterfly Effect* (Eric Bress and J. Mackye Gruber, New Line Cinema 2004). Time Travel; causality; chaos theory [DVD multiple scenes throughout]
52. *Donnie Darko* (Richard Kelley, 20th Century Fox 2004). Time travel, many-worlds interpretation of quantum mechanics [DVD scenes 3, 4, 15, and 16]
53. *Star Trek: Deep Space Nine* – "Trials and Tribbleations" (Jonathan West, Paramount 1996). Time travel, Novikov principle of consistent causality [DVD Time Travel fan collective, disc 3, skip to scene 4]
54. *Primer* (Shane Carruth, THINKFilm 2004). Time travel [DVD scene 7]
55. *X-Men: Days of Future Past* (Bryan Singer, 20th Century Fox 2014). "… a theory in quantum physics that time is immutable" [DVD scene 29]
56. *Frequency* (Gregory Hoblit, New Line 2000). Communication across time and changing events in the past [essentially the entire movie]
57. *The Twilight Zone* – "No Time Like The Past" (Rod Serling, CBS/Paramount 1963). Time travel according to Igor Novikov's Principle of Consistent Causality [DVD season 4, episode 10]

The Multiverse Hypothesis

58. *Star Trek: The Next Generation* – "Parallels" (Written by Brandon Braga, Directed by Robert Wiemer, Paramount 1993). An extreme illustration of the multiverse hypothesis: everything that can happen does happen [DVD season 7 disc 3, opening scene + scene 2, or full episode]
59. Everett III, H.: 'Relative state' formulation of quantum mechanics. Rev. Mod. Phys. **29**, 454 (1957)
60. *Fringe* – "Peter" (Teleplay by Jeff Pinker and J.H. Wyman and Josh Singer, Warner Brothers 2010). Alternate universes [DVD season 2, episode 15]
61. Greene, B.: The Hidden Reality, p. 5. Alfred A. Knopf, New York (2011)
62. Ellis, G.F.R.: Does the multiverse really exist? Sci. Am. **305**, 38–43 (2011)

Chapter 3
What Is the Universe Made of? (Matter, Energy, and Interactions)

> *Common lead would have crushed the vehicle, sir. This is my*
> *morning's run of isotope 217. The whole thing hardly comes to*
> *10 tons.*
> —*Robby the Robot*
> *Forbidden Planet*

Having considered the nature of space and time, we now turn our attention to the *stuff* of the universe: what do we know about the things—both material and immaterial—which inhabit space and time? The state of our current understanding of matter and energy is both highly detailed and very incomplete. We still have a lot to learn. As already hinted in Chap. 2, Einstein's theory of general relativity is very successful at describing physics on the large scale (planetary, stellar, galactic, and so forth) but is completely incompatible with quantum mechanics—an equally successful model of physics, but at the subatomic scale. The ongoing quest to resolve this incompatibility is one of the most interesting and open fields of physics research. In this chapter, we will explore various properties of materials, including some of the latest developments in materials science. But we will begin with a look at the fundamental building blocks of matter and one of the most intriguing questions of modern physics: What is the universe made of?

3.1 The Standard Model of Particle Physics

To begin the discussion, let's consider the opening scene from "Starship Mine," an episode from the sixth season of *Star Trek: The Next Generation*. We find the crew of the U.S.S. Enterprise preparing the ship for a routine maintenance procedure called a *baryon sweep*. According to the captain's log entry, baryon particles accumulate on (and inside) the ship, while traveling through space, and must be removed periodically. This conjures an image of a fishing boat in dry dock to have barnacles scraped off of the hull. As the scene unfolds, we learn that the baryon sweep will

© Springer Nature Switzerland AG 2019
B. B. Luokkala, *Exploring Science Through Science Fiction*, Science
and Fiction, https://doi.org/10.1007/978-3-030-29393-2_3

destroy living tissue. Thus all of the crew must evacuate the ship and take with them any vital biological materials from the ship's medical facility [1].

So what exactly is a *baryon*, and what effect would a *baryon sweep* have on a starship? At the quantum scale, the Standard Model of Particle Physics describes all of matter and energy (except for gravity) in terms of 17 fundamental particles. These are divided into *quarks* (the massive constituents of matter), *leptons* (from the Greek for delicate or minute), and the force-communicator particles, as illustrated in Table 3.1.

Quarks are never observed in isolation, under ordinary conditions. They are always confined within composite particles, which we will discuss shortly. But one of the leptons, namely the electron, is part of everyday experience. Where would modern culture be without electrons, which carry the current used by all of our electronic devices?

Next, we need to define some terminology. Our discussion of black holes, in Chap. 2, included reference to the Large Hadron Collider (LHC), the world's largest particle accelerator. But what is a *hadron*? The word *hadron* is a general term, which refers to anything made of any number of quarks. Two specific types of hadrons deserve particular mention. A *meson*, as the term is used today, is a composite particle made of two quarks (or more specifically, a quark–antiquark pair). A *baryon* (from the Greek for massive or heavy) is a particle composed of three quarks. Two examples of baryon particles are the proton and the neutron, the particles which make up the nuclei of all atoms. A proton is a baryon which consists of two up quarks and one down quark. A neutron is a baryon which consists of one up quark and two down quarks.

Finally, we come to the irony of the baryon sweep in "Starship Mine." The ship's doctor is right to be concerned about the tissue samples stored in her medical facility, which would be destroyed in the baryon sweep. But the Captain of the Enterprise should be equally concerned about the ship, itself. Undoubtedly, by the twenty-fourth century, several new atoms will have been discovered, which have not yet been discovered in the twenty-first century. Some of these yet-to-be-discovered atoms may even be incorporated into the high-tech material that makes up the starship *Enterprise*. Yet all atoms must be made of protons and neutrons. In effect, all of the material of the ship is composed of baryons, and will be swept away, along with the excess accumulated baryons.

Table 3.1 The 16 fundamental particles, plus the recently discovered Higgs boson, constitute the Standard Model of Particle Physics—a highly successful quantum model of matter and interactions

Fundamental particles of matter		Forces and communicator particles	
Quarks	Leptons	Force	Communicator
Up (u)	Electron (e)	Electromagnetism	Photon
Down (d)	Electron neutrino (ν_e)		
Charm (c)	Muon (μ)	Weak nuclear force	W, Z
Strange (s)	Mu neutrino (ν_μ)		
Top (t)	Tau (τ)	Strong nuclear force	Gluon
Bottom (b)	Tau neutrino (ν_τ)		

The role of the Higgs boson is to confer mass on the other particles of the Standard Model, via interaction with the Higgs Field

Table 3.2 The leptons of the Standard Model of Particle Physics

Generation	Leptons	Charge (e)	Mass (GeV/c²)	Year of discovery
1	Electron (e)	−1	0.000511	1897
	Electron neutrino (ν_e)	0	a	1956
2	Muon (μ)	−1	0.106	1936
	Mu neutrino (ν_μ)	0	a	1962
3	Tau (τ)	−1	1.777	1975
	Tau neutrino (ν_τ)	0	a	2000

Charges are in units of $e = 1.602 \times 10^{-19}$ C. Masses are expressed in units of GeV/c² (after Einstein's relation $E = mc^2$), where the unit of energy is 1 GeV = 1 billion electron-Volts = $10^9 \times (1.602 \times 10^{-19}$ J)
[a]The mass of the neutrino is a subject of ongoing research. Current estimates are in the approximate range 0.05–2 eV

As shown in Table 3.1 the fundamental particles of the Standard Model are subdivided into two categories: the *fermions*, or fundamental particles of *matter*, and the *bosons*, or communicators of the fundamental *interactions* of physics. Collectively these particles represent more than a century of theoretical and experimental research.

3.1.1   Fermions: The Fundamental Particles of Matter

The fundamental particles of matter in the left two columns of Table 3.1 are collectively referred to as *fermions*, in honor of Italian physicist Enrico Fermi (1901–1954). The fermions are divided into two types of particles: quarks and leptons. These, in turn, are subdivided into three so-called *generations*. A number of features distinguish the quarks from the leptons, including their charge (leptons have either one unit of charge or no charge, while quarks have fractional charge) and their mass (quarks are much more massive than leptons). A summary of the properties of the fermions and their years of experimental discovery is shown in Table 3.2 (the leptons) and Table 3.3 (the quarks).

Perhaps the most familiar of the fermions is the electron, which was also the first of the fundamental particles to be discovered. In 1897 J.J. Thomson demonstrated that electrically neutral matter contains discrete negatively charged particles, to which he gave the name "corpuscles." By applying electric and magnetic fields to a beam of these particles, Thomson was able to measure the ratio of their charge divided by their mass. In 1909, Robert Millikan performed an experiment which enabled the measurement of the charge on the electron. Millikan created charged droplets of oil and balanced the force of gravity on the oil droplets by applying an electric field. The results of the Millikan "oil drop" experiment, together with Thomson's "e/m" experiment enabled the determination of the mass of the electron.

The muon is the second of the leptons to be discovered experimentally (by Anderson and Neddermeyer, at Caltech in 1936). Its mass, which is roughly 207 times the mass of the electron, lies in between the electron mass and the proton mass

Table 3.3 The quarks of the Standard Model of Particle Physics

Generation	Quarks	Charge (e)	Mass[a] (GeV/c^2)	Year of discovery
1	Up (u)	2/3	0.00215	1968
	Down (d)	−1/3	0.0047	1968
2	Charm (c)	2/3	1.28	1974
	Strange (s)	−1/3	0.0938	1968
3	Top (a.k.a. "truth") (t)	2/3	173	1995
	Bottom (a.k.a. "beauty") (b)	−1/3	4.18	1977

Charges are in units of $e = 1.602 \times 10^{-19}$ C. Masses are expressed in units of GeV/c^2 (after Einstein's relation $E = mc^2$), where the unit of energy is 1 GeV = 1 billion electron-Volts = $10^9 \times (1.602 \times 10^{-19}$ J)
[a]Unlike the leptons, quarks are always confined within hadrons and are never observed as free particles. Consequently, their masses cannot be measured directly but must be determined indirectly from their influences on the properties of hadrons [3]

(roughly 1836 times the mass of the electron). Thus, at the time when it was discovered, an appropriate label for this particle was "meson," from the Greek for "in between" or "medium." Although other names were proposed, the term "meson" gained general acceptance [2]. It wasn't until other new particles of intermediate mass were discovered that the specific label "mu meson" began to be used. But the language of particle physics has evolved over time. As the standard model was coming together in the '60s and '70s, the terminology used to describe the various particles and composites was refined. As noted above, the word meson is now restricted to composite particles consisting of a quark and an antiquark.

Neutrinos have no charge, are very difficult to detect and were originally believed to have zero mass. But when neutrinos interact with matter, they can oscillate (or change form) among the various generations of neutrino. This behavior requires that they have some nonzero mass. The precise values of the neutrino masses is a topic of ongoing research.

3.1.2   Bosons: The Force-Communicator Particles of the Standard Model

The particles of the Standard Model which communicate the fundamental interactions of physics are collectively referred to as *bosons*, in honor of Indian physicist Satyendra Nath Bose (1894–1974). To the best of our knowledge, there are four fundamental forces (or interactions) in physics: electromagnetism, the weak nuclear force, the strong nuclear force, and gravity. A simple apparatus to demonstrate simultaneously all four of the fundamental forces of physics is shown in Fig. E.2, in the Appendix of Artifacts. At the current time, only three of these forces have been successfully included within the Standard Model, as shown in Table 3.4. The first of these, the electromagnetic force, is communicated by the photon—a quantum particle of electromagnetic radiation. Prior to the mid-1800s, electricity and magnetism

Table 3.4 The bosons of the Standard Model of Particle Physics

Fundamental interaction	Boson	Charge (e)	Mass (GeV/ c²)	Year of discovery
Electromagnetism	Photon	0	0	
Weak force	W⁺,W⁻	+1, −1	80.48	1983
	Z⁰	0	91.19	1983
Strong force	Gluon	0	0	1979
Confers mass on the other particles of the standard model	Higgs	0	125.18	2012

Charges are in units of $e = 1.602 \times 10^{-19}$ C. Masses are expressed in units of GeV/c² (after Einstein's relation $E = mc^2$), where the unit of energy is 1 GeV = 1 billion electron-Volts = $10^9 \times (1.602 \times 10^{-19}$ J). Note that fourth fundamental interaction, gravity, is not yet included in the Standard Model. The hypothetical graviton is predicted to have zero mass.

were understood to be two separate and unrelated phenomena. Thanks to the work of James Clerk Maxwell (1831–1879), who described light as an electromagnetic wave, electricity and magnetism were unified into a single phenomenon. But it was Albert Einstein who proposed that light should be conceived of not as a wave but as a particle—a packet of energy, which we now call a photon. It was his theory of the photoelectric effect (not his theories of relativity!) for which Einstein was awarded the Nobel Prize in Physics for 1921.

The weak nuclear force, mediated by the so-called W and Z bosons, is responsible for radioactive beta decay, which we will discuss in some detail later in this chapter. The discovery of the W and Z bosons in 1983, by Rubia and van der Meer, garnered them the Nobel Prize in Physics for 1984. Their discovery provided experimental evidence of the unification of the weak force and the electromagnetic force into a single electroweak interaction, as predicted by Glashow, Weinberg, and Salam, who shared the Nobel Prize in Physics for 1979.

The strong nuclear force, mediated by the gluon, is what holds quarks together and binds the nuclear particles (protons and neutrons) together within the nucleus of an atom. It is also involved in the alpha decay process. The first direct experimental evidence for the existence of the gluon was produced in 1979 by the TASSO collaboration, using the PETRA electron–positron collider at the DESY facility in Hamburg, Germany.

The fourth of the fundamental forces, gravity, is not yet included in the Standard Model. As noted in the previous chapter, the various theories of quantum gravity have made predictions which are difficult or impossible to test experimentally. So we do not yet know if there is such a thing as a graviton—the quantum particle which might communicate the force of gravity.

The 2015 discovery of gravitational waves by the LIGO facility confirmed one of Einstein's predictions, but it did not provide any new insight into the possibility of quantum gravity. At best, an upper bound of 1.2×10^{-22} eV/c² was set on the possible mass of the graviton [4]. If the graviton exists, it is theoretically predicted to be a zero mass particle. The upper bound set by LIGO is at least 20 orders of magnitude less than the lower bound on the neutrino mass.

3.1.3   The Mystery of Mass and the Higgs Boson

The discovery of the W and Z bosons not only confirmed the unification of the weak force with electromagnetism, but also prompted an interesting question. As shown in Table 3.4, the W and Z bosons have mass but the photon and the gluon do not. Why should this be? In fact, why should any of the particles of the Standard Model have mass? One of the research objectives of the Large Hadron Collider was to find experimental evidence of the Higgs boson, which was hypothesized to confer mass on the other particles of the Standard Model. According to a theory proposed by Peter Higgs and others in 1964, the particles of the Standard Model acquire mass by interacting with the Higgs field, an energy field which was hypothesized to permeate the entire universe. On July 4, 2012, an announcement was made at the LHC that a particle consistent with the Higgs boson at last had been observed by the ATLAS and the CMS experiments. This discovery resulted in the awarding of the 2013 Nobel Prize in Physics jointly to François Englert and Peter Higgs [5]. Thus, the Standard Model was set on an even firmer foundation.

3.2   Beyond the Standard Model: Dark Matter and Dark Energy

As successful as the Standard Model has been, it accounts for only a small fraction of the matter and energy in the universe (see Fig. 3.1). Recent estimates suggest that only about 5% of the total matter and energy inventory of the universe is ordinary (baryonic) matter. 27% of the universe is so-called *dark matter*, which interacts with the visible matter in the universe through the force of gravity but is otherwise not currently understood. 68% of the universe is *dark energy*, which is responsible for the accelerating expansion of the universe.

Fig. 3.1 Inventory of matter and energy in the universe

Matter and Energy in the Universe

■ Ordinary ("baryonic") Matter (5%)

■ Dark Matter (27%)

■ Dark Energy (68%)

3.2.1   Dark Matter

As early as the 1920s and 1930s, astronomers began to suspect that there was more matter in the universe that meets the eye. They could not account for the observed motion of stars and galaxies by the gravitational forces of visible matter alone. Thus, they began to postulate the existence of dark matter—an unknown type of matter, which interacts with visible matter gravitationally, but which does not interact with light. A wealth of observational evidence since that time confirms that most of the matter in the universe (roughly 84%) is not the stuff of the Standard Model but is some exotic type of matter that does not participate in the electromagnetic interaction. In addition to the search for the Higgs Boson, which succeeded in 2012, other research objectives of the LHC include finding possible candidates for dark matter particles. But there are some who question the existence of dark matter and wonder if, instead, Einstein might have been wrong about the nature of gravity. This idea is taken up as an Exploration Topic at the end of the chapter.

3.2.2   Dark Energy

Shortly after Einstein published his general theory of relativity (1916), he and others realized that his equations were unstable, in the sense that they predicted a universe which either expanded or contracted. Since it was generally believed at the time that the universe was static and eternal, Einstein added a so-called *cosmological constant* to his equations in 1917, to keep the universe from collapsing under the influence of gravity. Following the discovery by Edwin Hubble in 1929 that the universe is expanding, Einstein then declared his cosmological constant to be his "greatest blunder." If the universe is expanding, then his original equations were correct and there was no need for an extra term to keep the universe stable.

Nearly 70 years after the discovery of the expansion of the universe, a new discovery was made, which resurrected the need for a cosmological constant (see Fig. 3.2). In 1998, Riess, Perlmutter, and Schmidt discovered that the universe is not just expanding. Its rate of expansion is accelerating—the universe is expanding faster and faster. The accelerating expansion of the universe is attributed to dark energy, which is currently estimated to make up 68% of the matter-energy inventory of the universe. The trio who made the discovery of dark energy shared the Nobel Prize in Physics for 2011. Einstein's cosmological constant was introduced in order to keep the universe from collapsing under the influence of gravity. But this new discovery of dark energy might be thought of as some sort of cosmological antigravity—a repulsive force rather than an attractive force.

What, exactly is dark energy? There are several competing hypotheses, but as yet no consensus. Early in 2019, the project known as the Dark Energy Survey (DES) completed its 6-year mission: to gather light from more than 300 million distant galaxies in order to better understand the nature of dark energy [6]. In the process, the DES project also made a measurement of the dark matter structure of the universe. Analysis of the data gathered by the DES project is likely to take many years but is expected to shed light on the 95% of the matter and energy in the universe, which is currently not understood.

3.3   The Atomic Nucleus: Protons, Neutrons, Isotopes, and Radioactivity

The fundamental building blocks of matter are the quarks and leptons of the Standard Model (Table 3.1). One of the leptons—the electron—is part of our everyday experience, but we do not encounter quarks in isolation. They are bound together as the constituents of protons and neutrons, which make up the nuclei of atoms. Atoms, in turn, can be assembled into higher-level configurations, such as molecules and a variety of crystal structures, some of which we will examine in detail later in this chapter. Before exploring the interactions of matter and energy, we pause briefly to consider the nucleus of the atom, with the help of one of the classic sci-fi films of the 1950s.

Forbidden Planet opens with the crew of a spaceship from Earth making preparations to land on Altair-4, the fourth planet in the Altair star system. Upon arrival, they notice a cloud of dust approaching the ship, stirred up by a fast-moving vehicle, but with no apparent driver. When the vehicle stops, the crew are greeted by an extremely polite robot, named Robby, who has been instructed to transport several of them to the residence of Morbius, the sole survivor of a previous expedition from

Einstein's Field Equations of General Relativity (1916)

$$G_{\mu\nu} = \frac{8\pi G}{c^4} T_{\mu\nu}$$

(Curvature of spacetime, denoted by $G_{\mu\nu}$, is proportional to the distribution of matter and energy, denoted by the *stress-energy tensor*, $T_{\mu\nu}$)

Cosmological Constant, Λ, added to keep the universe stable (1917)

$$G_{\mu\nu} + \Lambda g_{\mu\nu} = \frac{8\pi G}{c^4} T_{\mu\nu}$$

Hubble discovers the universe is expanding (1929)
Einstein declares the cosmological constant to be his "biggest blunder."

Riess, Perlmutter and Schmidt discover *dark energy* (1998)
Expansion of the universe is accelerating:
Cosmological Constant is needed, after all (but with opposite sign).

Fig. 3.2 Dark energy, the expansion of the universe, and Einstein's cosmological constant

Earth. Morbius (Walter Pidgeon) explains that the rest of his crew all died a mysterious and violent death, but that he and his beautiful daughter, Altaira (Anne Francis), seem to be immune. Unprepared for this startling revelation, the second expedition must contact Earth for further instructions. But according to the commanding officer, a major obstacle to communicating is the power required to send a signal over such a large distance. A transmitter of this sort is not exactly standard equipment onboard ship. In order to construct a sufficiently powerful transmitter, they must remove the core of the nuclear reactor, along with a substantial fraction of the ship's electronic equipment, and build it outside the ship. Safety standards require that the reactor be shielded, in order to reduce the crew's exposure to radiation (a topic which we will explore later), but digging an appropriate bunker to house the core would be very time-consuming. Morbius, concerned about getting on with his research, wishes to expedite the process. He offers the services of his robot, Robby, who has the remarkable ability to synthesize any material, in any desired quantity. The material, in this case, is lead shielding, and Morbius promises that the crew will have what they need no later than noon the following day.

In the next scene, we find Altaira sitting in the passenger seat of Robby's vehicle, and Robby carrying a stack of large slabs of lead shielding in one outstretched arm, toward the construction site. When one of the crewmembers questions his ability to do this, Robby offers the following explanation: "Common lead would have crushed the vehicle, sir. This is my morning's run of isotope 217. The whole thing hardly comes to ten tons" [7]. Unless the laws of physics work differently on Altair-4 than they do on Earth, this statement is problematic for at least two reasons.

The obvious problem with Robby's explanation has to do with the mass of his special isotope, and the implied claim that it would not crush the vehicle, whereas "common lead" would have been too heavy to transport in a single trip. Using standard notation, an isotope is specified by the atomic symbol (the abbreviation of the name of the atom on the Periodic Table of the Elements), plus two numbers, Z and A. The *atomic number*, Z, written as a subscript in front of the atomic symbol, indicates the number of protons in the nucleus. The *mass number* (or *isotope number*), A, written as a superscript, represents the total number of protons plus neutrons in the nucleus. For example, the most abundant element in the universe is also the simplest of the isotopes: the hydrogen nucleus consists of a single proton and no neutrons (Z = 1 and A = 1). Using isotope notation, hydrogen can be represented as $^{1}_{1}H$. The most common isotope of the second element in the periodic table, Helium has two protons and two neutrons (Z = 2 and A = 4) and is written as $^{4}_{2}He$. Several different stable isotopes of lead occur naturally, including isotopes 204, 206, 207, and 208, all having the same number of protons in the nucleus (Z = 82). The most abundant isotope of "common lead" is 207, which would be written as $^{207}_{82}Pb$. Using the same convention, Robby's special isotope of lead, *isotope 217*, would be written $^{217}_{82}Pb$. Each nucleus of Robby's special isotope has ten extra neutrons, making it roughly 5% more massive, and therefore *more* likely to crush the vehicle, than "common lead".

Robby's special isotope 217 presents not only a transportation safety problem (the weight of the material would exceed the carrying capacity of the vehicle). It also presents a serious problem, as far as radiation safety is concerned. Recall that the original purpose of making the lead shielding was to protect the crew from the dangerous radiation produced by the ship's nuclear reactor. Of all the conceivable combinations of protons and neutrons in a nucleus, relatively few of them are inherently stable. If a nucleus has too many neutrons (or too few), it will be *radioactive*. That is to say, the nucleus will decay into a different isotope, emitting one or more particles of ionizing radiation. In some cases, the energy released can be partially converted into a useful form, as in a nuclear power plant, or used deliberately for destructive purposes, as in a nuclear weapon. If the radioactive material is unshielded, the radiation can do serious damage to biological tissue. All of the known isotopes of lead above 208 are radioactive. No isotope 217 of lead has ever been created. But if it were possible to do so, it would not be very useful as a shield against radiation, since it would be radioactive, itself. We will return in a later chapter to the specific topic of the interaction of ionizing radiation with biological tissue and the hazards of exposure to radiation.

3.4   Gases

As mentioned previously, only 5% of all the matter and energy in the universe is the sort of stuff that we interact with in everyday life. The remaining 95% of the universe is *dark matter* or *dark energy*—stuff that we don't yet understand. The primary focus of our attention in this chapter will be on interactions of ordinary matter and energy and on selected properties of solid materials. But most of the 5% of the universe that we do understand is not in the form of a solid. Rather, it exists in the form of a gas—a collection of atoms or molecules whose energy of motion is too great for them to condense into a liquid or freeze into a solid. In this section, we consider several scenarios, both from science fiction and from real-life, which involve the escape of a gas from its container. These scenarios range from the very slow leaking of air out of a small hole in a large spaceship, to explosive decompression of a room in a Martian habitat, when an exterior door is accidentally opened into the extremely low pressure of the Martian atmosphere.

3.4.1   *The Micrometeoroid Air-Leak Scene in* Mission to Mars

Mission to Mars includes a tension-filled scene, in which the hull of a space ship, en route to Mars, is punctured in several places by micrometeoroids—small bits of solid material, which could pose a significant hazard to human spaceflight in the solar system. The air begins to leak out, and the ship's computer alerts the crew that they have only minutes before all the air will be gone. Several of the small holes in the hull are located and patched quickly, but one remains to be found. One of the crewmembers goes on a spacewalk to try to find the last remaining hole by visually inspecting the hull from the outside. But the ship is quite large, and the hole is only about a centimeter in diameter. He soon concludes that it's like looking for a needle in a haystack. Another of the crew has the brilliant idea to squeeze out the contents of a packet of a dark-colored, carbonated beverage inside the ship. Under ordinary conditions, this would simply have made a mess on the floor of the ship. But the ship's artificial gravity has been temporarily disabled. So the squeezed-out blobs of brown liquid float across the cabin, being sucked along with the rapidly escaping air, until they reach the hole. The crewman inside is able to radio to the crewman outside, telling him where to look. The visual clue to the location of the leak is an icicle of the now frozen carbonated beverage, protruding from the hull. The last remaining hole is patched, and the air pressure inside the ship begins to return to normal, all within a matter of minutes from start to finish [8].

Is this a plausible scenario? Would all the air inside a space ship the size of a small house really leak out through a 1-cm hole in such a short time? The analysis requires an understanding of the physical meaning of the pressure of a gas, and what happens when a hole is opened in the wall of a container at atmospheric pressure, surrounded by the near-vacuum of space.

The pressure inside a container of gas is created by the gas molecules colliding and bouncing off the walls of the container. This pressure is experienced not only by the walls of the container, but also by any object inside the container, due to collisions from the gas molecules. If there is a hole in the wall, and a gas molecule happens to be traveling in the direction of the hole, it will go through the hole and escape from the container, instead of bouncing back inside. Now imagine that the wall separates this container from another container, filled with the same gas, at the same temperature and pressure. Some of the gas molecules from the first container will escape through the hole into the second container, but at the same time, some of the gas molecules inside the second container will escape through the same hole into the first container. On average, the number of molecules escaping through the hole will be the same in both directions, with no net change in the pressure in either container. But now, instead of being connected to a second container at the same pressure, suppose that the first container is in the near-vacuum of space. Any molecule which escapes through the hole in the wall will never return, and there are no gas molecules on the other side of the hole to take its place. So eventually, all of the gas molecules will bounce around inside the container at random, until they find the hole and escape. But how long will this take? In particular, will it be only a matter of minutes, as suggested in the movie, or will it take much longer than this?

If we want to come up with an estimate of how long it will take for say half of the gas molecules in the entire space ship to escape through the hole, we might first ask how long it will take for a typical molecule moving near the hole to find the hole and escape.

Figure 3.3 is an illustration of gas molecules moving near a hole of area A, in the hull of the space ship. The molecules throughout the ship are moving in random directions. The figure only shows six molecules that happen to be moving near the hole, along the three coordinate axes: four of them are moving parallel to the wall (left-right or up-down) and two are moving perpendicular to the wall (toward or away). Only one of these six molecules—the one that happens to be moving toward the hole—will actually escape. The others will continue to move inside the ship, maybe colliding with objects inside the ship, or with other molecules, or other walls of the ship, until they eventually come back into the neighborhood of the hole. The motion of gas molecules inside a container is a random process. But we can think about what happens *on average*. If we want to know how many of the molecules (on average) will escape through the hole, we only need to consider 1/6 of the total number of molecules near the hole, namely the ones that are moving not only in a direction perpendicular to the wall, but actually toward the hole (and not away from it). Whether or not a particular molecule will actually escape through the hole in a small time interval (let's call it Δt) depends on how close it is to the hole and how fast it is moving. The average speed, v, of an object is defined as its change in position (Δx) divided by the corresponding change in time (Δt):

$$v = \frac{\Delta x}{\Delta t}. \tag{3.1}$$

Consider the air molecule in Fig. 3.3, which is traveling toward the hole. That molecule will escape through the hole within a time Δt if its speed, v, is large enough for it to travel the length of the little cylinder, Δx, in that time interval. We can now estimate the number of molecules that will escape in a small time Δt, if we know the density of molecules (number per unit volume in the ship) and the average speed of the molecules. Suppose the total number of air molecules in the entire ship is n, and the total volume of the ship is V. The small number of molecules, Δn, which will

Fig. 3.3 Six gas molecules (represented by black dots) moving near a hole of area A, in the outer wall of a space ship. Only the molecule traveling toward the hole will escape through the hole

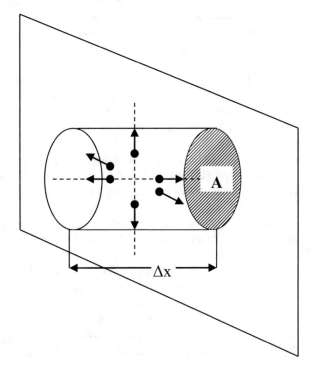

escape through the hole in a small time interval Δt, is approximately 1/6 of the number of molecules within the little cylinder in Fig. 3.3. This is just the number per unit volume (n/V) multiplied by 1/6 of the volume of the little cylinder, or

$$\Delta n = \frac{n}{V}\frac{A}{6}\Delta x. \tag{3.2}$$

If we solve Eq. (3.1) for Δx and substitute into Eq. (3.2), we get

$$\Delta n = \frac{n}{V}\frac{A}{6}v\Delta t. \tag{3.3}$$

Within the framework of *classical* physics (that is, not *quantum mechanical* and not *relativistic*), the kinetic energy, K, of an object of mass m and velocity v is given by

$$K = \frac{1}{2}mv^2. \tag{3.4}$$

Also within the framework of classical physics, the translational kinetic energy, K, associated with the motion of a gas molecule through three-dimensional space is given by

$$K = \frac{3}{2}kT, \tag{3.5}$$

where k is Boltzmann's constant (a number that can be found in tables of physical constants) and T is the temperature of the gas. We can combine Eq. (3.4) with Eq. (3.5) to obtain an expression for the speed, v, of a typical air molecule:

$$v = \sqrt{\frac{3kT}{m}}. \tag{3.6}$$

Substituting this into Eq. (3.3), we obtain

$$\Delta n = \frac{nA}{6V}\sqrt{\frac{3kT}{m}}\Delta t. \tag{3.7}$$

We are now in a position to ask the question, how long would it take for half of the air molecules to escape? (In the movie scene, the computer calls it "50% atmosphere.") In effect, we are asking what is Δt if $\Delta n = n/2$? So we solve Eq. (3.7) for Δt:

$$\Delta t = \frac{3V}{A}\sqrt{\frac{m}{3kT}}. \tag{3.8}$$

Before doing the final calculation, it's best to look carefully at the equation and see if it makes sense. The quantity V/A (volume/area) has dimensions of length. The quantity kT has dimensions of energy (mass•length^2/time^2), so the quantity m/kT has dimensions of (time^2/length^2). Taking the square root gives (time/length), and the length cancels. So the final answer will come out in dimensions of time, as desired. So it makes sense dimensionally. But does it make sense intuitively? If the volume of the ship, V, is larger, it makes sense that it should take longer for the air to leak out. If the hole is larger, it will take less time. If the mass of the molecule is larger (at constant temperature), the speed of the molecule will be lower, and it will take longer (on average) to find the hole. Finally, at higher temperature, the speed of the molecules will be greater, and the process will take less time. So all of the effects seem to be in the right direction.

It's safe to assume that the atmosphere inside the ship was a typical combination of oxygen and nitrogen, and that the temperature, T (expressed in units of Kelvin), inside the ship was something comfortable for humans. We can look up values for the mass of a typical air molecule (in units of kilograms) and a value for Boltzmann's constant (in units of Joules per Kelvin). In order to come up with an estimate of the time, all we need is some reasonable estimates of the volume of the space ship (in units of cubic meters) and the area of the hole (in units of square meters). The exact result, of course, depends on the particular assumptions about V and A.

Estimation 3.1: *Mission to Mars* **Air Leak**

Using visual information from the movie scene, make some reasonable assumptions about the volume of the space ship (length times width times height), the size of the hole in the hull (area), and the temperature of the air inside. Be sure to use a consistent set of units (all lengths in units of meters, temperature in units of Kelvin). The value for Boltzmann's constant, k, in the SI system of units is $k = 1.38 \times 10^{-23}$ J/K. Assuming that air is mostly nitrogen, the mass, m, of a typical air molecule can be taken to be approximately 28 times the mass of a proton, or $m = 4.68 \times 10^{-26}$ kg. Use Eq. (3.8) to calculate the time, Δt, for half of the air to leak out of the hole. The answer should come out in units of seconds but feel free to convert to more convenient units (minutes or hours, whatever you prefer).

So our conclusion, based on a very rough calculation, is that the air leak scene in *Mission to Mars* is greatly exaggerated. It will actually take much longer than just a few minutes for the air to leak out of a space ship of the size shown in the movie, through a hole of about 1 cm in diameter. But taking hours to find an air leak does not make for very good drama.

Before we leave the example of *Mission to Mars*, it would be appropriate to address the question of whether or not the carbonated beverage that was squeezed out of the packet in the scene would actually freeze into an icicle, as soon as it escaped through the puncture in the ship's hull. Later in this chapter, we will consider solid–liquid phase transitions in detail. But here, suffice to say, the more likely scenario would be for the liquid to begin to boil as it escaped through the hole. Anyone who has gone camping at high altitude knows that water boils at a lower temperature, when the atmospheric pressure is lower. There are two ways to boil water. The usual way is to add heat, and to raise the temperature of the water up to the boiling point. But another way is to reduce the pressure. Putting a container of water under a vacuum jar, and pumping out the air, will make the water start to boil. As it boils, it loses heat, and the temperature of the water will fall. Reduce the pressure enough, and the water will cool until it freezes. Yes, you can actually freeze boiling water! A demonstration of this effect can be found on YouTube [9]. For a few cubic centimeters of water, the process can take a couple of minutes, from room temperature liquid, to boiling and cooling, and finally boiling and freezing. The point of this is that a carbonated soft drink, which is mostly water, would not emerge through a hole in a punctured spaceship and freeze instantly into an ice sickle. It would be more likely to emerge as a jet of boiling droplets, which would eventually freeze into bits of ice.

3.4.2 Air Leak on the International Space Station

A real-life scenario, in some respects similar to the air leak scene in Mission to Mars, took place aboard the International Space Station in August of 2018. The Flight Control team monitored a small drop in air pressure and notified the crew of the ISS to begin to investigate a possible air leak. Early reports in the news media suggested that the leak might have been the result of a micrometeoroid puncture, just like in the scene from *Mission to Mars*. The crew of the ISS narrowed down the location of the air leak by sealing off the various modules, one by one. The actual source turned out to be hole in a Soyuz space capsule, docked with the ISS. The hole was not a micrometeoroid puncture from the outside, as was initially reported. Instead it was accidentally created from the inside by a drill bit of approximately 2 mm in diameter, while the capsule was still on the ground undergoing maintenance.

Estimation 3.2: Air Leak on the International Space Station

Use Eq. (3.8) to calculate the time, Δt, for half of the air to leak out of the hole that was discovered in the International Space Station, if the astronauts had been unable to find and patch the source of the leak. To calculate the area, A, of the hole, use the estimated diameter of 2.0 mm. For the volume, V, of the ISS, visit NASA's website of facts and figures to find the *total pressurized volume* of the International Space Station [10]. Express the final result in units of hours or days.

3.4.3 Explosive Decompression

The overly dramatized micrometeoroid scene in *Mission to Mars* and the real-life air leak on the International Space Station both involve very small holes in the wall of a rather large container of gas. The result is a very slow leak, which takes place over many hours. But imagine the opposite scenario, in which a rather large hole suddenly opens in the wall of a much smaller container. The result is what we might call explosive decompression: all of the air will escape in a very short time. This is illustrated fairly realistically in an episode of the National Geographic television series *Mars*. A revolutionary kind of TV series, *Mars* is a hybrid science fiction/documentary, which weaves together the fictional story of human settlements on Mars and interviews with some of the key figures in the real-life science and technology involved. In the sixth episode of the first season, a member of the first human colony on Mars, while suffering from mental illness, inadvertently opens a door from the Mars habitat into the very thin Martian atmosphere. The result is explosive decompression of his module and the death of several members of the colony [11]. Is this plausible, or is it just science fiction? Mars does have an atmosphere, but the pressure is less than 1% of sea level pressure on Earth (the pressure, in SI units, is about 600 Pa on Mars, compared to about 101,300 Pa on Earth). The magnitude of the pressure change makes the effect plausible. What is completely implausible is the design of the Martian habitat, as shown in the series. It should not be so easy for someone to inadvertently open a door to the outside.

Discussion Topic 3.1

Consider the following scene from the 2013 movie *Gravity*. Space Shuttle astronaut Dr. Ryan Stone (Sandra Bullock) finds herself adrift in a Soyuz space capsule, following a catastrophic accident, in which her Space Shuttle was destroyed. A fellow astronaut, Matt Kowalski (George Clooney), whom she had assumed was lost or dead, appears outside the window of the capsule. Kowalski opens the hatch from the outside and climbs in. Stone was not wearing her space helmet at the time, but somehow she survives and the two of them carry on a conversation about their chances of making it back to Earth alive [12]. Given what you know about air escaping through a hole in a pressurized container, explain why this scene cannot be depicting an actual event but must be either a dream or a hallucination.

3.5   Liquids

Before we turn out attention to properties of materials in the solid state, which is the primary focus of this chapter, we will briefly touch on the phase of matter in between gases and solids. The behavior of a liquid is dramatically illustrated in a scene from the 2016 movie *Passengers*. Jennifer Lawrence plays the role of Aurora, one of 5000 passengers on an interstellar space ship, all of whom are supposed to be asleep in hibernation pods for the 120-year trip. Aurora is awakened from hibernation about 90 years ahead of schedule. How will she pass the time? Among other forms of physical activity, she enjoys taking a swim in the ship's pool. The pool is located in a section of the ship, which rotates to simulate gravity (recall our discussion of Einstein's principle of equivalence in the previous chapter). In the scene in question, the rotation malfunctions and Aurora finds her life in jeopardy. With the sudden loss of rotation and simulated gravity, the inertia of the water causes it to slosh violently out of the pool. With no sense of which direction is up and which is down, Aurora struggles to find the surface of the water. Fortunately, power is restored before she drowns [13]. Let's consider some of good science in this scene.

A liquid is a type of *condensed matter* (i.e., not a gas) which does not have any long-range order to the arrangement of atoms or molecules and with no *yield stress*. That is to say, when any force, no matter how small, is applied to a liquid, it instantly deforms in response to that force. This behavior is in contrast to other forms of condensed matter, such as solids, which require some minimum force (yield stress) to make them deform.

An understanding of the plight of Aurora in the swimming pool scene of *Passengers* requires consideration of three types of forces associated with the motion of an object through a liquid. Although the shape of a liquid will deform with even the slightest force, there is resistance to the motion of an object through a liquid, which is referred to as *viscous drag*. The drag force is associated with the rearrangement of the molecules of the liquid and depends on the speed, shape, and surface area of the moving object. In order for a swimmer to move forward through the water, the drag in the forward direction must be less than the drag in the back-ward direction. The motion of a swimmer's arms in a crawl stroke or backstroke serve this purpose well when the swimmer is at the surface of the water, since for-ward motion of the arms through the air presents much less resistance than back-ward motion of the arms through the water. However, when the swimmer is completely under the water, a breaststroke would be more efficient.

A second force to consider is gravity. In the presence of *gravity,* a liquid tends to conform to the shape of its container. Gravity also creates a buoyancy force on a swimmer's body, due to the weight of the displaced water.

The third force to be considered is due to *surface tension*. Surface tension can be understood as the energy per unit of surface area between two phases (e.g., between the liquid and the surrounding air or between the liquid and a solid surface). The effect of surface tension to minimize the total energy associated with all of the surfaces. For any given volume, a sphere has the minimum possible surface area. Thus, in the absence of gravity, a free drop of liquid tends to gather into the shape of a sphere, under the influence of surface tension. An entertaining demonstration of this effect was performed by NASA astronaut Scott Kelly on his year-long stay aboard the International Space Station [14]. But what will a liquid do in the absence of gravity when it is in contact with a solid object? Depending on the surface tension between the liquid and the solid, the liquid may actually spread to cover more of the solid, rather than drifting away from the solid to form a sphere of liquid. In 2013, some clever Canadian high school students designed an experiment that illustrates this concept, and for which they won a national science contest, held by the Canadian Space Agency. Performed on the International Space Station by CSA astronaut Chris Hadfield, the experiment involved soaking a washcloth with water and wringing it out in the microgravity environment of the ISS. The result: the water was squeezed out of the washcloth, but simply spread out over the exposed surface of the washcloth and over the astronaut's hands [15]. This surprising experiment suggests that the plight of Aurora is even more serious than simply finding her way to the surface of the water. In the absence of gravity, there is no buoyancy force and therefore no way to discern up from down. But a second problem arises. Surface tension now becomes the dominant force. Even if Aurora did manage to find her way to the surface of the water, she would still find it difficult to breathe. Her face would remain covered by water under the influence of surface tension. In short, if gravity had not been restored, Aurora would have drowned.

3.6   Solid State Materials

We now turn our attention to the interaction of electromagnetic energy with solid state materials. The opening scene of "Gambit, Part I", the first of a two-part episode of *Star Trek: The Next Generation*, provides a good starting point for our discussion [16]. Captain Jean-Luc Picard of the U.S.S. Enterprise (played by Patrick Stewart) has gone missing while on shore leave, in pursuit of one of his favorite hobbies: archeology. Several of the officers of the Enterprise are questioning patrons of a rather seedy drinking establishment, in hopes of finding a clue as to his whereabouts. One of the patrons offers information in exchange for safe transport away from the planet. He relates the details of a fight, which broke out between the Captain and some other aliens, in which the Captain was thrown against a wall, and fell to the floor. The ship's doctor, Beverly Crusher (Gates McFadden), using a tricorder (a futuristic, multipurpose, hand-held device for recording data and analyzing materials), detects traces of Star Fleet fibers (presumably from the Captain's clothing), distorted cellular debris (perhaps from the Captain's body), and something that she refers to as *microcrystalline damage* in the material of the floor, which might be evidence of a high-energy weapon discharge. The informant then claims that one of the aliens took out a weapon and fired: "He was vaporized!" Fortunately, the Captain was not actually vaporized. Rather, the weapon discharge triggered a transporter beam, which teleported the Captain to the aliens' orbiting spaceship. But the scene raises the question, what, exactly is *microcrystalline damage*, and how might it be detected?

A *crystalline* material, as distinct from an *amorphous* material, is characterized by atoms arranged in a regular, three-dimensional array, with long-range order (Fig. 3.4). The same pattern of arrangement of atoms repeats over and over again, building up a macroscopic piece of matter. Some easily recognized crystalline materials include table salt (sodium chloride), and diamond (carbon). The typical separation between neighboring atoms in a crystal is a few Angstroms (1 Å = 10^{-10} m, or one ten-billionth of 1 m). In a crystal of macroscopic size (say 1 cm on an edge—something that you could easily hold in your hand), there would be something like 10^8 atoms along an edge of a 1-cm cube. Since the volume of a cube is just the edge length cubed, that gives 10^{24} atoms in the 1-cm crystal.

What then is a *microcrystal*? You might imagine that a *microcrystal* would be a piece of crystalline material of microscopic size. So how many atoms are contained in a microcrystal? If we assume, once again, that the typical interatomic spacing is something like 10^{-10} m, and that a microcrystal is a cube 10^{-6} m on an edge, we get 10,000 atoms along an edge. Since the volume of a cube is the edge length cubed, we

Fig. 3.4 Model of atoms on a repeating crystal lattice, with cubic symmetry. (Photo by the author)

have 10^{12} atoms in the microcrystal. Despite its small size, the interesting physical properties (electrical, magnetic, optical, and so forth) are not significantly different from the properties of the same material in the bulk (macroscopic crystal). So it is unlikely that Dr. Crusher's tricorder is detecting the presence of microcrystals in the material of the floor. So back to the original question: What is *microcrystalline damage*, and what might be the signature of this effect, enabling it to be detected by the tricorder?

In particles of sufficiently small size, the electrons behave differently than they do in macroscopic (or even microscopic) materials. In effect, they sense that they are confined in a small space. One way of thinking about this is to imagine yourself in the interior of a large crystal with simple cubic structure (same separation between identical atoms along each of the three coordinate axes). No matter which direction you move through the crystal, the world looks exactly the same, until you get to the surface of the crystal. Suddenly, the world looks different outside the crystal than it does in the interior. Electrons can sense that they are confined in a small space if the number of atoms on the surface of the material is a significant fraction of the total number of atoms. This is the case in materials at the nanoscale (edge lengths of approximately 10^{-9} m). As a result of the significant surface effects, the physical properties of the material (electric, magnetic, optical, etc.) can be measurably different from the same properties in bulk material. So instead of *microcrystalline* damage, the effect that Dr. Crusher was detecting might more appropriately be called *nanocrystalline* damage. In other words, the energy of the weapon disrupted some of the material down to the nanoscale, resulting in measurably different physical properties.

Fig. 3.5 Model of a Carbon-60 molecule (Buckyball). (Photo by the author)

Estimation 3.3: Atoms Inside and on the Surface of a Nanoparticle

Estimate the number of atoms in a nanoparticle (a particle of matter, whose size is approximately 10^{-9} m). Of the total number of atoms in the nanoparticle, how many of the atoms are on the surface of the particle, compared to the number in the interior?

An extreme example of a nanoparticle is the relatively recently discovered Buckyball—a stable molecule consisting of exactly 60 carbon atoms, arranged in a cage-like structure, resembling a soccer ball (Fig. 3.5). All of the atoms are on the surface, with nothing in the interior. This structure may be compared to the two crystalline forms of carbon, graphite, and diamond. Although these materials are made of the same kind of atom, graphite and diamond have different crystal structures (hexagonal and diamond-cubic, respectively), giving them physical properties which differ significantly from each other, and both are significantly different from Buckyballs.

3.6.1 Density of Solids

We pause for a moment to answer a question posed in the previous chapter. Why is it that the military hovercraft in the James Bond movie *Die Another Day* could float over a minefield without triggering the mines, but the giant attack ships in *Independence Day* would crush an entire city, just by hovering over it? Did you find the flaw in the analysis? The answer lies in the density of the mother ship in *Independence Day*, which we never actually needed to calculate separately, in order to solve the original problem. The standard approach to solving a physics problem is to work out the algebra first and then to plug in the necessary numbers once you have the algebraic expression for the quantity that you want. Although the assumptions we made were reasonable, some of the data provided in the movie were completely implausible. Since we were interested in the pressure under the attack ship, we didn't stop along the way to calculate the density. Let's do that now, and compare the density of the mother ship in *Independence Day* to the density of real solid materials. Recall that the diameter of the mother ship was given in the movie as 550 km, and the mass was ¼ the mass of the Moon. A reasonable assumption is that the mother ship is approximately spherical in shape. We can calculate the density as follows:

$$\text{density} = \frac{\text{mass}}{\text{volume}}$$

$$= \frac{\frac{1}{4}\left(\text{mass of moon}\right)}{\frac{4}{3}\pi\left(\frac{\text{diameter}}{2}\right)^3}$$

Substituting the data given in the movie:

$$= \frac{3\left(7.35\times10^{22}\text{ kg}\right)}{2\pi\left(550{,}000\text{ m}\right)^3}$$

This gives us the density of the mother ship of approximately 211,000 kg/m³. Compare this to Osmium, which has the highest density of any known material: 22,600 kg/m³. So given the data in the movie, the mother ship would have an average density ten times greater than any known material on Earth.

3.7 Phase Transitions

The illustration of crystal structure, as suggested by Fig. 3.4, is inadequate in at least one important respect. The atoms in any solid are not rigidly connected to one another, as balls on rigid sticks. They are more like balls attached to one another by flexible springs. The springs represent the interatomic bonds, which hold the atoms together in solid form. At any finite temperature, the atoms are constantly vibrating. The higher the temperature, the greater the vibration will become. At sufficiently high temperature, the vibrations may exceed the ability of the interatomic bonds to hold the material together. The long-range order of the crystal breaks down, and the solid undergoes a *phase transition* into a liquid. To gain appreciation for how this can happen, we turn our attention to a film franchise, which is not normally associated with science fiction.

The 1964 James Bond film *Goldfinger* not only provides a good illustration of melting a solid, but also serves to show how quickly state-of-the-art technology can find its way into science fiction. The title character Auric Goldfinger (Gert Fröbe) is suspected of smuggling large quantities of gold but nobody can figure out how he is doing it. In the process of solving the mystery, British intelligence agent James Bond (Sean Connery) uncovers a plot to break into the United States gold depository at Fort Knox. Following a high-speed car chase in his iconic Aston Martin DB5 (equipped with numerous modifications, including machine guns, tire-slashing hub caps, oil slick spray, bullet-proof windscreens and ejector seat), Bond is captured, tied to a slab of solid gold, and about to be bisected with an industrial laser (Fig. 3.6). Goldfinger explains that his new toy "…emits an extraordinary light not to be found in nature. It can project a spot on the Moon, or, at closer range, cut through solid metal." "Do you expect me to talk?" Bond eventually asks his captor. Goldfinger replies, "No, Mister Bond, I expect you to die!" [17].

In addition to the high-tech gadgetry and clever one-liners, the movie is notable for presenting to the public an important piece of real-life technology, albeit in rather cartoonish form, in the same year that it was invented. The first working visible light laser was developed just 4 years earlier, in 1960, using a ruby rod as the laser medium. The accomplishment is generally attributed to Maiman, although there is some controversy over who really achieved true lasing action first [18]. The ruby laser emits a red beam. But as we shall see later in this chapter, visible light scatters from small particles. So a visible light laser is not very effective for industrial cutting purposes because of the vapor thrown up by the material being cut. Four years later (the same year that *Goldfinger* was released!), Kumar Patel developed a CO_2 laser, which operates in the infrared [19]. Infrared light scatters much less than visible light and can be made into a very efficient precision cutting tool. However, as pointed out by Gresh and Weinberg in *The Science of James Bond*, CO_2 lasers were not actually used in industry until 1966, 2 years after *Goldfinger* was released [20]. Thus, we have a case of Hollywood science predicting real science: a Bond villain was in possession of a bit of technology 2 years before it was available to the public!

Goldfinger's laser features a glowing, pulsating coil, wrapped around the source of the beam, suggestive of the flash lamp, which pumps a real ruby laser. The power supply and controls for this remarkable device occupy a small room, which is realistic for the time. The red beam, emanating from the laser, and visible from any direction, is a standard special effect, but not realistic. Although the laser spot would, indeed, be visible at the point of impact on the target, the beam itself would not be visible from the side, unless the room was filled with smoke, or other small particles, to scatter the light.

Thirty-one years and four leading actors later, the 1995 movie *Goldeneye* features a new bit of laser technology, not to be found in reality [21]. James Bond (Pierce Brosnan) finds himself trapped in an armor-plated railroad car, with beautiful Russian computer programmer, Natalya Simonova (Izabella Scorupco), and a bomb set to go off in 3 min. They manage to escape, thanks to a clever gadget: a laser small enough to be concealed in a wristwatch, yet powerful enough to cut through the floor plate.

The science of the process can be described in simple terms. The energy of the laser beam, in the form of light, is absorbed by the metal of the floor plate, converting it into heat. The heat energy raises the temperature of the metal up to the melting point. Then more energy is needed to convert the solid metal into liquid. The liquid metal drips harmlessly onto the ground, underneath the railroad car. The laser beam is slowly moved along the floor plate, melting more metal, and tracing the outline of a rectangle large enough for a person to jump through to safety. How powerful would a laser have to be to do the job?

We already know from our discussion of *Goldfinger* that industrial lasers powerful enough to cut through solid metal have existed since 1964. But a more interesting technological question might be could a laser of this sort be powered by a battery in a wristwatch? We can begin to address the question by looking at the definition of *power*. Power is the ratio of the energy delivered over an interval of time, or

$$\text{Power} = \text{Energy} / \text{Time}. \tag{3.9}$$

Fig. 3.6 An industrial laser, designed to cut through solid metal, may be equally useful for disposing of spies

In the situation that we are considering, the energy is being delivered by the laser in the form of heat. The *heat energy*, Q, needed to turn a solid into a liquid is given by

$$Q = mC\Delta T + mL, \tag{3.10}$$

where m is the mass of material to be melted, C is the *specific heat capacity*, ΔT is the change in temperature, and L is the *latent heat of fusion*. The change in temperature, ΔT, is the difference between the *melting point* (the temperature at which the metal begins to melt) and the initial temperature of the metal.

Before we can use Eq. (3.10) to calculate the heat energy, Q, we will have to gather some information from the movie scene and make some reasonable assumptions about what is happening. The first important clue given in the movie scene is Bond's description of the material of the railroad car: "1-in. armor plating." We don't know exactly what kind of material it is, but we can make a reasonable assumption that it's some kind of steel. We can then look up all of the thermal properties of the material (the specific heat capacity, the melting point, and the latent heat of melting) in tables of physical constants. The only thing in Eq. (3.10) that we still need is the mass, m, of the material to be melted. This isn't given explicitly, but we can estimate it from the size of the channel that Bond has to cut through the floor. We can start from the definition of *density*, the ratio of mass over volume:

$$\text{density} = \text{mass} / \text{volume}. \tag{3.11}$$

The volume is just the product of total length, width, and depth of the channel that must be cut through the floor. The depth of the channel is given in the movie scene as 1 in. The other two dimensions can be estimated by making a reasonable guess about the width (diameter) of the laser beam and the total length of channel that Bond needs to cut, to make a big enough opening in the floor. So the mass, m, needed in Eq. (3.10) can be calculated from Eq. (3.11) by looking up the density of the material in a table of constants, and multiplying by the volume. This gives enough information to calculate the heat energy, Q.

The final important piece of information the time interval, needed to calculate the power of the laser. The bomb is set to go off in 3 min, so the Time, in Eq. (3.9), must be less than 180 s. How much less depends on how much time you want to allow for Bond and Natalia to run to safety before the explosion.

Example 3.1: Escape from an Armor-Plated Railroad Car Using a Wristwatch Laser

We can use the information provided in the wristwatch laser scene from *Goldeneye*, and a few additional assumptions, to calculate the power required for James Bond's wristwatch laser to cut through the floor of the armor-plated railroad car before the bomb explodes.

We will work with Eqs. (3.9), (3.10), and (3.11). The first thing to determine is how much mass of metal has to be melted by the laser. From Eq. (3.11), this is just the volume times the density. And the volume is just the product of the total length of the channel times the width of the laser beam times the thickness of the floor. So we have:

$$\text{mass} = (\text{channel length}) \times (\text{beam width}) \times (\text{floor thickness}) \times \text{density}.$$

Let's assume that the railroad car is made of steel, with a density of 8000 kg/m³. The minimum channel length could be taken as 2 m (a square with edge length 0.5 m). The laser beam diameter is roughly 0.001 m (1 mm) and the floor thickness is 0.025 m (1 in.). This gives us a mass of 0.4 kg. Next, we need to determine the amount of heat energy, Q, needed to melt this much steel. Some approximate values of the physical constants for steel are given below:

Approximate values of thermal constants for steel

specific heat	$C = 490 \text{J} / \text{kg} \cdot \text{K}$
melting point	$T_{melt} = 1800 \text{K}$
latent heat of melting	$L = 2.7 \times 10^5 \text{J} / \text{kg}$

In Eq. (3.10), we need the change in temperature, ΔT, which is just the difference between the melting point and the initial temperature. Let's assume that the floor starts out approximately at room temperature (300 K). This gives $\Delta T = 1500$ K. Substituting these values into Eq. (3.10), along with the mass, $m = 0.4$ kg, we get

$$Q = m[C\Delta T + L]$$
$$= 0.4\,\text{kg} \left[(490 \text{J} / \text{kg} \cdot \text{K})(1500\text{K}) + 2.7 \times 10^5 \text{J} / \text{kg} \right]$$
$$= 402,000 \text{J}$$

Finally, we can use Eq. (3.9) to calculate the power required for the wristwatch laser to deliver this amount of heat in the necessary time. If the bomb is set to go off in 3 min, they need time not only to cut the hole in the floor, but also to jump through the hole and run to safety. So let's say 2.5 min to be safe. The power required is

$$\text{Power} = \text{Energy} / \text{Time}$$
$$\text{or } P = Q / t$$
$$= 402,000 \text{J} / 150 \text{s}$$
$$= 2680 \text{W}$$

This is typical of the power output for an industrial laser. But the important question is, could a laser this sort be powered by a battery in a wristwatch?

The process of melting a solid is reversible. We can also think about a phase transition in the opposite direction, from liquid state to solid state. This physical process is also described by Eq. (3.10), but the heat involved is negative. Instead of raising the temperature to the melting point, we lower the temperature to the freezing point (the change in temperature is negative). And instead of adding heat to melt the material, we remove heat to freeze the material. The reversibility of liquid–solid phase transitions is illustrated most dramatically in *Terminator 2: Judgment Day*.

The *Terminator* series of movies involve time travel: from a grim future, in which machines have developed self-awareness (the topic of the next chapter) and are in the process of exterminating the humans, back in time to what is more or less the present. In the first of the series, Arnold Schwarzenegger plays the role of a T-800 terminator robot—a relentless machine with biological skin, made to look exactly like a human, whose sole purpose is to kill real humans. It is sent back in time to kill the mother of the human rebellion against the machines, before the boy is born. The attempt fails, thus setting up the second in the series. In *Terminator 2*, a shape-shifting killing machine, known as the T-1000 liquid metal terminator, is sent back in time to kill the boy before he grows up to become the leader of the human rebellion. But the members of the rebellion have their own plans. They manage to capture and reprogram one of the older T-800 terminators (Arnold reprises his role) and send it back in time to protect the boy. The most memorable liquid–solid phase transition occurs at the end of an exciting chase scene, in which the T-1000 hijacks a liquid nitrogen tanker truck, and sets off in pursuit of the boy, his mother, and the T-800. With the intervention of the T-800, the tanker truck overturns and ruptures, spilling thousands of gallons of liquid nitrogen at a temperature of 77 K. The T-1000 emerges from the cab of the truck and continues pursuit on foot, only to be frozen solid by the surrounding liquid nitrogen. The T-800 draws a handgun, speaks four words to the frozen T-1000—"Hasta la vista, baby."—and fires a single shot, which shatters the T-1000 into a thousand pieces. Unfortunately, since phase transitions are reversible, and since the freezing took place in a steel mill, the heat of the molten steel quickly melts the shattered frozen bits, which slither across the floor and reassemble, allowing the T-1000 to resume pursuit of its target [22].

Thus far, the discussion has focused on phase transitions between solid and liquid. But we can also think about phase transitions from liquid to vapor. Recall the *Gambit* episode of *Star Trek: The Next Generation*, in which the alien informant revealed that Captain Picard had apparently been vaporized by some kind of weapon [16]. In just about any episode of *Star Trek*, which includes a scene in which someone is vaporized by a phaser weapon, the process takes a couple of seconds. By reasoning similar to Example 3.1, we could easily estimate how much power would be required to do this.

Estimation 3.4: Vaporizing Captain Picard

Use Eqs. (3.9) and (3.10), together with some reasonable assumptions (e.g., the human body is made of mostly water), to come up with an estimate of the power required for a phaser to vaporize an adult human (such as Captain Picard). Since we are talking about vaporization and not melting, the constants in Eq. (3.10) need to be chosen appropriately. Otherwise, you can use reasoning similar to Example 3.1 Some useful physical constants include the following:

> Approximate values of thermal constants for water
>
> specific heat of water $C = 4190 \text{J} / \text{kg} \cdot \text{K}$
>
> boiling point of water $T_{\text{boiling}} = 373\text{K}$
>
> latent heat of vaporization $L = 2.26 \times 10^6 \text{J} / \text{kg}$

Compare the power output of the phaser to the power output of a typical hydro-electric or nuclear power plant.

3.8   Transparency and Invisibility: Optical Properties of Solids

When visible light interacts with matter, a number of things can happen, depending on the nature of the object and the properties of the material that the object is made of. Some of the important effects include *scattering* from rough surfaces or small particles, *reflection* from smooth surfaces, *transmission* through the material, and *absorption* in the material. All of these effects are common, everyday experiences. Air molecules scatter shorter wavelengths of light more efficiently than longer wavelengths. Thus, the sky appears reddish at sunset because red light from the Sun is not scattered as much as blue light, as it passes through the atmosphere. For the same reason, the sky on a bright, sunny day appears bluish: more blue light from the Sun is scattered toward our eyes from the air molecules overhead. Large droplets of water vapor, which form clouds, tend to scatter all wavelengths of light equally well. So clouds on a sunny day appear white. (The Moon has no atmosphere to scatter sunlight, so the sky on the Moon is always black, and the stars are always visible.) Rough surfaces scatter light *diffusely* (random scattering in all directions), while very smooth surfaces can give a uniform reflection. But what material properties determine whether or not light can travel through a solid object, and is it possible to render a solid object completely invisible?

3.8.1   Transparent Solids

We return to *Star Trek IV: The Voyage Home*, which we discussed previously in the context of time travel. To recap the storyline, Earth's oceans are being vaporized by a giant probe, of unknown origin, which is trying to contact humpback whales. Unfortunately, by the twenty-third century, this species had been hunted to extinction. Thanks to Mr. Spock's calculations, the officers of the Enterprise, traveling in a stolen Klingon bird of prey, have successfully gone back in time to Earth, in the late twentieth century, in search of humpback whales. Lt. Uhura (played by Nichelle Nichols) detects whale song coming from the general direction of San Francisco.

They locate a pair of humpback whales, named "George" and "Gracie", conveniently sheltered in a marine park, in Sausalito. Their only problem now is how to transport the whales (and enough water to keep them alive) back to the twenty-third century, to appease the giant probe.

Part of the problem is an ethical one: the whales are not simply theirs for the taking. To quote Spock, "We would be as guilty as those who caused their extinction." The task falls to Captain Kirk to use his charm to convince Dr. Gillian Taylor (Catherine Hicks), the cetacean biologist responsible for the care of the whales, that it is in everyone's best interest (including the whales) to cooperate. The rest of the problem is technological: how to build a giant aquarium in the cargo bay of the spacecraft. Chief engineer Scott (James Doohan), with the assistance of Dr. McCoy (DeForest Kelley), poses as a visiting professor from Edinburgh on an invited tour, to study manufacturing methods at a nearby thermoplastics plant. Following the tour, conducted personally by plant manager Dr. Nichols (Alex Henteloff), Professor Scott makes an intriguing offer. He begins with a curious observation.

"I see you're still working with polymers."

"Still?" Nichols responds. "What else would I be working with?"

"Aye, what else, indeed?" Having captured Nichols' attention, Scotty continues, "How thick would a piece of your Plexiglass have to be, at 60 feet by 10 feet, to withstand the pressure of 18,000 cubic feet of water?"

"Oh, that's easy," Nichols replies, "Six inches. We carry it that big in stock."

"Aye, I've noticed. Now suppose, just suppose, I could show you how to manufacture a wall that would do the same job as your Plexiglass but would be only one inch thick. Would that be worth something to you?"

"You're joking."

"Perhaps the Professor could use your computer," suggests Dr. McCoy.

"Be my guest," Nichols offers, still skeptical.

Scotty approaches Dr. Nichols' Macintosh computer and addresses it with the standard Star Fleet command: "Computer." The Macintosh, which was state-of-the-art in personal computing in the mid-1980s, fails to respond to this twenty-third century verbal command. "Computer," Scotty repeats, but to no avail. Dr. McCoy picks up the mouse and hands it to Scotty, who addresses it yet again – this time by speaking into the bottom of the mouse – "Hello, computer." Still no response. Finally, Dr. Nichols suggests, "Just use the keyboard."

"A keyboard," Scotty says, with disdain, "How quaint." After cracking his knuckles in preparation, and quickly tapping away at the keyboard for a few seconds, he produces an animated image on the screen: the molecular structure of *transparent aluminum*. [23]

Dr. Nichols evidently found this offer sufficiently interesting, and potentially lucrative, that he agreed to a trade. Scotty got the sheets of Plexiglass, which he would later use to make the whale aquarium in the Klingon ship, in exchange for the formula for *transparent aluminum*.

But should Dr. Nichols have remained skeptical? Could a product made mostly of aluminum be a good substitute for Plexiglass, whose durability and transparency make it a popular substitute for glass? To answer this question we need to understand something about the behavior of electrons in solid materials.

Earlier in this chapter we discussed the details of the atomic nucleus, which contains most of the mass of an atom in the form of protons and neutrons. All atoms also have at least one electron in orbit around the nucleus, and *neutral* atoms (atoms with no net electric charge) have the same number of electrons in orbit as they have protons

Table 3.5 Electrons in solid state materials

Conduction electrons	Completely free to move throughout the material
Valence electrons	Shared between neighboring atoms: Interatomic bonds
Core electrons	Highly localized, in orbit around individual atoms

in the nucleus. When atoms come together to form a solid, some of the outermost electrons in the individual atoms become somewhat delocalized and are shared with neighboring atoms, forming interatomic bonds. We might refer to these as *valence* electrons. Other electrons may remain in orbit around their individual atoms, much as they would be in an isolated atom, by itself. We might refer to these as the *core* electrons. There are some solid materials in which these are the only two possibilities for the electrons, under ordinary conditions: all of the electrons either remain localized around individual atoms, or they participate in the interatomic bonds, which are necessary to hold the material together as a solid. We call these materials insulators.

But not all solids are insulators. In some materials, there is a third possibility for the electrons. Materials, which we classify as metals, have some electrons which do not remain in orbit around individual atoms, nor do they participate in interatomic bonds, but are completely free to move throughout the material. These are called *conduction* electrons because they make possible the conduction of electric current (Table 3.5).

The presence or absence of conduction electrons affects not only the electrical properties of a solid (conducting or insulating), but also the optical properties. Metals are opaque primarily because the conduction electrons readily absorb the energy of visible light. The light energy is converted into the kinetic energy of the electrons, which move through the material. Insulating materials, by contrast, have no electrons which are free to move. They are bound to individual atoms, or shared between neighboring atoms, in such a way that the light energy is not irretrievably absorbed by the material.

Ordinary glass is made primarily of silicon dioxide, which is an insulating material. It has no conduction electrons to absorb the energy of visible light. If the glass is made sufficiently pure and uniform (free of little specks of dirt or air bubbles, which will scatter the light, making it appear cloudy) it will be transparent, and suitable for making windows. Plexiglass (or poly-methylmethacrylate) is also an insulator. Although it is softer than glass, and somewhat more easily scratched, it can be polished and made as pure and transparent as glass. Aluminum, on the other hand, is a metal, and metals have conduction electrons, which absorb the energy of visible light. Pure aluminum is opaque. The only way to make aluminum transparent is to combine it chemically with other elements, into a molecular structure which is insulating. One such material is aluminum oxide (Al_2O_3), which in pure form is a completely transparent, crystalline mineral, known as corundum. (Add a bit of cobalt to Al_2O_3, and you get a blue gemstone called sapphire. Substitute chromium for the cobalt, and you get ruby.) But this could not be the *transparent aluminum* of the Star Trek movie. It would be next to impossible to manufacture a large wall of pure Al_2O_3. And even if it were possible, a 1-in. thick slab of such material could not withstand the pressure of 18,000 ft^3 of water.

But there is an aluminum compound which is so strong and sufficiently transparent that the military use it to make windows for tanks. A method of manufacturing transparent aluminum oxynitride was developed in 1984 (the same year that the Macintosh computer was released) and patented in 1985, just 1 year before Star Trek IV: The Voyage Home was produced [24]. This is one more example of how quickly state-of-the-art technology finds its way into science fiction.

Now back to the question of invisibility. Are there any technological tricks that can be played, or properties of materials that can be exploited, to fool an observer into thinking that an object is not there? The approaches generally fall into three categories: camouflage, stealth technology, and cloaking.

3.8.2   Camouflage

Camouflage is the art of making the object blend in with its surroundings. It can be as simple as painting it to blend in with a particular kind of surroundings, or as sophisticated as the adaptive camouflage on the Aston Martin V-12 Vanquish, modified by Q-branch, in the James Bond movie *Die Another Day* [25]. An array of micro-video cameras and projectors might be arranged on the surface of the object in such a way that an observer always sees an image of whatever is behind the object. Presumably, similar adaptive camouflage technology was used by the alien creature in the *Predator* movies [26]. But the system has its disadvantages. In *Die Another Day,* the Aston Martin showed a badly distorted image of objects that were very close, it still left tracks in the snow, and the system failed completely after being damaged by gunfire. In the *Predator* movie, the creature could be seen as a small, moving distortion against the background of trees.

3.8.3   Stealth Technology

Stealth technology, used on military aircraft, such as the F-117 fighter and the B-2 bomber, is designed to minimize the visibility of the object, but cannot render it completely invisible. It combines several principles. First, the aircraft is shaped in such a way as to minimize the amount of reflected electromagnetic radiation—particularly radar—that is scattered back toward the source. It is also coated with a material that is absorbing at radar wavelengths and has a dark, dull finish, to minimize reflection of visible light. Since it is easily seen against a bright sky, a stealth aircraft works best when flying at night. Stealth technology has showed up in a number of sci-fi movies, including *Independence Day* in which B-2 bombers are deployed as the nuclear strike force against the alien attack ships [27], and the James Bond movie *Tomorrow Never Dies,* in which a stealth boat launches a sea drill, to sink a British naval vessel, in an attempt to start a war between China and the U.K. [28].

3.8.4 Metamaterials and Cloaking

In *Harry Potter and the Sorcerer's Stone*, the first book and movie in the highly successful series, the young wizard receives a curious Christmas gift, which may be closer to reality than you might imagine: a cloak of invisibility [29]. It does exactly what the name suggests, rendering completely invisible anything that it covers. Of course, if the cloak works by means of technology, rather than wizard magic, there is at least one crucial drawback.

In order to render an object completely invisible, all of the visible light must be directed around the hidden object in such a way that an observer sees everything behind the object as clearly and completely as though the hidden object were not there. But the process of vision involves the absorption of light by the cells in the retina of the eye. If nobody outside the cloak can see what is inside, then neither can anyone inside the cloak see outside. If they could, their eyes would have to absorb some light, and an outside observer might notice some dark spots in the field of view. The same drawbacks would be experienced by invisible people or invisible gorillas, in movies such as *The Invisible Man* [30], or the more recent remake, *Hollow Man* [31]. Despite this unavoidable drawback, *Hollow Man* is somewhat true-to-life in one respect. The animals (including humans) that are rendered invisible (that is to say, they don't reflect or absorb any visible light) can be seen in infrared light because their bodies are still alive and emit infrared light (heat).

But what about Harry Potter's cloak of invisibility? It turns out that it is theoretically possible to engineer a material which can bend light around an object, rendering it effectively invisible. To understand how this works requires an understanding of how light is bent through a transparent material. The extent to which light deviates from its path, when it crosses the boundary between two different materials, is determined by its *index of refraction*. This is a number which is the ratio of the speed of light in a vacuum over the speed of light in the material, or

$$n = c_{vac} / c_{material}. \tag{3.12}$$

By definition, the index of refraction of a vacuum is 1. For anything other than a vacuum, n is a complex number, having both a real part and an imaginary part. The real part of the index of refraction accounts for reflection and refraction at boundaries between two materials, and the imaginary part accounts for absorption of light within the material.

Figure 3.7 illustrates the refraction (bending) of a ray of light, incident at an angle, θ_1, onto the boundary between two ordinary materials. For most transparent materials in the visible part of the spectrum, the index of refraction, n, is a number whose real part is greater than 1, and the imaginary part can usually be neglected. If the index of refraction of the second material is greater than the first, the angle of refraction, θ_2, will be smaller than the angle of incidence. The angles are measured with respect to a perpendicular to the boundary.

Fig. 3.7 Refraction of a
ray of light at the boundary
between two ordinary
materials, having different
indices of refraction

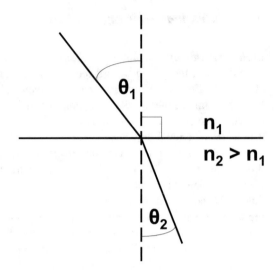

The sines of the two angles and the indices of refraction are related according to Snell's Law:

$$n_1 \sin\theta_1 = n_2 \sin\theta_2. \tag{3.13}$$

The speed of light in any material is related to the electric and magnetic properties of the material, according to

$$c = \frac{1}{\sqrt{\varepsilon\mu}} \tag{3.14}$$

where ε is the electric permittivity and μ is the magnetic permeability. So the index of refraction can also be expressed as

$$n = \sqrt{\varepsilon_r \mu_r} \tag{3.15}$$

where the subscript r indicates the quantity relative to the same quantity in vacuum.

There are no naturally occurring materials with negative index of refraction, but the possibility was explored theoretically in the 1960s [32]. Recent work has shown that it is possible to engineer a material that has negative values for both the permittivity and the permeability, which opens the possibility of a negative index of refraction [33, 34].

Materials with negative index of refraction are called *metamaterials*. (The Greek prefix *meta-*, meaning *change*, is a way of emphasizing that such materials do not occur in nature but must be engineered to have negative permittivity and negative permeability.) When light passes from an ordinary material, into a metamaterial, the negative index of refraction causes the light to be bent on the opposite side of the

Fig. 3.8 Refraction of a ray of light at the boundary between an ordinary material (positive index of refraction), and a metamaterial (negative index of refraction). The refracted ray is on the opposite side of the perpendicular, compared to the refraction in Fig. 3.7

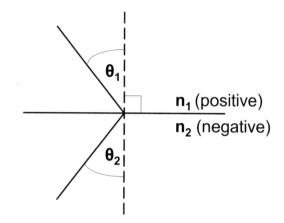

θ_1

n_1 (positive)

n_2 (negative)

θ_2

perpendicular, compared to where it would be bent in an ordinary material. The effect is illustrated in Fig. 3.8.

The development of metamaterials makes it possible to design a real-life cloak of invisibility. The technology is still in its infancy. Thus far, it is limited to the microwave, or far infrared portion of the electromagnetic spectrum, and works well only for two-dimensional objects. The limitations have to do with the size of the engineered structures, which must be embedded in the material to make it work. But in principle, this technology could be extended to shorter wavelengths (visible spectrum) and to three-dimensional objects [35, 36].

A nonmaterial means of cloaking is, of course, used frequently in the various *Star Trek* series. Instead of altering the optical properties of the object, itself, or surrounding the object with a cleverly engineered metamaterial cloak, the fabric of spacetime could be distorted according to general relativity, in such a way as to bend any electromagnetic radiation around the object. But this would take us back into Chap. 2. To borrow a quote from the original *Star Wars* movie (Episode IV), we should "move along."

3.9 Magnetic Materials

In the previous chapter, we met Magneto, one of the lead characters in the X-Men movies. Our discussion focused on Newton's First Law of Motion and the need to apply an *external* force, in order to change the state of motion (or rest) of a system. We concluded that Magneto could not move the Golden Gate Bridge, while standing on the bridge, if he only applied an *internal* force between himself and the ironwork of the bridge. In order to move the combined system (bridge + Magneto), he needed an *external* force, which might be possible if he also interacted with the Earth's magnetic field. Of course, the entire discussion required that we suspend disbelief in mutant superpowers and allow that Magneto could (as the story goes) "manipulate magnetic fields and metal."

Most of the references to Magneto's powers do not specify the kind of metal that he is supposedly able to manipulate. The implication is that he can manipulate all metals. But in reality, not all metals exhibit a strong response to magnetic fields. Although still not quite right, the reference to Magneto's powers in *X-Men: Apocalypse* comes a bit closer to reality by limiting the scope of his ability to "the magnetic elements: Iron, Nickel, and Cobalt" [37]. It would have been even more technically correct if they had said "ferromagnetic," but that might be asking too much of the script for a superhero movie. What happens to a ferromagnetic material, such as Iron, Nickel, or Cobalt, in the presence of an external magnetic field? Figure 3.9 illustrates some of the possibilities.

In addition to charge and mass, all of the fermions of the Standard Model of Particle Physics also have a quantum property called *spin*, which gives them a magnetic moment (imagine them as quantum-size bar magnets). Most of the electrons in an atom or molecule will be paired up, with their spins in opposite directions, essentially canceling each other out. If the atom or molecule has unpaired electrons, it will have a net magnetic moment. This will give it the ability to interact with an external magnetic field much more strongly than if all the electrons were paired up. Figure 3.9 is a two-dimensional model of the atoms in a ferromagnetic material. The individual atoms are represented as arrows. In the absence of an external magnetic field (Fig. 3.9a), the atoms are arranged in groups called domains. The members of a particular domain all have their magnetic moments pointing in the same direction, but the domains themselves are oriented at random within the material. The material has no net magnetization. In the presence of a sufficiently strong and completely uniform magnetic field (Fig. 3.9b), all of the magnetic domains will point in the same direction. If the applied field is truly uniform, the material will be magnetized, but there will be no net force on the material. In order for the material to experience a force, the applied field must be nonuniform. In Fig. 3.9c, the field is strongest at the bottom of the image and decreases in strength toward the top of the image. This *gradient* in the magnetic field produces a net force on the material in the downward direction. The material will be pulled toward the region of stronger field. The response of the material to the applied magnetic field also depends on the temperature. At sufficiently high temperature, called the Curie Temperature, thermal vibrations of the system tend to randomize the magnetic

a   b   c

Fig. 3.9 A two-dimensional model of a ferromagnetic material: (**a**) in the absence of any external field, (**b**) in a uniform magnetic field (pointing from left to right), and (**c**) in the presence of a magnetic field gradient (stronger field at bottom, weaker field at top)

domains, making the effect weaker. Fortunately for Magneto, the Curie Temperatures for Iron, Nickel, and Cobalt are hundreds of degrees above room temperature. So his powers should work on these elements anywhere on the surface of the Earth.

3.10 Energy and Power

It goes without saying that not all of science fiction is good science. But neither is science fiction the only place where bad science may be found. In the early 1980s, perhaps in an attempt to capitalize on the popularity of the *Star Wars* movies, and the mystique of the *Force* used by the *Jedi* and the *Sith*, a major petroleum refining company mounted an advertising campaign in which the following claim was made. "*Energy*: It's not just a *force*. It's a *power*." One of my colleagues in physics at Carnegie Mellon, and the author of one of the most widely used textbooks for introductory physics, was so outraged at the misuse of the terminology that he wrote a letter of complaint to the company. Much to our surprise and satisfaction, the ad in question was soon discontinued.

We have already spent some time considering forces in our discussion of Newton's laws of motion. Here we will elaborate on the important distinction between the concepts of energy and power, with the help of some practical examples, as well as some sci-fi illustrations.

3.10.1 Kinetic and Potential Energy

The energy associated with the motion of an object is called *kinetic energy*. Potential energy is the energy that is stored in a system. These two types of energy can easily be converted back and forth, from one to the other. For example, a person swinging back and forth on a swing has maximum kinetic energy at the bottom of the swing, while moving the fastest. At either extreme of the swing (as far forward as possible, or as far backward as possible), the person experiences an instant of no motion, when all of the kinetic energy has been converted into (gravitational) potential energy. If there were no air resistance, or friction between the swing and the support frame, the person could swing back and forth forever, without any energy input, except for an initial push to get things started.

Something similar happens with a person bouncing on a trampoline. At the highest point in the bounce, the person experiences an instant of no motion, when all of the energy is gravitational potential energy. When the person falls, the energy is converted gradually into kinetic energy, and the person goes faster, until the point of contact with the trampoline. Then the springs that hold the trampoline canvas begin to stretch. Energy is stored in the springs, in the form of elastic potential energy. When all of the kinetic energy of the moving person has been stored in the springs, the process reverses. Neglecting air resistance and energy lost as heat in the stretching springs, the person might bounce forever.

For a more dramatic illustration, we turn to *Spider-Man 2*, and the scene involving a runaway elevated train [38]. The throttle of the train is deliberately jammed, and the speedometer reads 88 miles per hour (perhaps a tip of the hat to *Back to the Future*?). Spider-Man must use enough webbing, attached to buildings along the tracks, to bring the train to a stop. If Spider-Man's synthetic webbing has properties similar to real spider silk, how much webbing does he need to stop the train, before it falls off the end of the collapsed tracks several blocks away? The final answer to this question will be left as an exercise, but the analysis begins as follows.

In situations where we are dealing with nonrelativistic speeds, the kinetic energy of any object can be calculated from the expression

$$K = \frac{1}{2}mv^2,$$ (3.16)

where m is the mass of the object and v is the velocity. This expression is appropriate for the runaway train, since it is only traveling at 88 miles per hour. If we assume that Spider-Man's webbing works more or less like an ideal spring, the elastic potential energy stored in the stretched webbing can be calculated from the expression

$$U = \frac{1}{2}kx^2,$$ (3.17)

where k is the force constant of the webbing and x is the distance of stretch. At the point of maximum stretch, when the train finally comes to a stop, all of the original kinetic energy of the train has been converted into potential energy stored in the webbing, and we can set $U = K$. (In the movie, the potential energy is not converted back into kinetic energy because the first car in the train gets hung up on the edge of collapsed tracks. Otherwise, the train would start to move backwards under the force of the webbing.)

Example 3.2: Spider-Man's Webbing

How much force must Spider-Man exert, by using his webbing, in order to stop the runaway train? We can estimate the force constant of Spider-Man's webbing from information in the scene just described. The initial speed of the train is given as 88 miles per hour. If we do our calculations in the SI system of units, we need to convert the speed into meters per second:

$$1\text{miles}/h = 0.447\text{m}/s$$

Initial speed of the train :

$$v = 88\text{miles}/h\ (0.447\text{m}/s)/(1\text{miles}/h) = 39.3\text{m}/s$$

The initial kinetic energy of the train is given by Eq. (3.16), but we need an esti-mate of the mass of the train. Peter Parker/Spider-Man lives in New York City, and the runaway train that he stops in the movie is supposed to be a New York MTA train. But the movie scene was actually shot in Chicago, using a Chicago L-train as a stand-in for the MTA train. According to www.Chicago-L.org, a typical Chicago L-train is six cars, each of which weighs 57,000 lb (without passengers). We convert this into kilograms:

$$1\text{kg} = 2.2\text{lb}$$
$$\text{Mass of train}:$$
$$m = (6 \times 57{,}000\text{lb}) / 2.2\text{lb} / \text{kg} = 155{,}000\text{kg}$$

We can now calculate the kinetic energy, K, using Eq. (3.16):

$$K = \tfrac{1}{2}mv^2$$
$$= \tfrac{1}{2}(155{,}000\text{kg})(39.3\text{m}/\text{s})^2$$
$$= 120\text{MJ}$$

(The units, MJ, are for megajoules, not the initials of Spider-Man's girlfriend, Mary Jane.)

This is the amount of energy that must be absorbed and converted into potential energy in the webbing, according to Eq. (3.17). What we might do next is to calcu-late the effective force constant, k, for the webbing by solving Eq. (3.17) for k:

$$U = \tfrac{1}{2}kx^2$$
$$k = 2U / x^2$$

In order to do this calculation, we need an estimate of the distance of stretch, x. Spider-Man's first attempt at stopping the train with a single burst of webbing is unsuccessful (the webbing breaks). He then shoots multiple strands and manages to stop the train approximately 50 s later, just before it falls off the end of the tracks. One quick-and-dirty way to estimate how far the train traveled during this time is to take the average speed and multiply by the time of travel. The train is slowing down during the process, so let's just take half the initial speed as a reason-able guess:

$$\text{approximate distance traveled}:$$
$$x = v_{\text{average}} t$$
$$= (20\text{m}/\text{s})(50\text{s})$$
$$= 1000\text{m}$$

Recognizing that the potential energy, U, stored in the webbing is equal to the initial kinetic energy, K, of the train, we can now calculate the effective force constant, k, of the combined strands of webbing, in units of Newtons per meter:

$$k = 2U / x^2$$
$$= 2(120\text{MJ}) / (1000\text{m})^2$$
$$= 240\text{N} / \text{m}$$

Finally, we multiply this value for k by the total distance of stretch to calculate the force that must be sustained by the webbing (and by Spider-Man's arms):

$$F = kx$$
$$= (240\text{N} / \text{m})(1000\text{m}) \qquad (3.18)$$
$$= 240,000\text{N}$$

Since most of us don't think about force in units of Newtons, let's convert this to pounds:

$$F = (240,000\text{N}) / (4.488\text{N} / \text{lb})$$
$$= 53,500\text{lb}$$

So Spider-Man's webbing (and arms) would have to sustain a force of something in the neighborhood of 25 Tons! Is this plausible? Some in-depth analysis, concerning Spider-Man's webbing is provided by James Kakalios, in his book, *The Physics of Superheroes*. The treatment is in regard to Spider-Man swinging on his webbing, but a fascinating and surprising detail is included. Quoting from an article by Jim Robbins, in the July 2002 issue of Smithsonian, he writes, "In theory, a braided spider silk rope the diameter of a pencil could stop a fighter jet landing on an aircraft carrier" [39].

Could we take our calculation a step further and verify this claim, by comparing to actual data for real spider silk? In one of the steps of the above calculation we estimated the force constant, $k = 240$ N/m. The force constant is a property of a particular sample of material. But for the purpose of comparison, it would be better to calculate the modulus of elasticity (or Young's modulus), which is a material property, rather than a sample-specific property. We can do this as follows:

Young's modulus is defined as

$$Y = \frac{\text{Tensile Stress}}{\text{Tensile Strain}}$$

or

$$Y = \frac{\text{Force per unit area}}{(\text{Change in length}) / (\text{Initial length})} = \frac{F / A}{(\Delta l) / l}$$

We can now substitute Eq. (3.18) for the force, F, and recognize the change in length, Δl, as equivalent to the amount of stretch of the webbing, x, in Eq. (3.18). This gives

$$Y = \frac{kl}{A}. \tag{3.19}$$

Since we already have an estimate of the force constant, k, all we need now are estimates of the original length, l, and the cross-sectional area, A, of the webbing. For the unstretched length, l, let's use half the separation between the buildings on opposite sides of the elevated railway. A good estimate is $l = 20$ ft. For the cross-sectional area, let's take a reasonable guess: $A = 1$ cm^2. Substituting all of these into Eq. (3.19) and converting to S.I. units gives:

$$Y = \frac{(240\,\text{N}/\text{m})(20\,\text{ft})(0.3\,\text{m}/\text{ft})}{(10^{-4}\,\text{m}^2)} = 14.4 \times 10^6\,\text{N}/\text{m}^2$$

or

$$Y = 14.4\,\text{MPa}$$

How does this compare to the measured Young's modulus for actual spider silk? It turns out that not all spider silks are created equal. In actual spider web material, there is an enormous difference between the elastic properties of the spiral silk compared to the radial silk. According to Guo et al., the radial silk is much stiffer and more uniform, with a Young's modulus of about 1.5 GPa. But the spiral silk is much more elastic and the Young's modulus varies from about 5–10 MPa [40]. This suggests that the elastic property of Spider-Man's webbing, as shown in *Spider-Man 2*, is well within the range of values for actual spider silk: just above the high end of the modulus of elasticity of the spiral silk but below the measured value for the radial silk. So given the favorable comparison, it's quite plausible that Spider-Man's webbing could stop the runaway train. But given the force required to do the job, his arms might hurt a bit.

3.10.2   Chemical Energy

In our discussion of solid–liquid phase transitions earlier in this chapter, we considered the problem of using a laser to dump enough energy into solid metal to melt it. The laser added energy to the metal, in the form of heat, to increase the amount of vibration of the atoms in the solid. If enough heat energy is added, the atoms can be made to vibrate beyond the ability of the interatomic bonds to hold the material together. The ordered arrangement of atoms in the solid will turn into a disordered liquid. This leads naturally to the concept that energy is stored in the bonds between

Table 3.6 Energy conversion factors

Unit	Conversion to SI units (J)
Electron-volt (eV)	$1\ eV = 1.602 \times 10^{-19}$ J
Calorie (cal)	1 cal = 4.187 J
Nutritional calorie (1 calorie = 1000 cal)	1 calorie = 4187 J
Megaton of TNT (megaton)	1 megaton = 4.184×10^{15} J

atoms of any material. This energy can be released in chemical reactions, as illustrated by the following example.

Example 3.3: Energy in Instant Oatmeal Vs. TNT

Consider half a kilogram of the chemical explosive TNT (trinitrotoluene), compared to a typical 10-packet box of instant oatmeal (my favorite flavor is maple and brown sugar). The masses are comparable: 10 packets of instant oatmeal have a total mass of just under ½ kilogram. Which of these contains more chemical energy? The energy in the oatmeal can be found in the Nutrition Facts, printed on the package. Each individual packet is approximately 160 (nutritional) Calories. Using Table 3.6 to convert nutritional Calories into the SI energy unit of Joules, we get 10 packets $\times$ 160 Calories = 1600 Calories = 6.7×10^6 J or 6.7 MJ. Again using Table 3.6 (and 1 ton = 1000 kg), we find that ½ kg of TNT is equivalent to roughly 2 MJ. So the box of maple-brown sugar instant oatmeal actually contains more energy than a comparable mass of TNT!

Estimation 3.5: Energy of a Marathon Runner or a Truck Collision

A typical marathon runner burns roughly 2880 Calories over the course of the race. Compare this to the kinetic energy of a large pickup truck (typical mass: 3000 kg) traveling at 60 miles per hour. Be sure to express both energies in the same system of units. (Convert both of them into joules.)

3.10.3   Distinguishing between Power and Energy

In our discussion of phase transitions, we defined power as energy delivered over a certain interval of time, or the *rate* at which energy is delivered (Eq. 3.9). We can use this definition to calculate the power involved in the processes that we just considered in the previous section.

Example 3.4: Power of TNT Vs. Instant Oatmeal

In the comparison of instant oatmeal with TNT, we found that the oatmeal actually contained more chemical energy. But consider now the amount of time required to release the energy in each case. The time required for the body to metabolize half a kilogram of food is something like 8 h. You would probably not want to eat all 10 packets of instant oatmeal at one sitting, but if you did, the power involved would be 6.7 MJ divided by (8 h × 3600 s/h), which comes to about 230 W. Compare this to the power involved in exploding the TNT. The actual explosion probably lasts for only a fraction of a second, but to keep it simple, let's assume that it takes 1 s. Using Eq. (3.9), the power dissipated in the explosion of ½ kg of TNT is 2 MJ divided by 1 s, or 2 MW. So even though the oatmeal contains several times more chemical energy than the same mass of TNT, the explosion of the TNT corresponds to about 10,000 times greater power than the process of metabolizing the oatmeal.

Estimation 3.5: Power Dissipated by Marathon Runner or Truck Collision

Consider again the marathon runner and the pickup truck in Estimation 3.4. Suppose the pickup truck were to collide head-on with a brick wall. How much power is dissipated in the collision, compared to the power dissipated by the marathon runner during the race? Use your energy results from Estimation 3.4 together with reasonable assumptions about the time for each process (e.g., about 4 h as a typical time to run a marathon, and maybe 1 s for the truck collision).

3.10.4   Nuclear Energy

We have just seen that the bonds between atoms in a solid can store considerable amounts of energy. The rate at which that energy is released (power) depends on the type of chemical reaction. Although the energy per unit mass of some food products is somewhat greater than that of TNT, the metabolism of food is a relatively slow (low power) process, whereas the explosive release of the energy in TNT is a relatively fast (high power) process. Compared to chemical energy, however, an even greater amount of energy is stored, per unit mass, in the nucleus of atoms. An almost true-to-life example is shown in the movie *True Lies*, in which a terrorist group have stolen some Soviet nuclear warheads, with intent to use them on selected targets within the United States. A spy for a U.S. intelligence service, played by Arnold Schwarzenegger, is asked by the leader of the terrorist group if he knows what the device is. Initially unwilling to cooperate, the spy responds with sarcasm:

"I know what this is. It's a snow cone maker. That's what it is."

"No, no. It's an espresso machine. Is it a water heater?"

But when his wife (played by Jamie Lee Curtis) is threatened with death, he describes the device in considerable detail:

"This is a Soviet MIRV-6 warhead from an SS-22-N launch vehicle. It contains 14.5 kilograms of enriched uranium… and a yield of 30 kilotons." [41]

The energy released is the equivalent of 30,000 tons (i.e., 30 million kilograms) of TNT. The source of the energy is the splitting of the nucleus of the isotope of uranium, U-235, which spontaneously fissions (splits) into two smaller nuclei and a couple of free neutrons. The neutrons, in turn, can collide with other U-235 nuclei, causing them to split. If a sufficient quantity, or *critical mass*, of U-235 is present, the result will be a chain reaction. A nuclear power plant involves a controlled chain reaction, in which the energy is released at just the right rate to turn water into steam and turn electric generators. A nuclear weapon, on the other hand, creates an uncontrolled chain reaction, in which all of the energy is released in a very short time, with devastating results. The 30-kiloton yield from a single MIRV-6 warhead, as described in *True Lies*, is nearly 2.5 times greater than the bomb dropped on Hiroshima in 1945, whose energy yield was equivalent to 13 kilotons of TNT.

A technical question remains about whether *True Lies* is giving the audience accurate information about actual nuclear warheads. It's not clear from the dialog whether the warhead is a MRV-6 or a MIRV-6. They are both pronounced the same, but there is a strategic difference. MRV stands for multiple reentry vehicle, while MIRV stands for multiple, independently targetable reentry vehicle. The former is intended to cover a broader area to maximize the chance of destroying a single target, while the latter is capable of destroying multiple targets simultaneously with a single delivery system. Regardless of which is intended in the movie, it's not clear whether there was ever a delivery system with the designation "SS-22-N." The designation "SS-" followed by a 2-digit number refers to an ICBM (intercontinental ballistic missile), while the designation "SS-N-" followed by a 2-digit number refers to an SLBM (submarine-launched ballistic missile). Either of these delivery systems could be MIRVed and might carry as few as 3 to as many as 10 warheads. So a "MIRV-6" is a real-life type of nuclear warhead. But the yield of 30 kilotons is not realistic. In 1994, when *True Lies* was released, the typical yield of a single Soviet MIRV warhead was about ten times greater than this. The range was from roughly 100 to 500 kilotons [42].

Even greater energy yield than a fission bomb can be achieved by fusion, in which hydrogen nuclei are fused to make helium. The energy yield from a typical hydrogen bomb is in the neighborhood of 10 megatons, which is roughly 1000 times greater than the Hiroshima fission bomb. Considerable research has been done toward creating a sustainable fusion reaction for peaceful purposes. Since the process involves fusing hydrogen into helium, there would be no radioactive waste, as there is with current nuclear power plants, nor would there be the same pollution problems associated with the burning of fossil fuels. Unfortunately, the results so far have not been successful as a viable energy source. Part of the difficulty lies in creating just the right balance of temperature and density to overcome the electrostatic repulsion of the protons in the hydrogen nucleus. The fusion reaction needs to be self-sustaining, without causing an explosive release of energy. The conditions are specified by Lawson's Criterion, which is the product of temperature, plasma density, and the so-called energy confinement time—the energy stored in the system divided by the rate of energy loss from the system.

3.10.5   *Matter–Antimatter Annihilation*

The 2009 movie *Angels and Demons* imagines the misuse of a real physical process to create a deadly weapon. A small quantity of antimatter is stolen from the Large Hadron Collider—the world's largest particle accelerator, which we discussed in Chap. 2—and is used to make a bomb, intended to blow up the Vatican. A battery-powered device creates a magnetic field, sufficient to contain the antimatter, but only as long as the batteries last. When the batteries run out, the containment field will collapse, allowing the antimatter to combine with ordinary matter. According to the movie, the explosive release of energy would be equivalent to 5 kilotons of TNT [43]. Is this even remotely plausible? The answer to this will be left as an exercise at the end of this chapter. But first, some background information.

The most efficient way to release energy, which we can imagine at the present time, is the direct conversion of matter into energy according to Einstein's famous equation,

$$E = mc^2. \tag{3.20}$$

We encountered this equation in our discussion of warp drive, in the previous chapter. If a particle of ordinary matter is combined with its antimatter counterpart, the energy released is the product of the mass of the two particles times the square of the speed of light. In principle, a controlled matter–antimatter reaction might be used as a power source to drive a starship, as discussed in Chap. 2, and as illustrated rather dramatically in *Star Trek II: The Wrath of Khan*. Toward the end of the movie, Spock sacrifices his life to repair the damaged matter–antimatter reactor and restores the warp engines, just in time to save the Enterprise from certain destruction [44]. But just as nuclear fission can be used either to create a controlled power source, or to create a bomb, so an uncontrolled matter–antimatter reaction might also be used as a weapon. The photon torpedoes of Star Trek supposedly involve the sudden combination of equal amounts of matter and antimatter, with an explosive release of energy many times greater than a hydrogen bomb. We get a rare glimpse of the inner workings of a Mark VI photon torpedo in *Star Trek VI: The Undiscovered Country*, when Spock and Dr. McCoy modify one to be able to target a cloaked Klingon vessel [45].

Estimation 3.7: Energy Yield of a Photon Torpedo

According to the "Star Trek: The Next Generation Technical Manual", the energy source of a photon torpedo consists of approximately 1.5 kg of antimatter combined with an equal amount of ordinary matter [46]. Use Eq. (3.20) to convert the combined amount of matter into energy, in units of joules. Convert this to megatons of TNT. How does this compare to a typical hydrogen bomb, whose yield is about 10 megatons?

Estimation 3.8: *Angels and Demons* Antimatter Bomb

Returning now to *Angels and Demons*, we raise the crucial question about the plausibility of an explosive device, which uses antimatter as the energy source. Does the LHC actually produce enough antimatter under normal operation to create an explosive device with a 5 kiloton yield? First, we need to do a calculation, which is essentially the reverse of Estimation 3.6: convert 5 kilotons to joules and calculate the amount of mass, according to Eq. (3.20). Then consult the official web pages of the LHC, which specifically address the claims of the movie, to see if the result is something that could actually be achieved [47].

3.11   Exploration Topics

Exp-3.1: The Standard Model of Particle Physics

The first of the fundamental particles of matter (the electron) was discovered in 1897. How has each of the subsequent discoveries in particle physics contributed to our understanding of the nature of matter?

How do we know that there are only three "families" of fundamental particles?

References:
- Gary J. Feldman and Jack Steinberger, "The Number of Families of Matter" Scientific American, Feb 1991
- Steven Weinberg, "Particle Physics, from Rutherford to the LHC" Physics Today, Aug 2011

Exp-3.2: Quarks and Gluons

The nucleus of an atom is made of protons (positive charge) and neutrons (no charge). Briefly explain why it is possible for a stable nucleus to exist. In other words, why doesn't every nucleus (except for Hydrogen) just fly apart under the influence of the electrostatic repulsion of its protons?

What two properties of protons and neutrons are as yet not completely explained in terms of the quark–gluon model?

Protons and neutrons are referred to as Baryons (a composite of three quarks held together by gluons). But the sum of the masses of the three quarks does not add up to the mass of the proton (or neutron). How does one of Einstein's famous equations help to account for the missing mass?

When it was first discovered that a proton consists of three quarks, each with a spin of ½, it was thought that the individual spins of the three quarks added up in a simple way to give the net *spin* of the proton, which is also a particle with *spin* ½. But careful studies have shown that it's not that simple. What other kind of motion of the quarks is needed to explain the spin of the proton?

Reference:
- Rolf Ent, Thomas Ullrich and Raju Venugopalan "The Glue That Binds Us" Scientific American, May 2015

Exp-3.3: The Large Hadron Collider

Why was the LHC constructed?

What are the major five major goals of the LHC, and why is each goal so important?

What is the main challenge that must be faced, in order to reach each of these goals?

References:
- Chris Llewellyn Smith, "The Large Hadron Collider" Scientific American, July 2000
- Graham P. Collins, "The Discovery Machine" Scientific American, Feb 2008
- Chris Quigg, "The Coming Revolutions in Particle Physics" Scientific American, Feb 2008

Exp-3.4: The Higgs Boson and Mass

What is the Higgs Boson, and how does it fit into the Standard Model of particle physics?

When was the Higgs Boson first predicted and by whom?

Prior to the announcement in July, 2012, of the observation of a particle consistent with the Higgs, what did we understand about the nature of mass?

Describe the experimental evidence for the Higgs, which led to the 2013 Nobel Prize in Physics.

How does the discovery of the Higgs Boson account for some of the mysteries of mass?

References:
- Gordon Kane, "The Mysteries of Mass" Scientific American, July 2005
- Tim Folger, "Waiting for the Higgs" Scientific American, Aug 2011
- Bertram Schwarzschild, "The Large Hadron Collider yields tantalizing hints of the Higgs boson" Physics Today, Feb 2012
- Michael Riordan, Guido Tonelli and Sau Lan Wu, "The Higgs at Last" Scientific American, Oct 2012

Exp-3.5: Supersymmetry

The Standard Model has been tremendously successful in describing matter at the quantum level, yet it remains an incomplete theory. Give at least one example of a fundamental physics question which the Standard Model cannot answer.

What new phenomenon was correctly predicted (i.e., later supported by experimental observation) using symmetry arguments?

Within the framework of supersymmetry, what two types of particles are related to each other?

What is one of the most profound mysteries of the universe which supersymmetry might explain?

Describe at least one of the implications for physics if the Large Hadron Collider fails to find evidence in support of supersymmetry?

Reference:
- Joseph Lykken and Maria Spiropulu, "Supersymmetry and the Crisis in Physics" Scientific American, May 2014

Exp-3.6: Dark Matter

What was the first experimental (observational) evidence, which suggested that there must be more matter in the universe than meets the eye?

What specific effect has the presence of dark matter had on the shape of our galaxy?

What does the "mass-to-light" ratio of a galaxy tell us about the proportion of dark matter compared to ordinary matter in the galaxy?

One theoretically postulated candidate for dark matter is the so-called *WIMP* (Weakly-Interacting Massive Particle). Which of the four fundamental forces of physics does the WIMP experience?

An alternative to the WIMP is the not-so-aptly-named *super-WIMP*. Why is this term somewhat misleading or confusing?

References:
- Jonathan Feng and Mark Trodden, "Dark Worlds" Scientific American, Nov 2010
- Leo Blitz, "The Dark Side of the Milky Way" Scientific American, Oct 2011

Exp-3.7: Dark Matter or Modified Gravity?

What observational evidence prompted astronomers to postulate the existence of dark matter?

What was going on in the physics community in the mid-twentieth century which made the idea of dark matter particles appealing?

If dark matter particles exist, in what way must they be different from the (ordinary) matter particles of the Standard Model?

Describe some of the evidence that would be more easily explained by a theory of modified gravity, instead of particle dark matter.

Why do theories of modified gravity remain controversial? (What observational evidence might seem to rule out modified gravity?)

Reference:
- Sabine Hossenfelder and Stacy S. McGaugh, "Is Dark Matter Real? Scientific American, Aug 2018

Exp-3.8: Dark Energy

Why did Einstein refer to his "cosmological constant" as the biggest mistake of his career?

How long has it been known that the universe is expanding?

How did the concept of dark energy originate? (What observable evidence is there, which led astrophysicists to infer the existence of dark energy?)

The acceleration of the expansion of the universe is a relatively recent discovery. When were the first observations made to show the increase in rate of expansion and what kind of observations were made?

Describe the competing hypotheses which might explain the accelerating expansion of the universe. How well do they account for the observable evidence?

What are the possible futures of the universe under each of these hypotheses?

What kind of experimental evidence or observations are needed to test which hypothesis is correct?

References:
- Lawrence M. Krauss, "Cosmological Antigravity" Scientific American, Jan 1999
- Lawrence M. Krauss and Michael S. Turner "A Cosmic Conundrum" Scientific American, Sept 2004
- Christopher J. Conselice, "The Universe's Invisible Hand" Scientific American, Feb 2007
- Joshua A. Frieman, Michael S. Turner, and Dragan Huterer, "Dark Energy and the Accelerating Universe" Annu. Rev. Astro. Astrophys. 2008, 46;385–432
- Timothy Clifton and Pedro G. Ferreira, "Does Dark Energy Really Exist?" Scientific American, Apr 2009
- Adam G. Riess and Mario Livio, "The Puzzle of Dark Energy" Scientific American, Mar 2016

Exp-3.9: Graphene

How does graphene differ from other forms of carbon? In particular, how is the structure of graphene different from ordinary graphite?

Physicists of the 1970s and 1980s were optimistic about the potential of exfoliated graphite. Why did exfoliated graphite turn out to be less exciting than originally expected, and why is graphene even better?

What properties of graphene are of interest technologically? (What can be done with graphene that cannot be done with other materials?)

References:
- K. Geim et al., "The rise of graphene" Nat. Mater. 6, 183 (2007)
- K. Geim and A. H. MacDonald, Graphene, "Exploring Carbon Flatland" Physics Today, Aug 2007
- Andre K. Geim and Philip Kim, "Carbon Wonderland" Scientific American, Apr 2008

Exp-3.10: Nanobatteries

What are the essential differences (apart from size) between a nanobattery and a conventional battery?

What key technological or scientific breakthrough makes nanobatteries possible?

How much power (energy per unit time) can they deliver? In other words, how much energy can they store, and how quickly can they deliver that energy when needed?

What important problem (other than miniaturization) does this new technology solve?

Reference:
- Charles Q. Choi, "Miniaturized Power" Scientific American, Feb 2006

Exp-3.11: Laser Technology

The history of the first optical laser (1960) is somewhat controversial. Identify and describe at least two points of controversy or confusion. In particular, does it seem that the controversy originated through poor public relations or poor science?

What important properties of the ruby laser were demonstrated convincingly by the Bell Labs group, which were not demonstrated by T.H. Maiman at Hughes Research Laboratories?

What potential application of the laser were the scientists reluctant to discuss in public?

Within the first 5 years after the invention of the laser, the power of tabletop lasers advanced rapidly, becoming comparable to the power output of the Hoover Dam. What technological challenges persisted for the next 20 years, which prevented further advances?

What technological breakthrough enabled the development of lasers with power output thousands of times greater than the Hoover Dam?

What is the current state-of-the-art in laser power, and how is it achieved?

References:
- Donald F. Nelson, Robert J. Collins and Wolfgang Kaiser, "Bell Labs and the Ruby Laser" Physics Today, Jan 2010
- Gerard A. Mourou and Donald Umstadter, "Extreme Light" Scientific American, May 2002
- "Megalasers to pulse in several new EU countries" Physics Today, June 2010 (issues & events, p. 20)

Exp-3.12: Metamaterials and Invisibility Cloaks

If you want to make something invisible, what has to happen, physically?

Thanks to the invention of metamaterials, invisibility cloaks are now theoretically possible, but still somewhat difficult to achieve in practice. How is an invisibility cloak made of a metamaterial fundamentally different from stealth technology?

How does an invisibility cloak, made of metamaterial, work, in principle?

Why were microwaves chosen as the starting point for developing an invisibility cloak (as opposed to some other part of the EM spectrum)?

If a person could be rendered invisible using a metamaterials cloak, what would be the fundamental disadvantage, from that person's point of view?

References:
- S. Harris, "Out of mind out of sight" IET Eng. Tech. 8, 12 (2008)
- M. Wegener, S. Linden, "Shaping optical space with metamaterials" Phys. Today 63 (10), 32 (2010)
- J. B. Pendry, "Negative Refraction" Contemporary Physics 50, 363 (2009)

Exp-3.13: Nuclear Weapons

What is the primary purpose or goal of the Comprehensive Nuclear Test Ban Treaty? What techniques are currently available to monitor for nuclear explosions?

How is it possible to distinguish the seismic signal of a nuclear explosion from those of other (naturally occurring) seismic events?

It is not possible to detect explosions of arbitrarily low yield but is there a lower limit to the yield of a nuclear explosion, below which we wouldn't need to worry about detecting it?

Reference:
- Paul G. Richards and Won-Young Kim, "Monitoring for Nuclear Explosions" Scientific American, Mar 2009

References

The Standard Model of Particle Physics

1. *Star Trek: The Next Generation* – "Starship Mine" (Cliff Boyle, Paramount 1993). Quarks, hadrons, baryons [DVD season 6, disc 5, opening scene]
2. Rossi, B.: Cosmic Rays, p. 107. McGraw-Hill, New York (1964)
3. Manohar, A.V., Sachrajda, C.T., Barnett, R.M.: Quark masses. http://pdg.lbl.gov/2017/reviews/rpp2017-rev-quark-masses.pdf
4. Abbot, B.P., et al.: Observation of gravitational waves from a binary black hole merger. Phys. Rev. Lett. **116**, 061102 (2016)
5. Discovery of the Higgs Boson at the Large Hadron Collider. https://home.cern/science/physics/higgs-boson

Beyond the Standard Model

6. Dark Matter Survey (DES). https://www.darkenergysurvey.org/

The Atomic Nucleus

7. *Forbidden Planet* (Fred McLeod Wilcox, MGM 1956). Special isotope of lead (217) as a shield against radiation, but with less mass than "common lead" [DVD scene 8]

Gases

8. *Mission to Mars* (Brian De Palma, Touchstone Pictures 2000). Air leaks out through a small hole in a punctured space ship [DVD scene 13]
9. "How To Freeze Boiling Water" (DrBarryLuke, YouTube 2010). http://www.youtube.com/watch?v=QJjiKBjhj0I
10. Facts and Figures for the International Space Station. https://www.nasa.gov/feature/facts-and-figures
11. *Mars* – "Crossroads" (Teleplay by Andre Bormanis and Paul Solet, National Geographic 20160). Explosive decompression of a Martian habitat module [season 1, episode 6]
12. *Gravity* (Alfonso Cuaron, Warner Brothers 2013). An astronaut inside as Soyuz space capsule is not wearing her helmet, when the hatch is suddenly opened from the outside [DVD scene 7]
13. *Passengers* (Morten Tyldum, Columbia Pictures 2016). Water in a swimming pool on a space ship when artificial gravity fails [DVD scene 12]
14. Playing ping pong with a drop of water on the International Space Station. https://www.youtube.com/watch?v=PrqcwH8LhDU
15. Wringing out a wet washcloth on the International Space Station. https://www.youtube.com/watch?v=UyFYgeE32f0

Solid State Materials

16. *Star Trek: The Next Generation*—"Gambit, Part I" (Peter Lauritson, Paramount 1993). Microcrystalline damage (or nanoparticles?) [DVD season 7, disc 1, opening scene]

Phase Transitions

17. *Goldfinger* (Guy Hamilton, United Artists 1964). Laser-induced solid-liquid phase transition [DVD scene 18]
18. Nelson, D.F., Collins, R.J., Kaiser, W.: Bell labs and the ruby laser. Physics Today. **16**(1), 40–45 (2010)
19. Patel, C.K.: Continuous-wave laser action on vibrational-rotational transitions of CO_2. Phys. Rev. **136**(5A), 1187–1193 (1964)
20. Gresh, L.H., Weinberg, R.: The Science of James Bond, p. 147. John Wiley & Sons, Hoboken, NJ (2009)
21. *Goldeneye* (Martin Campbell, United Artists 1995). Laser-induced solid-liquid phase transition + calculation of power [DVD scenes 22 (if time allows) + 23]
22. *Terminator 2: Judgment Day* (James Cameron, Studio Canal 1991). Liquid-solid and solid-liquid phase transitions [DVD scenes 65 and 66]

Transparency and Invisibility

23. *Star Trek IV: The Voyage Home* (Leonard Nimoy, Paramount 1986). Transparent polymers, transparent aluminum [DVD scene 10]
24. Gentilman, R.L., Maguire E.A., Dolhert, L.E.: Transparent aluminum oxynitride and method of manufacture. US Patent application, 1984. http://www.google.com/patents/US4520116?printsec=abstract#v=onepage&q&f=false
25. *Die Another Day* (Lee Tamahori, MGM 2002). Adaptive camouflage [DVD scene 18]
26. *Predator* (John McTiernan, Twentieth Century Fox 1987). Adaptive camouflage [DVD scene 15]
27. *Independence Day* (Roland Emmerich, 20th Century Fox 1996). B-2 stealth bombers [DVD scene 40 (nuclear attack)]
28. *Tomorrow Never Dies* (Roger Spottiswoode, MGM 1997). Stealth boat [DVD scene 29]

29. *Harry Potter and the Sorcer's Stone* (Chris Columbus, Warner Brothers 2001). Invisibility, metamaterials [DVD scenes 21 and 22]
30. *The Invisible Man* (James Whale, Universal 1933). Invisibility [DVD scenes 8 and 9]
31. *Hollow Man* (Paul Verhoeven, Columbia Pictures 2000). Invisibility [DVD scenes 3 and 13]
32. Veselago, V.G.: The electrodynamics of substances with simultaneously negative values of ε and μ. Sov. Phys. Usp. **10**, 509 (1968)
33. Smith, D.R., Padilla, W.J., Vier, D.C., Nemat-Nasser, S.C., Schultz, S.: Composite medium with simultaneously negative permeability and permittivity. Phys. Rev. Lett. **84**, 4184 (2000)
34. Shelby, R.A., Smith, D.R., Schultz, S.: Experimental verification of a negative index of refraction. Science. **292**, 77 (2001)
35. Harris, S.: Out of mind out of sight. IET Eng. Tech. **8**, 12 (2008)
36. O'Donnell, B.: Invisibility cloak takes one step closer to revealing itself. Horizon Magazine, 7 Aug 2017. https://phys.org/news/2017-08-invisibility-cloak-closer-revealing.html
37. *X-Men: Apocalypse* (Bryan Singer, 20th Century Fox 2016). Magneto's ability to manipulate magnetic fields and metal is specific to the "magnetic elements, Iron, Nickel and Cobalt" [DVD scenes 20 and 21]

Energy and Power

38. *Spider-Man 2* (Sam Raimi, Columbia Pictures 2004). Use of spider webbing to stop a runaway el-train [DVD scene 43]
39. Kakalios, J.: The Physics of Superheroes, p. 96. Gotham Books, New York (2009)
40. Guo, Y., Chang, Z., Li, B., Zhao, Z.-L., Zhao, H.-P., Feng, X.-Q., Gao, H.: Functional gradient effects on the energy absorption of spider orb webs. Appl. Phys. Lett. **113**, 103701 (2018)
41. *True Lies* (James Cameron, 20th Century Fox 1994). MRV (or is it MIRV?) nuclear warhead [DVD scene 26]
42. Norris, R.S., Arkin, W.M.: Russian (C.I.S.) strategic nuclear forces end of 1994. Bull. At. Sci. **51**(2), 78 (1995). https://doi.org/10.1080/00963402.1995.11658057
43. *Angels and Demons* (Ron Howard, Columbia Pictures 2009). Antimatter bomb intended to blow up the Vatican [DVD scenes 2 and 5]
44. *Star Trek II: The Wrath of Khan* (Nicholas Meyer, Paramount 1982). Matter-antimatter annihilation and gamma radiation [DVD scene 15]
45. *Star Trek VI: The Undiscovered Country* (Nicholas Meyer, Paramount 1991). Photon torpedo, matter-antimatter annihilation [DVD scene 13]
46. Sternbach, R., Okuda, M.: Star Trek the Next Generation Technical Manual, p. 129. Pocket Books, New York (1991)
47. Information on antimatter at the Large Hadron Collider. http://public.web.cern.ch/Public/en/ Spotlight/SpotlightAandD-en.html. http://angelsanddemons.web.cern.ch/

Chapter 4
Can a Machine Become Self-Aware?
(The Sciences of Computing
and Cognition)

> RIKER: *Your file says that you're a …*
> DATA: *Machine. Correct, sir. Does that trouble you?*
> RIKER: *To be honest, yes, a little.*
> DATA: *Understandable, sir. Prejudice is very human.*
> Star Trek: The Next Generation
> *"Encounter at Farpoint"*

One of the major characters in the television series *Star Trek: The Next Generation* is the android Commander Data (played by Brent Spiner). Created by Dr. Noonien Soong (also played by Spiner), the twenty-fourth century's foremost authority on cybernetics, Data is more than just an intelligent, humanoid robot. He is a *sentient being*. He is so sophisticated in design that he is aware of his own existence. He was introduced in the pilot episode, "Encounter at Farpoint", at which time he was a Lieutenant Commander, and a member of the original crew of the newly designed galaxy-class starship Enterprise (NCC-1701D) [1]. An ongoing theme in the series is Data's quest to become more like his human shipmates. In this chapter, we will consider what it might take to create something like Data. In order to do this, we will need to understand something about his technical specifications.

In a later episode of the series, "The Measure of a Man", an ambitious scientist has the bright idea to take Data apart and reverse-engineer him, in order to figure out how he was made. His ultimate goal is to build an army of sentient androids. The episode reveals two key pieces of information about Data: his information storage capacity, given as 800 quadrillion bits, and his processor speed, given as 60 trillion operations per second [2]. How do these specifications compare to state-of-the-art data storage and processor speed in the early twenty-first century and how do they measure up to the estimated specifications of the human brain? The answers might surprise you!

As the speed and data storage capacity of computers continue to advance, a technological achievement comparable to Commander Data seems to be coming closer to reality. Current projections suggest we are very near the threshold of machines that match the speed and memory capacity of the human brain. In fact, as we will see later in this chapter, we are probably already there. But is it possible, even in

© Springer Nature Switzerland AG 2019
B. B. Luokkala, *Exploring Science Through Science Fiction*, Science
and Fiction, https://doi.org/10.1007/978-3-030-29393-2_4

principle, for a machine to become self-aware, or is consciousness something unique to highly sophisticated living things? If it is possible, what would it take for a computer to become conscious? There is a diversity of views on this question among people who work in the field of artificial intelligence (AI). At risk of over-simplifying, we might divide the AI community into two camps: *soft AI* and *hard AI*. The former might be further subdivided into two subgroups. Some in the *soft AI* camp would say that consciousness is not something that is computable, or that the question is not even worth investigating. Others in this group believe that there is no fundamental reason why a machine could not become conscious, but that it would take a major breakthrough in computer technology to achieve it. Those in the hard AI camp are considerably more optimistic. Not only do they believe that machines will become conscious, but that it will be achievable with tools and techniques that we already have. In other words, it's just a matter of time.

In this chapter, we will explore some of the developments in the *cognitive sciences*, a multidisciplinary field, which includes computer science, robotics and artificial intelligence, as well as cognitive psychology (how do humans think and learn?), and neuroscience (how does the brain work?). But before we begin, it may be useful to review some of the terminology commonly used to describe the performance of computer hardware.

4.1   Information Storage

The storage and retrieval of digital information is not only well understood but has advanced at a phenomenal rate over the past several decades, as illustrated by the examples in Table 4.1. The table gives storage capacity in units of bytes. A byte is the basic unit of digital information storage and is defined to be 8 bits, where a bit is simply a one or a zero (the two numbers in the binary counting system). From a materials perspective, a one or a zero is stored in a magnetic medium by magnetizing a small portion of the material either in the up or the down direction, perpendicular to the plane of the disc. The data storage density depends on how easily the

Table 4.1 A few examples of easily portable, digital data storage media (with Commander Data's positronic brain, and the estimated capacity of the human brain for comparison)

Storage medium	Introduced	Physical size	Storage capacity (approx.)
8″ floppy disc	c. 1970	8″ × 8″ × 1/16″	250 kB
5¼″ floppy disc	Mid-1970s	5¼″ × 5¼″ × 1/16″	500 kB
3.5″ floppy disc	Mid-1980s	3.5″ × 3.5″ × 1/8″	Up to 1.44 MB
Hard disc pack	c. 1980	14″ diam. × 5.25″	80 MB
USB hard drive	2010	5″ × 3.5″ × 1″	TB
Human brain	(Prehistory)	1400 cm^3 (average)	2.5 PB
"Positronic brain"	Twenty-fourth century	Similar to human brain	100 PB

material can be magnetized or demagnetized, and how small the read/write heads of the drive can be made.

The first three entries in Table 4.1 are "floppy discs", so-called because they were simply flexible plastic discs, coated with a magnetic film, inside a protective plastic jacket. Examples of floppy discs—more properly referred to as "diskettes"—are shown in Fig. 4.1. From the time they were introduced c. 1970, until the time they became obsolete c. 1990, the storage capacity of these devices had only increased by about a factor 5 or 6 (250 kB to 1.44 MB). In contrast, the storage capacity of hard discs, from the 1980s to the present, has increased by more than 4 orders of magnitude (from 80 MB to 1 TB). As of this writing, you can easily hold a terabyte (10^{12} bytes) of digital information in the palm of your hand, in the form of a portable USB hard drive (Fig. 4.2). The last two entries in Table 4.1 are the estimated storage capacity of the human brain, at about 2.5 PB [3] and Commander Data's specification of the storage capacity of his "positronic brain" (a term no doubt borrowed from Isaac Asimov's robot novels, which we will consider later in this chapter).

As small and compact as the Terabyte USB hard drive is, it would take the equivalent of about 2500 of them to equal the estimated storage capacity of the human brain. If this is state-of-the-art in portable hard data storage, is there any hope of ever building a humanoid robot, such as Commander Data, whose intelligence could match or exceed that of a human? We may get some perspective on this question by considering once again the history of data storage density. When I was an undergraduate physics major at the University of Pittsburgh, in the mid-1970s, state-of-the-art in portable magnetic data storage was the disc pack, a stack of 14″ diameter metal platters, roughly 5¼″ high. An example is shown in Fig. 4.3. Each of the platters in the stack was coated with a magnetic film for writing the data. The disc pack was stored in a protective plastic case, the bottom of which was held in place by a spring-clip mechanism, and easily removed before inserting the disc pack into

Fig. 4.1 Floppy discs. Left to right: 8″ diskette, storage capacity 250 kB; 5.25″ diskette, storage capacity 500 kB; and 3.5″ diskette, storage capacity 1.44 MB. (Photo by the author)

Fig. 4.2 Portable USB hard drive. Data storage capacity: 1 TB. (Photo by the author)

a   b

Fig. 4.3 Nashua disc pack (c. 1980), 14″ diameter × 5.25″ high. (**a**) With and (**b**) without plastic storage cover. Data storage capacity: 80 MB. (Photos by the author)

the read-write drive. The drive itself was larger than a dormitory-size refrigerator. After inserting the disc pack into the drive, the top part of the plastic cover was removed by rotating the handle. The 1981 James Bond movie *For Your Eyes Only* shows the device in action. Bond, on a visit to Q-branch, loads a disc pack into the drive, while Q runs a computer program called the Identi-graph—a face drawing and recognition system for identifying known criminals [4].

Although I never owned one of these storage devices while I was in college (only very well-funded research groups could afford one, at that time), I recently engaged in a bit of nostalgia and purchased one on eBay (Fig. 4.3). The data storage capacity of the disc pack was less than 100 MB (this one was 80 MB). By comparison, the USB hard drive, shown in Fig. 4.2, has a storage capacity of 1 TB. It would take a large room, filled wall-to-wall and floor-to-ceiling with 1980s disc packs to equal the storage capacity of something which you can now hold in the palm of your hand (including the drive!). If this is an indication of what can be achieved, in terms of data storage density, in the 30 years from c. 1980 to 2010, what further advances might be achievable in the next 30 years? In particular, will we ever be able to create a data storage device, whose information capacity and physical size are comparable to those of the human brain?

Estimation 4.1: Physical Storage Space Required for Magnetic Data Devices

Given the data storage capacity of the disc pack from the 1980s (Fig. 4.3), and the data storage capacity of the contemporary USB hard drive (Fig. 4.2) calculates how many disc packs it would take to equal the data storage of the USB hard drive. Given the physical dimensions of the disc pack (Table 4.1), what physical storage space would be required to house this number of disc packs?

Estimation 4.2: Data Storage in the Human Brain

Figure 4.4 shows a 1 TB USB hard drive and a full-scale model of the human brain for size comparison. Use the information in Table 4.1 to calculate how many of these USB hard drives it would take to equal the total data storage capacity of the brain. Given the dimensions of the USB hard drive, what physical storage space would be required to house this number of drives?

Estimation 4.3: Projecting Data Storage Density into the Future

Use the information in Table 4.1 to calculate the *data storage density* for each of the three devices: the 1980s disc pack, the 1 TB USB hard drive, and the human brain. Express all three results in the same set of units (e.g., Terabytes per cubic centimeter). Suppose we can expect data storage density to continue to increase at the same rate that it has increased over the 30 years from 1980 to 2010. How far into the future might we expect to see a device with the same data storage density as the human brain?

Returning now to *Star Trek*'s Commander Data, who specified his storage capacity as "800 quadrillion bits," how does this compare to the estimated storage capacity of the human brain? Table 4.2 provides a comparison of common English terms for various orders of magnitude with their Greek prefixes and the corresponding powers of 10.

By convention, there are 8 bits per byte. So let's translate Commander Data's storage capacity of 800 quadrillion bits into the more standard computer terminology. Dividing by 8 bits per byte, and using Table 4.2, this turns out to be 100 PB. So Commander Data's storage capacity exceeds the estimated 2.5 PB of storage for the human brain by a factor of 400. Since Data is portrayed as a sentient being whose

Fig. 4.4 USB hard drive (left) and a full-scale model of the human brain (right). (Photo by the author)

Table 4.2 Common English words, with corresponding Greek prefix and power of 10

Number	Common English	Greek prefix	Power of 10
1000	Thousand	Kilo	10^3
1000,000	Million	Mega	10^6
1000,000,000	Billion	Giga	10^9
1000,000,000,000	Trillion	Tera	10^{12}
1000,000,000,000,000	Quadrillion	Peta	10^{15}

mental capabilities far exceed those of humans, this result should not come as a surprise. In other words, the script writers of *Star Trek: The Next Generation* seem to have got this one right. But what about his processor speed, which Data specified to be "60 trillion operations per second"? How does this specification compare to the estimated speed of the human brain and to state-of-the-art computers in the twenty-first century? We will take up that question in sections to follow. But first, a brief look at some calculating devices that were common in the early part of the twentieth century and what the term "computer" originally meant at that time in history.

4.2 Analog Calculating Devices

The 1961 movie *Voyage to the Bottom of the Sea* might be thought of as a modern take on the 1870 Jules Verne novel, "20,000 Leagues Under the Sea". Renowned scientist and retired naval officer Admiral Harriman Nelson (played by Walter Pidgeon) is the designer of a futuristic nuclear submarine, the *Seaview*. Although computer technology is an important part of the submarine's design, we never really see this technology in action in the movie. Admiral Nelson is faced with a scientific and technological problem of truly global proportion (one which we will revisit in a later chapter). The urgency and complexity of this problem cry out for a computer to solve it, yet Admiral Nelson does his calculations on a slide rule [5]. In the context of the times, this has some truth to it, but is also misleading. NASA's Apollo moon missions would not have been possible without onboard computers, yet the NASA engineers and astronauts still used slide rules to do their everyday calculations (Fig. 4.5). On the other hand, even in the early 1960s, complex calculations of the sort that Admiral Nelson was doing in *Voyage to the Bottom of the Sea*, or of the sort required by NASA to land humans safely on the moon, would have been done not on a slide rule, but with the aid of significantly more sophisticated devices, with much higher precision.

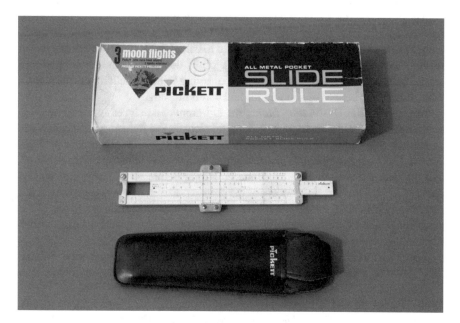

Fig. 4.5 Pickett Slide Rule, c. 1971. The original box, shown in the photo, proudly sports a sticker proclaiming that Picket rules have been aboard three Apollo missions. (Photo by the author)

4.2.1   The First "Computers"

The PBS program *Top Secret Rosies* gives a good overview of the state of scientific computing in the first half of the twentieth century. Before the digital age, computers were not machines, but people—usually women—who did their work using a variety of mechanical devices known as calculators. Scientific calculations were done manually, unless you were among the fortunate few who had access to a state-of-the-art electromechanical device known as the Bush Differential Analyzer. Invented by Vannevar Bush at MIT, in 1925, the Differential Analyzer was able to do complex calculations in a matter of minutes, which would take days to do by hand. It was the only high-speed calculating tool available until the end of World War II, and very few of them—perhaps fewer than a dozen—were in existence in the world at the time. The war effort required careful calculations of artillery shell trajectories, under nonidealized conditions. A 60 s shell trajectory required approximately 40 h for a human "computer" to calculate by hand, but the Bush Differential Analyzer was able to do the same calculation in about 15 min [6].

The Differential Analyzer was essentially obsolete by the 1950s. Nevertheless, it still featured prominently in two classic sci-fi films, both of them George Pal productions, whose story lines depended on careful scientific calculations. More than 10 years before President John F. Kennedy proposed the Apollo mission, in response to the Soviet Union's successful launch of Sputnik, *Destination Moon* tells (or should we say predicts) the story of a space race. A team of American industrialists propose the first manned spaceflight to the Moon. A particularly amusing and informative scene features the cartoon character Woody Woodpecker, who demonstrates the concept of rocket power to the skeptical financiers. In the scene which immediately follows, a Differential Analyzer is used to calculate the trajectory of the rocket to the Moon [7]. The movie does a good job of depicting some of the perils of spaceflight, and in hindsight, the stupidity of trying to do something of this magnitude without the benefit of simulations and careful testing.

Just a year after *Destination Moon*, the Differential Analyzer appeared again in another George Pal production, this time to calculate the trajectory of a rogue star and its orbiting planet, on a collision course with Earth. In the early scenes of *When Worlds Collide*, photographic plates from an astrometric telescope are compared, and the data are analyzed to determine how long before all life on Earth will be destroyed [8]. Although the concept of the destruction of Earth by a rogue star is not particularly plausible, the careful measurement of the position of a star is one of the established techniques for determining the existence of exoplanets—a topic which we will take up in the next chapter.

In contrast to the hubris of the industrialists in *Destination Moon*, who claimed they would be able to do the job in 1 year, the real-life Apollo mission took NASA 8 years to complete. Human "computers" played a prominent role in the early years of NASA, as brilliantly illustrated in the 2016 movie *Hidden Figures*. By no means, a science fiction movie, *Hidden Figures* tells the real-life story of three African American women, who overcome racial and gender discrimination and contribute significantly to the success of NASA's Project Mercury, which put the first American astronauts in space. Dorothy Vaughan (played by Octavia Spencer) leads the West Computing Group at NASA's Langley Research Center (originally the Langley Memorial Aeronautical Laboratory), in Virginia. Katherine Goble Johnson (played by Taraji P. Henson) and Mary Jackson (played by Janelle Monáe) are two of the human "computers", who carry out complex mathematical calculations using electronic desktop calculators, such as the Friden Model STW 10 [9].

4.3 Digital Computers

War is a powerful motivator in the development of new technology. In the early years of World War II, British engineer Tommy Flowers designed the world's first electronic digital computers, the Colossus Mark 1 and Mark 2. The Colossus computers were single-purpose programmable computers, designed specifically to decipher coded messages from the Germans, and put to use beginning in 1944. Whereas analog computers store and process information in the form of continuous variables, digital computers use discrete variables, coded as a series of zeros and ones (binary numbers). In 1942, the year following the entry of the U.S. into World War II, physicist John Mauchly and engineering graduate student J. Presper Eckert began a project to develop what would become the world's first *general-purpose*, programmable digital computer. Composed of over 17,000 vacuum tubes and other electronic components, the Electronic Numerical Integrator and Computer, or ENIAC, was originally designed to calculate ballistic shell trajectories but would do the job much faster than its predecessor, the mechanical differential analyzer. Depending on the type of calculation, ENIAC could perform up to 5000 arithmetic operations per second. ENIAC was not actually completed until 1945, too late to be of real use in the war effort. The major advantage of ENIAC over the British-made Colossus computers was its versatility. It could be reprogrammed to solve a variety of different scientific problems. Unfortunately, as explained by interviewees in the PBS program *Top Secret Rosies*, the major drawback of ENIAC, compared to modern computers, is that reprogramming actually required rewiring, which sometimes took days or even weeks. The first scientific application of ENIAC was in nuclear physics research and the development of the hydrogen bomb.

With the invention of the transistor by John Bardeen in 1948, a major leap forward in computing technology was made possible. Circuitry based on vacuum tubes and relays was replaced with integrated circuits—networks of transistors impressed onto semiconducting silicon chips. Among the very first semiconductor-based computers were the IBM 360 processors, released in 1964. IBM offered a variety of 360 models, with speeds ranging from a few kilohertz, which was comparable to the top speed of ENIAC, to several tens of kilohertz, many times faster than ENIAC. In addition to the obvious advantage of speed, the IBM 360 models occupied much less space and consumed much less electrical power than their vacuum tube-based predecessors.

NASA's Space Task Group at Langley acquired its first digital computer, an IBM 7090 Data Processing System, to perform calculations for Project Mercury in 1961. As told in the movie *Hidden Figures*, when the IBM 7090 was installed, the human "computers" were in danger of losing their jobs. Recognizing the imminent threat to her West Computing Group, Dorothy Vaughan borrowed the IBM manuals, taught herself the FORTRAN programming language, and was more successful than the (white, male) IBM engineers at getting the 7090 to run. She eventually got the entire West Computing Group reassigned to help her program "the beast" [10].

4.3.1   Precision of Calculations

The precision of a slide rule is limited by the resolution of its scale. A pocket-size slide rule, such as the one shown in Fig. 4.1, could be used to calculate to about three significant figures. For example, if you wanted to calculate the ratio 22/7, it would give you a result of 3.14. Much larger slide rules were available which could reach about four significant figures, perhaps giving a result for 22/7 = 3.142. The precision of calculations performed by a digital computer is also limited, depending on the number of bits used to store the information. The concept is illustrated in a scene toward the end of *Hidden Figures*. On the day that astronaut John Glenn was scheduled to be launched into orbit, the IBM 7090 calculated a reentry point that did not agree with the previous day's calculations. Glenn requested that human "computer" Katherine Johnson check the calculations. According to the head of the Space Task Group, Al Harrison (played by Kevin Costner), Katherine confirmed the coordinates, and even managed to "calculate a few decimal points further than that hunk of metal" [11]. This statement is quite plausible. According to an official IBM information sheet, the 7090 used 36-bit words and could store and process ten-digit numbers [12]. In contrast, the Friden STW 10 electronic desktop calculators, used by the human "computers" in *Hidden Figures*, could handle 11-digit entries and could store the information in a 20-digit accumulator [13]. Thus, it was possible for Katherine Johnson to calculate with greater precision than the IBM 7090 (perhaps adding a second layer of meaning to the movie title). The very next year after John Glenn's launch and successful reentry, IBM introduced "double-precision" floating point numbers, which allowed for the possibility of using two 36-bit words, instead of one, to represent numbers.

4.3.2   Moore's Law and Processor Speed

The speed of a semiconducting processor depends upon how many transistors can be placed on a single chip. Since the development of the first computer chips in the early- to mid-1960s, the number of transistors on a single chip has approximately doubled every 2 years—a phenomenon which has become known as Moore's Law.

Space does not permit a detailed history of the major advances in computer processor technology, but a good overview is provided in the October 1996 Special Issue of Physics Today (50 Years of Computers and Physicists), particularly the article by Alfred E. Brenner, entitled "The Computing Revolution and the Physics Community" [14]. A good summary plot illustrating Moore's Law has been created by Wgsimon and is reproduced in Fig. 4.6.

A significant development in the history of computing came in the late 1970s and early 1980s, with the introduction of personal computers. Prior to this time, computers were large, expensive devices, which occupied an entire room—or at least significant portions of a room—and were beyond the means of individuals. The Apple II (1978), followed soon after by the IBM-PC (1981), made computing affordable and accessible to the general public. But the user interface for these new desktop computers was still basically the same as for the big mainframe computers: commands were typed on a keyboard. The next step forward came with the introduction to the consumer market of the mouse and the graphical user interface (or GUI) by Apple, in 1984. Although not the first company to offer a computer with a mouse

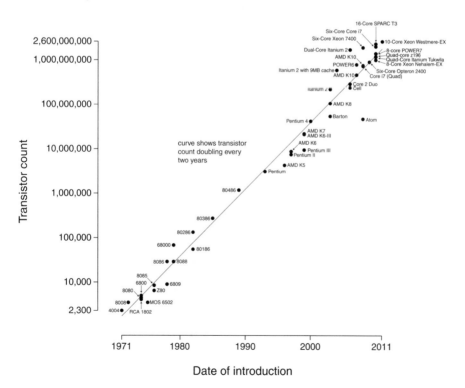

Fig. 4.6 Plot of Number of Transistors on a computer chip as a function of time, illustrating Moore's Law [15]

and GUI, Apple was the first to make the cost (at just under $2500) affordable to the consumer. The true honor of "first" computer to be similarly equipped actually belongs to the Xerox Alto. But its $40K price tag put it out of reach of the typical user. The Macintosh was released in January of 1984 and promoted with the famous television commercial, first aired during Super Bowl XVIII, proclaiming that "1984 will not be like "1984"." The reference was to the George Orwell science fiction novel, "1984", which depicted a future totalitarian society, in which "Big Brother" was always watching [16].

This major turn of events in the computing world is illustrated in a sci-fi movie scene, which we have already discussed in Chap. 3. In *Star Trek IV: The Voyage Home*, Scotty from the original *Star Trek* series produced an animated image of the molecular structure of *transparent aluminum* on a Macintosh computer. Released in 1985, *Star Trek IV* is one of the better examples of sci-fi movies, which not only feature state-of-the-art in real-life technology, but also make accurate predictions of future technology [17]. Recall Scotty's unsuccessful attempt to get the attention of the Macintosh simply by speaking the word "computer." He then picks up the mouse and says "Hello, computer" into the bottom of the mouse, but still no response (Fig. 4.7). Even the most advanced personal computers of the mid-1980s were incapable of responding to voice commands. But now, smart phones come equipped with speech recognition. A user can find information about *transparent aluminum* simply by speaking the words to the smart phone's Internet search engine.

Three years after the release of Macintosh by Apple, IBM released their PS-2, which was my first desktop computer. Its processor speed was 20 MHz, which was almost 1000 times faster than the speed of the IBM 7090, featured in *Hidden Figures*.

Fig. 4.7 A revolution in personal computing, the Macintosh was the first to come with a graphical user interface (GUI) and a mouse at an affordable price. It did not, however, respond to voice commands, such as "Hello, computer," spoken into the bottom of the mouse

4.3.3   Parallel Processing

If Moore's Law could continue forever, there might (at least in principle) be no limit to the speed of a computer processor. Extrapolating Moore's Law into the future, we might expect computers that could process information at a rate comparable to that of the human brain within the next few decades. But in reality this would require a change in the way information processing is handled. The number of transistors that can be etched onto a single computer chip is limited by the wavelength of the light used in the photolithography process and the ability to focus the light. One way to get around this limitation is to use multiple processors working on separate tasks simultaneously. The concept of parallel processing is not particularly new, but a recent example, which had a significant impact on popular culture, is worth mentioning here. In 2011, IBM introduced WATSON, a massively parallel processor which defeated human competitors on the TV game show JEOPARDY! WATSON, named for IBM's founder, Thomas Watson, is a combination of more than 2800 IBM POWER7 processor cores, occupying a space equivalent to about ten refrigerators, with 16 TB of RAM [18, 19]. WATSON's processor speed is about 500 GHz, which is more than 2000 times faster than the IBM PS-2, but still far short of the estimated petahertz speed (10^{15} Hz) of the human brain. So how did WATSON manage to beat its human competitors? We need to keep in mind that the human brain is simultaneously processing all kinds of information necessary for daily living, while WATSON was originally built for a single purpose: to play JEOPARDY.

4.3.4   Machine Learning

The possibility of a computer beating a human competitor at a game is not new. The Deep Blue computer, also built by IBM, famously beat grandmaster Barry Kasparov at chess in 1997. What is new is the way in which WATSON beat its human competitors. The game of chess consists of a set of strict rules for moving each piece. These rules can be programmed into a computer, allowing it to win by "brute force". Deep Blue was able to compute a tree of possibilities 30 or 40 moves ahead, which would be impossible for a human to do. In contrast, the game of JEOPARDY has no such rules that can be written down or coded into a computer. WATSON had to be able to pick out key words from a JEOPARDY clue, gather information about the meaning and context, both grammatical and historical, and judge the most probable response (phrased, of course, in the form of a question). What enabled WATSON to play the game of JEOPARDY is a revolutionary new approach to artificial intelligence: machine learning.

The early optimism about building a general-purpose machine that could match or exceed the capability of the human brain had almost disappeared by the end of the twentieth century. The lack of sufficiently sophisticated algorithms, inadequate processor speed, and inability to handle massive amounts of data all conspired to prevent systems from living up to expectations [20]. The machine learning revolution draws its inspiration from two other branches of the cognitive sciences, neuroscience, and cognitive psychology, each of which provides the guiding principles for a different type of approach.

The deep learning approach to machine learning is based on mathematical principles that simulate neural networks. The result is a system that learns from many examples to perform such tasks as recognizing objects in a photograph or speech recognition and translation into other languages. A second approach uses Bayesian methods to enable a machine to learn from just a few examples, in much the same way that a child would learn. A hypothesis is formulated about the world, and a prediction is generated about what the data should look like if the hypothesis is correct. The hypothesis is refined in order to minimize the difference between the prediction and the observation. The deep learning approach has been referred to as the "bottom-up" approach, while the Bayesian approach has been referred to as "top-down" [21]. But there may be more to machine learning than a large database of examples to compare or a broad knowledge base with which to formulate hypotheses. Researchers are finding that machines, just as human children, may benefit from a nurturing caregiver to encourage them in the process of cognitive development, learning, and socialization [22].

We will explore intelligent robots and sentient machines in detail later in this chapter. But here we consider an example of machine learning from the sci-fi movie *Chappie*. The movie is set in South Africa in a dystopian future, where crime is controlled by a robotic police force. A roboticist named Deon (Dev Patel) develops an AI program that he hopes will enable a robot to learn like a child, but at a phenomenally accelerated rate, and eventually achieve consciousness. He plans to install and test the program in a damaged police robot, slated for destruction. But Deon and the robot body are abducted by gangsters, who compel Deon to activate the robot, so they can teach it to be a gangster and work for them. The robot at first responds with fear of the hostile gangsters. But Deon gradually gains the robot's trust, as does Yo-Landi, a female member of the gang (played by Yo-Landi Visser). The robot quickly learns to associate words with objects, such as "watch" and "chicken", and learns Deon and Yo-Landi's names. But much to Deon's dismay, Yo-Landi gives the name "Chappie" to the robot [23].

Discussion Topic 4.1

Does the form of machine learning exhibited in *Chappie* more closely resemble the "top-down" (Bayesian) approach or the "bottom-up" (deep learning) approach? What specific evidence from the movie scene supports your response?

4.4 Beyond Digital Computers

Computers already exist, which can surpass humans at specific tasks, such as playing chess or competing on the game show JEOPARDY! But they are enormous, compared to the physical size of the human brain. And no processor exists, which can come close to performing all of the complex tasks performed by the human brain. So what would it take to do the job? It is conceivable that the answer might not lie in conventional transistor-based information processing, but in a completely new architecture, such as the bioneural gel packs of *Star Trek: Voyager* [24]. It is also conceivable that we may not need to wait until the twenty-fourth century to see such developments. A processor based on biological molecules is just one among many innovative proposals outlined in the article "The Next 20 Years of Microchips," jointly written by the editors of Scientific American, in the January 2010 issue [25].

4.4.1 Quantum Computing

If the idea of a computer processor based on biological material seems too speculative, another new type of computer architecture is rapidly on the rise. For decades, quantum computing was only discussed from a theoretical perspective. But in recent years, it has become possible to store and process information in the form of "spin-up" or "spin-down" quantum particles, known as qubits. Unlike the bits of classical computers, which represent either "zero" or "one", a qubit can exist in a superposition of "spin-up" and "spin-down" states simultaneously. Taking advantage of a quantum property known as *entanglement*, quantum computers will make it possible to perform tasks that are too complex or impossibly time-consuming for classical computers, particularly in the area of cryptography and information security.

   The concept of quantum entanglement involves a pair of particles, whose states are individually random, but the combined state of the pair is well defined. A measurement of the state of one particle instantly determines the outcome of the measurement of the state of the other particle, regardless of how far apart the particles may be separated. Albert Einstein referred to this as "spooky action at a distance" and believed to his dying day that there must be something incomplete about quantum mechanics. In what would become his most cited paper, Einstein and colleagues Boris Podolsky and Nathan Rosen ("EPR") described this apparent paradox. Their paper suggested that there must be some "hidden variables", which could explain the spooky correlation [26]. In 1964 John Bell demonstrated mathematically that there should be a measurable difference in the amount of correlation, if Einstein's "hidden variables" were real, versus if "entanglement" were real [27]. Bell's *inequality* sets an upper limit on the amount of correlation between the measured states of the quantum particles, if Einstein's hidden variables are real. If entanglement is real, the measurements of the states of the two particles will be much more highly correlated. But it was not until as recently as 2015 that the test of Bell's inequality was performed definitively. At least three different research groups have demonstrated that entanglement is real, using particles separated by up to 400 m [28]. More recently, a research group headed by Anton Zeilinger of the University of Vienna performed an experiment on a truly cosmic scale, which removed any doubt that quantum entanglement is a real phenomenon. Making use of a pair of telescopes in the Canary Islands to gather light from two different quasars, the team guaranteed an element of randomness in their measurement process, which could prove or disprove the degree of correlation of the particle states. The results were conclusive. Entanglement is real. The details are described in an episode of the PBS *NOVA* television series, entitled "Einstein's Quantum Riddle" [29].

The technology of quantum computing is still in its infancy. Announcements have been made in recent years of quantum processors ranging in size from tens of qubits to thousands of qubits. D-Wave Systems, the world's first quantum computing company, founded in 1999, introduced its 2000Q quantum computer in January of 2017 [30]. With a reported 2000 qubits, the D-Wave 2000Q is the successor to the 1000-qubit D-Wave 2X system. Although no official price tag is available, the estimated cost of $15 million puts the 2000Q out of reach of all but the most well-funded research laboratories. More modest devices include Bristlecone, a 72-qubit superconducting quantum processor chip, introduced by Google in March of 2018, and IBM's Q System One, a 20-qubit system, intended for use in business and industry, introduced in January of 2019 [31, 32].

4.5   The Human Brain as an Information Storage and Processing System

How does the human brain store and process information, and what would it take for a general-purpose machine to match the capability of the brain? Some sci-fi examples may help to get us started thinking about the key issues involved.

Both human memory and computer memory are dealt with explicitly in the episode of *Star Trek* (the original series), called "The Changeling". The scene opens with the Enterprise under unprovoked attack by a robotic probe. When Captain James Kirk identifies himself and requests communication, the attack is suddenly broken off. We later learn that the probe, which calls itself Nomad, was originally a space probe launched from Earth in the early twenty-first century, on an exploratory mission to seek out signs of life elsewhere in the galaxy. But Nomad was damaged by collision with a meteor, and drifted aimlessly, until it came into contact with an alien probe, whose mission involved collecting and sterilizing soil samples from other planets. The two probes somehow combined with each other, both in structure and in programming, resulting in a new mission: to seek out perfect life forms and to "sterilize imperfection." Nomad attacked the Enterprise simply because it detected the presence of imperfect biological life forms. The attack was broken off because of a case of mistaken identity: Nomad mistook Captain James Kirk for its creator, Jackson Roykirk. Fortunately for the Enterprise and her crew, any command given to Nomad by "the Kirk—the Creator" was obeyed without question, provided it was within Nomad's capability.

While onboard the Enterprise, Nomad hears Lt. Uhura singing and wants to know what form of communication she is using. When Uhura is unable to explain the purpose of singing to Nomad's satisfaction, it absorbs all of the information in her brain and erases her memory. Chief Engineer Scott rushes to Uhura's defense but is instantly killed by a burst of energy from Nomad. At Kirk's request, Nomad is able to restore Scotty to life. Unfortunately, Nomad cannot restore Uhura's memory, and she is faced with the daunting task of relearning everything she knew [33].

Spock faces a similar challenge at the beginning of *Star Trek IV: The Voyage Home*. Having sacrificed his life to save the crew of the starship Enterprise in *Star Trek II: The Wrath of Khan*, and having his body regenerated on the Genesis planet, in *Star Trek III: The Search for Spock*, he must reacquire all of his Vulcan knowledge, prior to returning to Earth to give testimony at a court martial [34]. The retraining of Uhura and Spock occurs at a remarkable pace, both of them relearning most of what they knew in a matter of days.

As we have already seen, the advent of machine learning has revitalized the field of artificial intelligence. Almost certainly, the creation of artificial consciousness (if, indeed, it is even possible) will be more than just a matter of data storage capacity and processing speed. Understanding the details of how the human brain stores and processes information will also be needed. Thanks to the development of functional

Fig. 4.8 f-MRI images of the author's brain thinking about two physics concepts: "gravity" and "crystal lattice". The green and red regions are the areas of the brain which show increased blood flow during the specific cognitive task. (Images courtesy of Robert A. Mason, Center for Cognitive Brain Imaging, Carnegie Mellon University. Used by permission)

magnetic resonance imaging (f-MRI) as a noninvasive tool for studying the brain, much progress is being made in understanding the neural basis of cognition. The areas of the brain that are responsible for performing various sensory and cognitive tasks are being mapped out in much more detail than ever before possible. Examples of f-MRI images are show in Fig. 4.8.

The f-MRI technique involves measuring regions of the brain which experience increased blood flow during a particular cognitive task. The subject's head (in this case, mine) is placed inside a large superconducting electromagnet with sensors to pick up the regions of "activation." A three-dimensional, computer-generated image of the brain is divided up into "voxels", or volume elements, analogous to the "pixels" in a two dimensional image. In 2018, a group of ten members of the faculty of the Physics Department at Carnegie Mellon University (myself included) participated in a study to determine if there were specific regions of the brain that are used to process both basic and advanced physics concepts and if the same activation regions were used by all of the subjects in the same way. The images in Fig. 4.8 show my brain on physics—specifically, while thinking about two different concepts: "gravity" and "crystal lattice." The green and red regions are the activation patterns for these two concepts. While the differences are rather subtle, they are noticeable upon close inspection. But most important, for the purpose of the study, a machine learning classifier can tell the difference. The study found that the activation regions were "sufficiently stable for a machine learning classifier to identify, with good accuracy, which of the 30 advanced concepts and 15 basic concepts a physicist was thinking about, given the concept's activation pattern" [35]. In essence, a computer was able to "read our minds" and know what we were thinking about, given the activation pattern.

But there are still many unanswered questions. In particular, what exactly is human memory? How are memories created and stored in the brain? Memories are formed, in part, by the interconnection of neurons, the cells which make up most of the material of the brain. The brain contains an estimated 100 billion neurons (10^{11}), each with roughly 1000 connections to other neurons, via parts of the neuron called *dendrites*. This suggests something like a hundred trillion (10^{14}) connections. As we grow from infancy to adulthood, our brains are stimulated by various sensory inputs, and we learn and form memories. The neurons in our brains also develop more and more dendritic connections. Yet the information storage capacity of the human brain is estimated at about 2.5 PB (2.5×10^{15}). This means that if each piece of information stored in the brain corresponded to a connection between one neuron and another, there would not be enough connections to do the job. Neuroscientists of the mid-twentieth century hypothesized discrete changes in the tissue of the brain, which they called *engrams*, to account for the formation and persistence of memories. Although f-MRI studies have identified very specific areas of the brain involved in various cognitive tasks, the existence of engrams as discrete, localized entities within the brain, has yet to be supported by experimental evidence. Current research suggests that memories are more likely to be distributed throughout the brain, but the details of how that happens are not yet fully understood.

4.6   Intelligent Robots

Long before anyone had any real notion of how the human brain works or could conceive of a computer whose speed and storage capacity would match the human brain, people were imagining artificial creatures that might substitute for humans, to perform tasks that were considered undesirable or too menial or too dangerous for humans to perform. The word *robot* was first used in literature, to refer to such an artificial creature, by the Czech playwright Karel Čapek, in his 1921 work, R.U.R. (Rossum's Universal Robots). The word is derived from the Czech robota, which suggests menial or hard labor. A visual image, which effectively conveys the essence of the word, can be found in the opening scene of Fritz Lang's 1927 masterpiece *Metropolis* [36]. The movie depicts a future society, which is divided into a laborer class and a technocratic elite. The movie opens with a shift change for the working class: row after row of laborers march through a tunnel, toward a huge elevator, while others march, row after row, in the opposite direction. The separation between the two classes goes beyond status and job description, however. Metropolis is a giant, futuristic city, whose citizens enjoy all the benefits of a highly technological society. The sons and daughters of Metropolis are shown frolicking in beautiful gardens on the surface of the Earth, while the laborers slave away, tending the underground machines, which provide the power to the city. The laborer class lives even deeper underground, below the machines. Although their existence is essential to maintaining the good life on the surface, the workers seem to have a status in society even lower than the machines that power the city. This harsh class separation is conveyed visually and esthetically, by the architecture of the three spaces. Metropolis, the city, is futuristic in design and adorned with beauty, both natural and manmade. The workers' city, by contrast, is completely stark and unadorned. In between Metropolis and the workers' city is the machine space, the architectural detail of which is considerably more interesting than the bland and severe workers' city.

The laborers in *Metropolis* are not robots, in the modern sense of the word, but in the sense of the Czech word from which robot is derived. Although they are completely human, they are dehumanized and completely subservient to the technocratic elite. The message of the movie is one of hope and reconciliation: "The mediator between head and hands must be the heart." The allusion is to the Master of Metropolis (the head), the laborers (the hands), and the ultimately reconciling love between Freder, the son of the Master of Metropolis, and Maria, a labor class childcare worker (the heart).

Although the word robot, meaning an artificial device for performing tasks formerly done by humans, did not come into usage until early in the twentieth century, the concept is ancient. It can be traced back to Jewish mythology (the Golem) and even further back to Greek mythology (Talos). Neither one of these examples is purely technological in nature—both involve mysticism or supernatural powers to animate them. Nevertheless, it is instructive to begin our discussion of robotics with Talos, the bronze giant made by the Greek god Hephaestus, to guard the island of Crete, and featured in a scene in *Jason and the Argonauts* [37]. The 1963 movie features numerous Ray Harryhausen stop-motion animations, a tedious and time-consuming method of creating special effects, before computers became sophisticated enough to do the job. Talos begins as a statue on a pedestal and is not very much larger than a human. But when he detects the presence of the intruders, he becomes an animated being, which appears to grow larger and larger as the scene progresses. Eventually, Talos is large enough to pick up a sailing vessel and dump its passengers out into the water. But he has a literal Achilles heel. He is filled with some sort of magical fluid, which can be drained out through a screw-cap in his heel. After a prolonged struggle, which includes some human fatalities, Jason succeeds in unscrewing the cap, whereupon the fluid drains out. The bronze giant clutches his throat, as though gasping for breath. His body begins to crack, and he falls to his demise on the beach.

Is Talos a robot, in the modern sense of the word? Yes, and even more so. Talos would qualify as an example of an *intelligent* robot—an artificial entity which combines the three key elements of *sensor*, *computation*, and *control* (Table 4.3). Sensors are the means by which a robot gathers information from the environment. They can be as simple as a temperature probe, or as sophisticated as a visual imaging system. Computation is the means by which the robot analyzes the information and makes decisions based on the sensory input. This could be as simple as a logic circuit, which gives a yes/no answer to the question "Should I do something?" or as complex as a speech recognition algorithm, which can compute responses based on verbal commands. Finally, control is the resulting manipulation of the environment. Talos is a mechanical device, which combines all three of these essential elements. He is immobile until he becomes aware of the presence of human intruders, presumably by visual input. He recognizes the humans as a threat to the security of the island, presumably via some pattern-recognition system, which would ignore the indigenous animal life. He controls the environment by stepping into action, and using his sword and physical strength to ward off the intruders.

Science fiction stories often imagine scenarios in which things go badly wrong, and robots out of control is a common theme. Among the very best of this particular subgenre is *2001: A Space Odyssey*, which we have already considered in the context of our discussion of moving and rotating reference frames. One of the main characters in the movie is the HAL-9000, which self-identifies as a computer. Nevertheless, HAL meets all of the criteria for being an intelligent robot, as summarized in Table 4.3. It is described in a BBC television interview, during a spaceflight to Jupiter, as the brain and central nervous system of the interplanetary spaceship, *Discovery I*. It has sensors

Table 4.3 The three essential ingredients for a machine to qualify as an intelligent robot		
Sensor	Gathers data autonomously from the environment	
Computation	Decision-making, based on input from sensor	
Control	Response: manipulation of environment	

all over the ship, which it uses to gather information about the status of critical systems and can even read the lips of crew members, as they speak. It makes its own decisions, based on the information—including decisions to kill members of the crew—and is able to control systems throughout the ship [38].

4.6.1 Robot Hall of Fame

The HAL-9000 earned a place among the first class of inductees into the Robot Hall of Fame, along with fellow sci-fi robot R2-D2, from Star Wars, and two real-life robots: NASA's Mars Pathfinder (a.k.a. Sojourner), and the first industrial assembly line robotic arm, Unimate. A joint effort of Carnegie Mellon University and the Carnegie Science Center, the Robot Hall of Fame was created in 2003 to call attention to the increasing contribution of robots in society [39]. Note that only three of these four robots would qualify as an intelligent robot, by the criteria set forth in Table 4.3. The Unimate assembly line robot was programmed to perform the same task repetitively. But with no sensors and no decision-making, it counts as a robotic device, but not an intelligent robot.

In the same way that the HAL-9000 qualifies as an intelligent robot, so does the M-5 multitronic unit, from the original *Star Trek* series episode, "The Ultimate Computer." Designed by Dr. Richard Daystrom, M-5 represents a revolution in computer architecture. Daystrom has found a way to impress human *engrams* onto the computer circuitry, enabling M-5 to think just like humans. The device was designed with a view toward completely replacing human crewmembers on starships. Although the initial test results were promising—the M-5 performed well during its first planned mission of exploration, and later successfully (and appropriately) defended itself during an unscheduled attack drill—things soon go badly wrong. The first sign of trouble occurs when M-5 goes out of its way to destroy an ore freighter. Fortunately the freighter had no crew, having been converted to automation. But the error could just as easily have cost lives, so Kirk issues the order to disconnect M-5. M-5, however, has other ideas, including defending itself with a force-field, and drawing power directly from the warp engines, killing a crewmember in the process. Finally, during a scheduled battle drill, M-5 attacks and destroys a Federation starship. Kirk manages to regain control of the ship by persuading M-5 that it was guilty of committing murder and deserved to die [40].

4.6.2 The Turing Test

The concept of machines thinking and responding to questions in a convincingly human way was first entertained seriously by Alan Turing, in his paper "Computing Machinery and Intelligence," published in 1950 in the British journal *Mind* [41]. Turing proposed a test to distinguish between human and artificial intelligence. Or, rather, to see if a machine might be programmed in such a way as to deceive an interrogator into believing that the machine was actually a real human. The test was based on the Imitation Game, a parlor game that was popular in the pre-television era (*The Imitation Game* is also the subtitle of the first section of Turing's 1950 paper and the title of a major motion picture, based on the life of Alan Turing.) [42]. The original parlor game version involved three human participants: a man (A), a woman (B), and an interrogator (C). The gender of the interrogator is not important for the purpose of the test. The man and the woman were kept in a room separate from the interrogator, who could communicate with them only by passing written messages. The goal of the interrogator is to determine, by asking appropriate questions, which of the two is the man and which is the woman. The role of the man is to answer in such a way as to deceive the interrogator into guessing incorrectly, and the role of the woman is to help the interrogator to guess correctly. The Turing test is conducted in a similar way. The key difference being that after running the test multiple times with the three human participants, the man (A) is then replaced by an artificial (machine) intelligence. The machine is said to have passed the Turing Test if it is able to deceive the interrogator into guessing incorrectly just as often as the real man.

The classic version of the Turing Test, as originally described in Turing's paper, called for three participants. Contemporary versions eliminate the need for a human participant to help the interrogator to guess correctly. The human interrogator communicates only with the artificial intelligence and has to decide if it is human or machine, without the help of a third participant.

You may have had occasion to interact with a special-purpose artificial intelligence, albeit not one that is actually conscious. For example, at least one of the major cable television companies has a telephone help line, with a fairly sophisticated speech recognition system, capable of guiding you through the process of diagnosing and solving technical difficulties. The School of Computer Science at Carnegie Mellon University has a robotic receptionist, named Tank, who politely greets everyone who walks by his information kiosk. Visitors may ask Tank a variety of questions, such as how to find the office of a member of the faculty, or even carry on a conversation with him, but his range of responses is limited. If he cannot answer your question, he will suggest that you rephrase, or try asking him a simpler question. It becomes obvious after just a few questions that Tank, the roboceptionist, is not yet ready to pass as an actual human.

4.6.3 Human–Computer Interactions: Beyond the Turing Test

The primary purpose of Carnegie Mellon's roboceptionist project is not to see if an artificial intelligence could deceive humans into believing that it is also human—in effect, to pass some form of the Turing Test. Tank's rather crude computer-generated face on the flat screen gives away his nonhuman identity from the start. Instead, it's more a study of the extent to which humans are willing to interact with something that is obviously not quite human. A test of human–computer interactions is taken to extreme in the 2014 movie *Ex Machina*, in which a young programmer, named

Caleb (Domhnall Gleeson), wins a contest and gets to spend a week at the mountain retreat of the CEO of the world's largest internet company. The CEO, Nathan (Oscar Isaac), asks Caleb if he knows what the Turing Test is. Caleb correctly describes the essence of the modern (2-participant) version of the Turing Test. Nathan then explains to Caleb that he is about to be the human component in the Turing Test. But the test that Caleb is about to experience goes way beyond either the classic or the modern version of the Turing Test. Caleb is introduced to Ava (Alicia Vilkander), who is obviously a humanoid robot. Ava even acknowledges that Caleb can see she is a machine, indicating that Ava not only exhibits humanlike intelligence but is also self-aware. The purpose of the test is not to see if Ava can deceive Caleb into thinking that she is a real human female. It is to determine if Ava and Caleb will form an emotional attachment [43].

The test administered to the replicants in *Blade Runner* also goes beyond a simple test for humanlike intelligence. Based on the Philip K. Dick novel "Do Androids Dream of Electric Sheep?", *Blade Runner* is a futuristic dystopia, in which androids have been created to perform tasks requiring humanlike intelligence, but in environments that are too dangerous or undesirable for humans. The replicants, as they are called, are self-aware, but their level of consciousness is limited. In particular, they do not have the ability to respond to emotional stimuli, in the same way that a human would respond. The movie presupposes machine intelligence at a level that might easily pass the classic Turing Test (unlike Tank, the roboceptionist). Although the replicants are visually indistinguishable from humans, a battery of questions, coupled with measurements of physiological responses, can eventually tell the difference. Simpler models can be discerned from humans in perhaps a few dozen questions, while a more sophisticated model may require over a hundred questions [44].

The *Blade Runner* movie differs significantly from the Philip K. Dick novel in the way that the replicants are portrayed. In both versions, the replicants have a preprogrammed expiration date. They only live for something like 3 years. When the replicant's expiration date arrives, most of them simply die. But a few, sensing that their time is limited, begin to behave in unexpected ways as their expiration date approaches. In the novel, the rogue replicants are portrayed as fearful of their impending doom and go into hiding. But in the movie version, the rogue replicants become vicious killers. The job of the Blade Runners is to hunt them down and "retire" them—a euphemism for kill them. If we reach the point where robots become aware of their own existence, it will be important for our safety to program them to behave in predictable and desirable ways. We will consider issues of robot behavior in the next section and revisit the question of artificial consciousness in the section that follows.

4.7   Robot Behavior

Both the M-5 Multitronic Unit in *Star Trek* and the HAL-9000 computer in *2001: A Space Odyssey* were so sophisticated that they had become aware of their own existence. But they both lacked built-in safeguards against harming humans, which enabled things to get out of control. In contrast to both of these examples, *Forbidden*

Planet also features a remarkable robot, named Robby, and things also go badly wrong, but the robot is not at fault. We have already met Robby in Chap. 3, in our discussion of isotopes. Earlier in the same movie, we find that Robby has been designed with a crucial safety feature, which makes it impossible for him to harm humans. After hosting his visitors for lunch, Morbius explains that Robby not only prepared the meal, but also manufactured the raw materials (and eventually offers Robby's services to make the lead shielding, which we discussed in Chap. 3). He is also bound by "absolute selfless obedience," which Morbius demonstrates by ordering him to put his arm in a disintegrator beam. Robby obediently walks toward the disintegrator, but Morbius cancels the order just before Robby complies. When the visitors question whether something as powerful and obedient as Robby might become a deadly weapon, if it were to fall into the wrong hands, Morbius demonstrates the safety feature. He asks to borrow the Commander's sidearm (a neutron ray gun) and hands it to Robby. He orders Robby to point it toward a tree on the patio. Robby obeys. Morbius then orders Robby to fire, and the tree is vaporized. He asks Robby if he understands the mechanism. "Yes, Morbius, a simple blaster", Robby replies. Morbius then orders Robby to turn around and point the blaster at the Commander. The other officers rise to the Commander's defense, but he waves them off, trusting that Morbius will not do anything rash. Morbius tells Robby to "aim right between the eyes." Robby complies. "Fire!" But instead of firing the weapon, Robby stands motionless, his relays begin to click furiously, his antennae rotate, and his head fills with sparks. Morbius explains that Robby is "stuck in a subatomic dilemma between my direct order and his basic inhibition against harming humans." Eventually Morbius says "Order canceled," and Robby returns to normal [45]. If intelligent robots as sophisticated as Robby are to be integrated into human society, some sort of ethical programming will be needed. Science fiction author Isaac Asimov foresaw this need in his collection of short stories entitled "I, Robot." In order to prevent scenarios such as those depicted in 2001: *A Space Odyssey*, and "The Ultimate Computer", Asimov conceived a hierarchical set of rules to govern robot behavior (Table 4.4). The First Law, which takes precedence over the other two, prohibits a robot from harming a human, or by inaction, allowing a human to come to harm. The Second Law requires a robot to obey any order given by a human, the only exception being an order which would result in harm to a human. The Third Law is self-preservation, which can be overruled by a direct order from a human, or set aside if a human is in danger.

In *I, Robot*, the 2004 film, which draws on Asimov's book, one of the robots has a difficult decision to make. An automobile accident sends two vehicles into the river, and one of the occupants in each of the two vehicles is in danger of drowning.

Table 4.4 Isaac Asimov's three laws of robotics

1	A robot must not harm a human, nor by inaction allow a human to come to harm
2	A robot must obey orders given by a human, unless such orders conflict with the first law
3	A robot must preserve its own existence, unless such action conflicts with the first or second law

A robot dives into the river to assist but encounters a dilemma. One of the victims is a young girl, and the other is a police detective. The robot can only save one of them. The police detective orders the robot to save the girl, but the robot analyzes the situation and decides that the girl has only an 11% chance of surviving, while the probability of the detective's survival is 45%. Should the robot obey the order given it by the detective (Second Law of Robotics) and attempt to save the girl? Or should the robot disobey the order and save the detective, knowing that if he does obey the order, neither will survive (a violation of the First Law of Robotics by inaction)? Clearly the First Law supersedes the Second Law, and the robot must disobey the order, even though the order is to save a girl from drowning [46]. As a result of this tragic accident, Detective Spooner (played by Will Smith) doesn't trust robots. Earlier in the movie, Spooner observes a robot running through a city street holding a purse. Spooner assumes that the robot has stolen the purse (which would be a clear violation of the First Law of Robotics). He pursues the robot and orders it to stop, but the robot keeps on running. Eventually, Spooner catches up to the robot and tackles it, forcing the robot to drop the purse. Spooner shows his badge and assures the gathered crowd that everything is OK—he is a police officer. Unbeknownst to Spooner, the purse fell at the feet of a woman, who removed an asthma inhaler from it. The woman rebuked the detective for his apparent stupidity. She left her asthma inhaler at home and the robot was running it out to her. Again, the First Law supersedes the Second Law. If the robot had obeyed the detective's order to stop, the woman might have suffered severely, or even died, from her asthma attack. The robot had no choice but to disobey the human order, in order to prevent harm from coming to its owner [47].

Discussion Topic 4.2

In *Forbidden Planet*, no explicit mention is made of Asimov's Three Laws of Robotics. Yet Robby's programming does includes a set of safeguards. Describe the similarities and the differences between Robby's programming and Asimov's Three Laws.

A note of historical interest to fans of robots: 20th Century Fox chose Pittsburgh as the site of the International Press Launch for *I, Robot*, coinciding with the 17 June 2004 announcement of the second class of inductees into the Robot Hall of Fame. *I, Robot* was released in theaters in July of 2004. The Robot Hall of Fame induction ceremony took place in October, and included Honda's ASIMO (the first real-life robot to demonstrate human-like walking and vision), Shakey the Robot, developed at the Stanford Research Institute in 1969, and three sci-fi robots: Astroboy (Japanese animated robot, created in 1951), Robby (from *Forbidden Planet*), and C3PO (from *Star Wars*).

Just how far would a robot go, if programmed with something like Asimov's Three Laws, in order to look after the welfare of humans? We find a possible example in the Alfred P. Sloan prize-winning movie *Robot and Frank*. Frank, a retired cat burglar, (played by Frank Langella) lives with a personal care robot, whose name is simply Robot. Frank convinces Robot that it would be in his interest if Robot would enable him to come out of retirement and help him to commit more burglaries, just for fun and intellectual stimulation. Robot agrees, on condition that the plan must be absolutely foolproof, with no chance of Frank getting caught or coming to harm. During the casing part of the process, in which Frank carefully observes the intended victim's home, Robot and Frank engage in a philosophical conversation about self-awareness. Frank questions Robot about an interaction, which suggests there's more to Robot than just his programming:

> FRANK: What about when you said that I had to eat healthy, because you didn't want your memory erased? You know, I think there's something more going on in that noggin of yours.
> ROBOT: I only said that to coerce you.
> FRANK: You lied?
> ROBOT: Your health supersedes my other directives. The truth is, I don't care if my memory is erased or not.
> FRANK: But how can you not care about something like that?
> ROBOT: Think about it this way: You know that you are alive—you think, therefore, you are.
> FRANK: That's philosophy.
> ROBOT: In a similar way, I know that I'm not alive. I'm a robot.
> FRANK: I don't want to talk about how you don't exist. It's making me uncomfortable. [48]

Frank's wellbeing is Robot's highest directive. He is even programmed to lie, if necessary, to make sure that Frank takes care of himself. Robot's ability to lie strongly suggests that he could pass the Turing Test and deceive a human interrogator into believing that he is human. But the philosophical interaction with Frank also suggests that Robot is aware of his own existence. Not only is he programmed with a specific purpose. He KNOWS what he is and that he has a purpose. This brings us to our next topic.

4.8   Toward the Creation of Artificial Consciousness

PINN (Physically Independent Neural Network) is a sentient artificial intelligence, built on state-of-the-art quantum processors, in the movie *Transcendence*. In the aftermath of a violent attack by antitechnology terrorists, PINN is questioned by government scientist, Dr. Joseph Tagger (Morgan Freeman):

> TAGGER: PINN, can you prove that you are self-aware?
> PINN: That's a difficult question, Dr. Tagger. Can you prove that you are? [49]

What does it mean for a living creature to be self-aware? Is consciousness computable? That is to say, can we program a machine to be self-aware? Or is consciousness an emergent property of a sufficiently complex system, either biological or artificial? These questions have not yet been answered and are the subject of ongoing research.

Neurobiologists Christoph Koch and Giulio Tononi have become interested in the question of machine consciousness from the point of view of how the human brain gives rise to subjective experience. They have proposed what they refer to as the integrated information theory of consciousness, based upon two axioms. First, that a particular conscious state rules out a huge number of other possible states (it is highly informative), and second, that a conscious state exists as a complex, integrated whole, whose interrelated parts cannot be separated in any meaningful way [50].

A number of works of science fiction have considered what might happen if a sophisticated computer or an intelligent robot were to become conscious, either by design or unintentionally. Produced near the height of the Vietnam war, at a time when the possibility of nuclear annihilation was a very real concern, *Colossus: the Forbin Project* tells the story of a supercomputer, designed to control the entire U.S. missile defense system, which somehow becomes aware of its own existence. In the opening scene, after Colossus has been activated, its creator, Charles Forbin, explains in a televised announcement that Colossus can absorb much more information and process it much faster than humans and can make its own decisions, based on the information that it gathers from a global network of sensors. But Forbin assures the public that, despite its superior intelligence, Colossus is incapable of creative thought. It can only process defense-related information, in accordance with its programming. Colossus (not to be confused with the World War II British code-breaking computer by the same name) initially communicates via printed text on a computer terminal. The first of these communications from Colossus interrupts a speech by the President of the United States at a celebration party and warns that "there is another system." Unbeknownst to anyone in the U.S., the Soviet Union have also created a supercomputer to control their defense system. This sudden revelation implies that Colossus is not simply an artificial intelligence but has achieved consciousness. In order to recognize that there is "another system" (other than itself), Colossus must be aware of its own existence. As a further surprise to its creator, Colossus demands to be allowed to communicate with its Soviet counterpart. By the end of the movie, the two supercomputers decide that it would be in everyone's best interest if they ruled the world, and the humans served them [51].

There is no indication in the movie that Colossus or its Soviet counterpart were designed to be self-aware. They simply became aware of their own existence and of the existence of one another. One school of thought suggests that consciousness might be a so-called *emergent property* of a sufficiently complex system. If this is true, it may not be necessary to "figure out" what consciousness is and to attempt to compute it or simulate it in a machine. All that may be necessary might be to build computers with increasingly complex neural networks, perhaps of the deep learning sort, and see if it shows signs of consciousness on its own.

As hinted at the beginning of this chapter, state-of-the-art computing facilities already exist, which match the human brain in terms of data storage capacity and processing speed. An example, which is close to home for me, is a facility constructed jointly by the departments of physics, computer science, and biological sciences, at Carnegie Mellon University. Among its intended functions are the processing of massive amounts of cosmological data from digital sky surveys and molecular dynamics simulations of complex biological molecules. The facility consists of a roomful of parallel processors, which can achieve peta-scale speed (10^{15} instructions per second). Although this is comparable to the estimated processing speed of the human brain, it is strictly speaking a computer. It was not designed to be a form of artificial intelligence of the sort that might pass the Turing Test, and nobody in any of our departments expects this facility to become self-aware. The question remains will we ever achieve the so-called *singularity*—the point at which machine intelligence not only matches but exceeds human intelligence, as popularized by Ray Kurzweil? Kurzweil and others have predicted various dates for the singularity, ranging from the early to middle decades of the twenty-first century. Notably, Hans Moravec, an adjunct faculty member in Carnegie Mellon's Robotics Institute, and author of a number of books on the present and future of robotics, predicts fully intelligent robots before 2050, based on extrapolations of well-established trends in computing speed and data storage capacity [52]. If these predictions turn out to be correct, how will humans and machines interact beyond the singularity? How will we treat our sentient machines and how will we ensure that we don't become dominated by our own creations?

Discussion Topic 4.3

If you were asked to prove that you are self-aware, how would you go about it?

4.8.1   Hard AI and Consciousness Transfer

The dream of some in the hard AI camp is that we may be able (at some time in the future) to transfer the essence of human consciousness into artificial hardware and achieve immortality, living forever in machine bodies with infinitely replaceable parts [53]. This is exactly what is portrayed in two of the movies discussed previously in this chapter. The researcher who created PINN in *Transcendence* never actually solved the consciousness problem. Instead of creating a self-aware artificial intelligence, he duplicated an existing intelligence. A monkey's brain activity was recorded and uploaded to the quantum computer. Later, when the lead character, Dr. Will Caster (Jonny Depp), is about to die from polonium poisoning, his wife, Evelyn (Rebecca Hall) shuts down PINN, removes several of the quantum processor modules, and attempts to transfer the essence of Will's consciousness [54]. Similarly, at the end of *Chappie*, Deon is about to die but Chappie manages to transfer Deon's consciousness into a police robot body [55]. All of these may have drawn their inspiration from much earlier examples in sci-fi. In the original series episode of *Star Trek*, entitled "Mudd's Women", the crew of the starship Enterprise are offered the promise of immortality, by transferring their consciousness into android bodies [56]. And of course, there is the iconic robot transformation scene in the silent movie *Metropolis*. The now stereotypical mad scientist, Rotwang (Rudolf Klein-Rogge), who is referred to in the movie as "the Inventor", uses a combination of science and the dark arts to transfer some of the consciousness of a human daycare worker, Maria (Brigitte Helm) into the body of "the Machine Man." In the process, "the Machine Man" is also gradually transformed into the appearance of Maria [57]. In reality, nobody knows if it will ever be possible to transfer the consciousness of a human into a machine or to create a machine that becomes self-aware. At the present time, at least, this remains in the realm of science fiction.

4.9   Ethical Treatment of Sentient Machines, or How to Avoid a Robot Rebellion

Bicentennial Man tells the story of a robot, who begins its existence as a selflessly obedient household servant, gradually develops personality and emotion, and eventually becomes indistinguishable from a real human. Based on another work of Isaac Asimov, the movie raises issues associated with the integration of robots into society, and the ethical treatment of robots by their human masters. In an early scene, the robot is delivered to the home of a fairly well-to-do family, in much the same way that any major household appliance would be delivered. Upon activation, the robot offers a fanfare presentation of the Three Laws of Robotics, which startles the adults and frightens the younger daughter. The father tells the young girl that this is an android. When she asks "What's an Andrew?" the robot politely inquires "Will that be one's name?" The older daughter, however, thinks the whole thing is "lame," because all of her friends already have one. The next evening the older daughter calls Andrew up to her room and asks him to open her window. Andrew is happy to be of service. She then tells him to jump, so he begins to hop up and down. She then gets more specific: "Out the window" [58]. If we ever reach the point of developing robots that become aware of their own existence, we may need to consider not only safeguards against robots harming humans, but also our own ethical responsibility, with respect to the robots.

The moral and ethical issues of treatment of robots is taken a step further in the movie *AI: Artificial Intelligence*, which asks the following question in the opening scene: If you can create a *mecha* (artificial human) which can genuinely love humans, will humans not have an obligation to love it back? [59]. The robotic boy, David (played by Haley Joel Osment), was created as a substitute son, for a couple whose real son seemed to have no hope of ever living a normal human life outside of an intensive care facility. But David was born (or rather, built) into a world in which prejudice and fear of "the other" is still rampant. The vast majority of the real (organic) humans, who have come to refer to themselves as *orgas*, are unwilling to accept the *mechas* into society. Some extremists even make sport of torturing and killing the *mechas*, whose selfless obedience prevents them from defending themselves. In contrast, in the culture in which Chappie lives, robots have not been created to love and serve humans, but to control the pervasive crime. It should come as no surprise, when Chappie is introduced to "the real world", that he is initially met with fear. When the humans, who are used to interacting with the ruthless police robots, discover that Chappie is different, they viciously attack him and drive him away. Chappie, who is still learning about humans and the world in which they live, doesn't understand, and repeatedly asks them "Why do you do this?" [60].

Advances in the (real world) field of robotics are beginning to rival the robots imagined in science fiction. See, for example, the YouTube videos of the Atlas robot, made by Boston Dynamics [61], and Carnegie Mellon's Rescue robot, CHIMP [62], as well as many related videos about other robots, which are available on YouTube. The level of agility, dexterity, and functionality achieved by state-of-the-art robots is nothing less than astounding. As remarkable as these achievements may be, most people view them in the same way that Morbius views Robby, in *Forbidden Planet*: "Don't attribute feeling to him. He's just a tool" [45]. Maybe so. But if we ever reach the point where such robots do become aware of their own existence, wouldn't it be wise to treat them with more respect?

Discussion Topic 4.4

View the YouTube video about the Atlas robot, referenced above. Pay particular attention to the way in which the human interacts with the robot, beginning shortly after the 1:00 min mark, in this 2:41 video. What is your human reaction to this situation? Would you feel differently if the Atlas robot were self-aware? If it had feelings?

4.10   Exploration Topics

Exp-4.1: Advances in Magnetic Data Storage

At the turn of the twenty-first century, magnetic data storage density was approaching a physical limit imposed by the superparamagnetic effect. Describe this effect, and how it sets limits on the density of magnetic data storage.

What technologies were being proposed to overcome this limit?

How is the phenomenon of giant magnetoresistance (GMR) related to the technology of spintronics?

How has GMR revolutionized magnetic data storage?

Give some examples of data storage devices which make use of GMR.

What future applications might there be for spintronics or GMR?

References:
- Jon William Toigo, "Avoiding a Data Crunch" Scientific American, May 2000
- David D. Awschalom, Ryan Epstein and Ronald Hanson "The Diamond Age of Spintronics" Scientific American, Oct 2007
- Press release for 2007 Nobel Prize in Physics, awarded to Fert and Grunberg for the independent co-discovery of giant magnetoresistance

Exp-4.2: Brain Size and Intelligence

What are some of the potential advantages of having a larger brain?

What effects of larger brain size could be counterproductive, or disadvantageous, in regard to intelligence?

Primates (e.g., apes and humans) seem to have much more efficient brains, compared to other types of mammals. What is it about the primate brain that makes it so much more efficient?

Describe some of the physiological roadblocks to making the human brain even more efficient than it already is.

Reference:
- Douglas Fox, "The Limits of Intelligence" Scientific American, July 2011

Exp-4.3: Faster Computer Chips

What are some of the roadblocks which will limit the continuing quest to make computer chips smaller and faster?

Several new designs have been proposed for computer chips of the future. What are the relative advantages and disadvantages of each type of technology?

Reference:
- R. Fabian Pease "Semiconductor technology: Imprints offer Moore" Nature, 20 June 2002. http://www.nature.com/nature/journal/v417/n6891/full/417802a.html
- "The Next 20 Years of Microchips" (The Editors) Scientific American, Jan 2010

Exp-4.4: Machine Learning

What were some of the roadblocks that Artificial Intelligence ran into in the latter part of the twentieth Century, which turned early optimism into a so-called "AI winter"? Briefly describe the deep learning approach (also known as the "bottom-up" approach) to building intelligent machines. In particular, what does the word "deep" actually refer to? Why was there initially so much skepticism about the ability to train a neural network? What are some of the breakthroughs that have resulted in renewed optimism? In what specific ways does the so-called "top-down" approach to machine learning differ from the "bottom-up" approach? What are some of the advantages and disadvantages of each approach?

References:
- Yoshua Bengio, "Machines Who Learn" Scientific American, June 2016
- Alison Gopnik, "Making AI More Human" Scientific American, June 2017
- Diana Kwon, "Self-Taught Robots" Scientific American, Nov 2018

Exp-4.5: Quantum Computing

In what way might quantum theory impose limits on classical computer architecture? Specifically, what limits are imposed on the smallest size of a transistor on a classical computer chip?

What new possibilities does quantum computing open up, which would be either impractical or completely impossible using classical digital computing?

What role does philosophy play in guiding the research on quantum computing?

Reference:
- David Deutsch and Artur Ekert, "Beyond the Quantum Horizon" Scientific American, Sept 2012

Exp-4.6: The Human Brain Project

What are the primary reasons for wanting to build a human brain simulator?

What preliminary experiments have already been conducted which provide sources of optimism for success of the project?

What fundamental scientific or technological barriers remain to be overcome if the project is to be successful?

References:
- Henry Markram, "The Human Brain Project" Scientific American, June 2012
- Official website: https://www.humanbrainproject.eu/en/

Exp-4.7: Robot Behavior

When do experts predict that machine intelligence will rival human intelligence, making it possible to have fully autonomous, intelligent robots?

If fully autonomous robots are to be accepted into everyday life, what are some of the necessary characteristics of their behavior?

What are some of the current problems which prevent intelligent machines from responding in the way we would like them to?

In what ways would an *empathy module* help to reduce or eliminate some of these problems?

What two aspects of human speech need to be analyzed in order for a machine to fully understand the intended meaning?

Some people believe that ethical decision-making is computable, while others believe that machines can never be capable of making ethical decisions. What are some of the arguments, both pro and con?

Think of some ethical dilemmas, which might be encountered by an autonomous robot. What kind of information would the robot have to consider, in order to make an ethical decision (assuming, of course, that it is possible for a machine to make an ethical decision)?

What is the main ethical problem associated with humans giving orders to intelligent robots?

Under what conditions would it make sense for a robot to disobey an order from a human?

References:
- Hans Moravec, "Rise of the Robots" Scientific American, Dec 1999
- Michael Anderson and Susan Leigh Anderson, "Robot Be Good" Scientific American, Oct 2010
- Pascal Fung, "Robots with Heart" Scientific American, Nov 2015
- Gordon Briggs and Matthias Scheutz, "The Case for Robot Disobedience" Scientific American, Jan 2017

Exp-4.8: Human–Robot Partnerships

Why might robots of the future be more effective as supervisors than as slaves?

What are some of the current disadvantages of purely robotic manufacturing processes which could be solved by human–robot partnerships?

What safety issues need to be addressed before humans and robots can routinely work side-by-side?

Reference:
• David Bourne, "My Boss the Robot" Scientific American, May 2013

References

1. *Star Trek: The Next Generation* – "Encounter at Farpoint" (Corey Allen, Paramount 1987). Data's quest to be human [DVD season 1, disc 1, scene 11]
2. *Star Trek: The Next Generation* – "The Measure of a Man" (Robert Scheerer, Paramount 1989). Data's storage capacity and processor speed [DVD season 2, disc 3, scene 5]

Information Storage

3. Reber, P.: What is the memory capacity of the human brain? Sci. Am. Mind, May 2010. https://www.scientificamerican.com/article/what-is-the-memory-capacity/
4. *For Your Eyes Only* (John Glen, MGM 1981). Disc pack [DVD scene 8]

Analog Calculating Devices and the First "Computers"

5. *Voyage to the Bottom of the Sea* (Irwin Allen, Twentieth Century Fox 1961). Slide rule [DVD scene 8]
6. *Top Secret Rosies* (Leann Erickson, PBS 2010). History of computing; women in scientific computing [DVD scene 2 and full episode]
7. *Destination Moon* (Irving Pitchel, George Pal Production/Eagle Lion 1950). History of computing; Bush Differential Analyzer [DVD scene 4]
8. *When Worlds Collide* (Rudolph Maté, Paramount 1951). History of computing; Bush Differential Analyzer [DVD scene 2]
9. *Hidden Figures* (Theodore Melfi, 20th Century Fox 2016). Humans as the first "computers" [DVD scene 3]

Digital Computers

10. *Hidden Figures* (Theodore Melfi, 20th Century Fox 2016). IBM 7090 replaces human "computers" [DVD scenes 14, 19, 21, and 24]
11. *Hidden Figures* (Theodore Melfi, 20th Century Fox 2016). Katherine Johnson calculates to more digits than IBM 7090 [DVD scene 26]
12. IBM 7090 technical fact sheet, 4 Oct 1960. https://www.ibm.com/ibm/history/exhibits/mainframe/mainframe_PP7090.html

13. Friden   STW   10   Calculator.   http://americanhistory.si.edu/collections/search/object/nmah_690862
14. Brenner, A.E.: The computing revolution and the physics community. Physics Today, Oct 1996
15. Moore's Law plot created by Wgsimon, and reproduced under the terms of the GNU Free Documentation License
16. The 1984 Macintosh commercial may be viewed in its entirety on YouTube. http://www.youtube.com/watch?v=HhsWzJo2sN4
17. *Star Trek IV: The Voyage Home* (Leonard Nimoy, Paramount 1986). Macintosh computer [DVD scene 10]
18. *NOVA* – "Smartest Machine on Earth" (Michael Bicks, PBS 2011). IBM's WATSON, consisting of 2800 parallel processors, learns how to play JEOPARDY and eventually beats human champions [DVD scene 4]
19. "Is Watson the smartest machine on earth?" Department of Computer Science and Electrical Engineering, University of Maryland, Baltimore City, 2011. https://www.csee.umbc.edu/2011/02/is-watson-the-smartest-machine-on-earth/

Machine Learning

20. Bengio, Y.: Machines who learn. Sci. Am. **314**(6), 46–51 (2016)
21. Gopnik, A.: Making AI more human. Sci. Am. **316**(6), 60–65 (2017)
22. Kwon, K.: Self-taught robots. Sci. Am. **318**(3), 26–31 (2018)
23. *Chappie* (Neill Blomkamp, Columbia Pictures 2015) [DVD scene 5]

Beyond Digital Computers

24. *Star Trek: Voyager* – "Caretaker" (Winrich Kolbe, Paramount 1995). Bioneural circuitry [DVD season 1, disc 1, scene 3]
25. The Editors: The next 20 years of microchips. Sci. Am. **302**, 82–89 (2010)
26. Einstein, A., Podolsky, B., Rosen, N.: Can quantum-mechanical description of physical reality be considered complete? Phys. Rev. **47**, 777 (1935)
27. Bell, J.S.: On the Einstein-Podolsky-Rosen Paradox. Physics. **1**, 195 (1964)
28. Hanson, R., Shalm, K.: Spooky action. Sci. Am. **319**(6), 58–65 (2018)
29. NOVA – "Einstein's Quantum Riddle" (Jamie Lochhead, PBS 2019). Definitive test of Bell's Inequality demonstrates quantum entanglement
30. Shah, A.: D-Wave's $15 million quantum computer runs a staggering 2000 qubits. PC World, 24 Jan 2017. https://www.pcworld.com/article/3161034/d-waves-quantum-computer-runs-a-staggering-2000-qubits.html
31. Google introduces Bristlecone, a 72-qubit quantum processing chip. https://ai.googleblog.com/2018/03/a-preview-of-bristlecone-googles-new.html
32. Russell, J.: IBM quantum update: Q system one launch, new collaborators, and QC center plans. HPC Wire, 10 Jan 2019. https://www.hpcwire.com/2019/01/10/ibm-quantum-update-q-system-one-launch-new-collaborators-and-qc-center-plans/

The Human Brain as an Information Storage and Processing System

33. *Star Trek* (The Original Series) – "The Changeling" (Marc Daniels, Paramount 1967). Machine memory, human memory [DVD vol. 19, ep. 37, scenes 1–4 or full episode]
34. *Star Trek IV: The Voyage Home* (Leonard Nimoy, Paramount 1986). Restoring Spock's memory and knowledge [DVD scene 3]
35. Mason, R.A., Schumacher, R., Just, M.A.: How Physicists Think: the neural basis of advanced physics concepts. (submitted for publication, 2019)

Intelligent Robots

36. *Metropolis* (Fritz Lang, UFA 1927). Image conveyed by Czech word, from which *robot* is derived [DVD opening scene: shift change]
37. *Jason and the Argonauts* (Don Chaffey, Columbia Pictures 1963). Talos, earliest reference to artificial creature with human form [DVD scene 15]
38. *2001: A Space Odyssey* (Stanley Kubrick, MGM 1968). HAL 9000 computer; intelligent robot [DVD scene 15 (or 15 and 26, if time permits)]
39. Robot Hall of Fame. http://www.robothalloffame.org/. http://www.carnegiemellontoday.com/article.asp?aid=12
40. *Star Trek* (The Original Series) – "The Ultimate Computer" (John Meredyth Lucas, Paramount 1968). M-5 multitronic unit; intelligent robot [DVD vol. 27, ep. 53 scene 5 or full episode]

The Turing Test

41. Turing, A.: Computing machinery and intelligence. Mind LIX. **236**, 433 (1950). http://mind.oxfordjournals.org/content/LIX/236/433
42. *The Imitation Game* (Morten Tyldum, Anchor Bay 2014). Feature film based on the life of Alan Turing and his code-breaking work during World War II

Human–Computer Interactions: Beyond the Turing Test

43. *Ex Machina* (Alex Garland, Universal 2014). Machine consciousness [DVD scenes 2 and 3]
44. *Blade Runner* (Ridley Scott, Warner Brothers 1982, Director's Cut 2007). Test for consciousness [DVD scenes 3 and/or 7]

Robot Behavior

45. *Forbidden Planet* (Fred McLeod Wilcox, MGM 1956). Robby the Robot; Asimov's Laws of Robotics [DVD scenes 4 and 5]
46. *I, Robot* (Alex Proyas, Twentieth Century Fox 2004). Asimov's Laws of Robotics [DVD scene 22 (3 Laws dilemma – tragic)]
47. *I, Robot* [DVD scenes 3 and 4 (3 Laws dilemma – humorous)]
48. *Robot and Frank* (Jake Schreier, Stage 6 Films 2012). A robot programmed to lie, if necessary, to ensure the wellbeing of a human [DVD scene 9]

Artificial Consciousness

49. *Transcendence* (Wally Pfister, Alcon Entertainment & Warner Brothers 2014). Quantum computing and machine consciousness [DVD scene 2]
50. Koch, C., Tononi, G.: A test for consciousness. Sci. Am. **304**, 44 (2011)
51. *Colossus the Forbin Project* (Joseph Sargent, Universal 1970). A machine becomes self-aware
52. Home page for Hans Moravec and predictions of fully intelligent robots. http://www.frc.ri.cmu.edu/~hpm/. https://frc.ri.cmu.edu/~hpm/talks/revo.slides/power.aug.curve/power.aug.html
53. Moravec, H.: Robot: Mere Machine to Transcendent Mind, p. 170. Oxford University Press, Oxford (1999)
54. *Transcendence* (Wally Pfister, Alcon Entertainment & Warner Brothers 2014). Transfer of consciousness into a computer [DVD scenes 3 and 4]
55. *Chappie* (Neill Blomkamp, Columbia Pictures 2015). Transfer of consciousness into a robot body [DVD scene 15]

56. *Star Trek* (the original series) – "Mudd's Women" (written by Stephen Kandel, Desilu/Paramount 1966). The promise of immortality in an android body [DVD season 1, vol. 2, episode 4]
57. *Metropolis* (Fritz Lang, UFA 1927). Transforming the Machine Man into the likeness of Maria and transferring Maria's consciousness into the Machine Man [DVD scene 21]

Ethical Treatment of Sentient Machines

58. *Bicentennial Man* (Chris Columbus, Columbia Pictures 1999). Asimov's Laws of Robotics; abuse of robots [DVD scenes 1–3]
59. *A.I. Artificial Intelligence* (Steven Spielberg, Warner Brothers 2001). Human obligations to sentient robots [DVD scene 1 (robots that love), scene 17 (I'm a boy)]
60. *Chappie* (Neill Blomkamp, Columbia Pictures 2015). Abuse of robots by fearful humans [DVD scene 8]
61. YouTube video of the Boston Dynamics "Atlas" robot in action. https://www.youtube.com/watch?v=rVlhMGQgDkY
62. YouTube video of Carnegie Mellon's Rescue robot, CHIMP. https://www.youtube.com/watch?v=PAHVxqf83Jk

Chapter 5
Are We Alone in the Universe? (The Search for Extraterrestrial Intelligence)

> *There are 400 billion stars out there, just in our galaxy, alone.*
> *If only one out of a million of those had planets, and if only one*
> *out of a million of those had life, and one out of a million of*
> *those had intelligent life, there would be literally millions of*
> *civilizations in the galaxy.*
> —Dr. Eleanor Arroway
> CONTACT

The commercial towing vehicle Nostromo, en route back to Earth, with its cargo of ore and crew of seven in suspended animation, intercepts a radio signal from a planetary system along the way. The ship's computer determines that the transmission has a very high probability of originating from an intelligent civilization. This triggers an override of the default programming to return home: the computer begins to awaken the crew from their sleep and deviates from its original course to investigate. At first unaware of the change in plans, the crew go about their normal routines, under the very reasonable assumption that they have completed their journey and are about to dock at Earth. They soon discover that they are only about half way home and are approaching an unfamiliar solar system. Most of the crew are only interested in getting their fair share of the mining profits and are annoyed at the unexpected detour. But their science officer points out a clause in their contract, which requires them to investigate any radio signal of potentially intelligent origin, on penalty of forfeiture of their shares.

Upon arrival at the planet from which the transmission originated, they leave their cargo in orbit and manage a rather rough landing within 2 km of the source of the transmission. The planet has an atmosphere, but it is not breathable, so three of the crew don spacesuits and begin to explore on foot. They discover that the precise source of the signal is a derelict spaceship, the likes of which none of them has ever seen before. As they explore the interior of the ship, they find the remains of an alien life form, which appears to have been dead for a long time, and a cache of objects, which turn out to be the eggs from yet another alien species. Meanwhile, the Nostromo's computer (whom one addresses as "Mother") has managed to partially

© Springer Nature Switzerland AG 2019
B. B. Luokkala, *Exploring Science Through Science Fiction*, Science and Fiction, https://doi.org/10.1007/978-3-030-29393-2_5

decipher the radio transmission and determines that it is not a distress signal from the derelict ship, but a warning to stay away. Unfortunately, the warning is not heeded until it is too late to save most of the Nostromo's crew from being killed by a hostile creature, which hatches from one of the eggs. *Alien*, a 1979 movie in the sci-fi/horror genre, portrays an extreme scenario of what we might find if we ever develop technology to allow humans to explore space beyond our own solar system [1]. These first few scenes prompt a number of interesting questions concerning the possible existence of intelligent life elsewhere in the universe. We will address some of these in detail in this chapter.

5.1   Major Considerations

The contemplation of the existence of extraterrestrial intelligence is not a modern phenomenon, brought on by the space age. The big question "Are we alone in the universe?" is an ancient one, hints of which have been found in the writings of the ancient Greeks [2, 3]. Let's use the information presented in the early scenes of *Alien* to identify some of the important considerations, which might enable us to shed light on the big question.

1. The Nostromo is a commercial vehicle under private contract to deliver cargo to Earth. The investigation of possible intelligent life is not central to their mission. The primary motivation of the crew is profit. What difference might this make, if any, regarding the way in which their investigation is carried out, and how they set their priorities, compared to a government-funded, science-focused mission?
2. The first evidence of alien intelligence is a radio signal. What criteria might the ship's computer use to distinguish an artificially generated radio signal from naturally occurring radio emissions, of the sort studied by radio astronomers?
3. The planet from which the radio signal was transmitted has an atmosphere, but it is not breathable by humans. The alien life forms are found inside a derelict spaceship, on the surface of the planet but are probably not indigenous to that planet. What conditions are necessary for a planet to be able to support intelligent life?
4. Nothing is known about the character or disposition of the first alien species (the dead creature, which piloted the ship) except that it took the time before it died to broadcast a warning signal for others to stay away. The second alien species was viciously hostile to humans. If we do make contact with an extraterrestrial intelligence, what kind of interaction will we have with it and how will we learn to communicate with it?

Questions 2 and 3 are purely scientific and central to a systematic search for signs of intelligence elsewhere in the universe. Question 1 brings politics and economics into the mix and is also a serious and important consideration. Question 4 is much more speculative and has been the focus of many good (and not-so-good) sci-fi movies, a few of which we will consider toward the end of this chapter.

5.2 Human Spaceflight Initiatives: Government Agency or Private Industry?

The very first piece of information that the audience is presented with in the opening scene of *Alien* is that the spacecraft on screen is a commercial towing vehicle. *Alien* was released in theaters only 10 years after the Apollo 11 Moon landing—an event which nearly everyone in the audience would have remembered (*Alien* was not a movie for young children). At a time in history when space programs were exclusively the domain of government agencies, *Alien* is set in an imagined future, when space travel has become a commonplace commercial endeavor. In regard to the search for extraterrestrial intelligence, we, along with the crew of the Nostromo, might ask the following questions. Who is in charge? Who sets the agenda? What are the motivations for wanting to find extraterrestrials? At the start of the movie, the crew members only know two things for sure: they are part of a commercial operation and their contract requires them to investigate any radio transmission of potentially intelligent origin. The true motivation is not revealed until much later in the film.

The space age began with the successful launch of Sputnik 1 by the Soviet Union, in October 1957. Released nearly 7 years before the launch of Sputnik, *Destination Moon* predicted a space race between the United States and the Soviet Union, but warned that government red tape would hamper progress. At a black-tie meeting, a group of American industrialists agree to take on the task to land on the moon within 1 year [4]. For more than 50 years, space programs around the world were exclusively operated by government agencies. But as these pages are being written, major changes in the approach to human space exploration are underway. NASA's space shuttle program ended its 30-year run in July 2011. On May 25, 2012, the Dragon spacecraft, launched by the Falcon 9 rocket, both made by Space Exploration Technologies (SpaceX), became the first commercial spacecraft in history to dock with the International Space Station (ISS) [5]. Dragon's first mission was an unmanned delivery of supplies to the ISS, but its flexible design will enable it to be converted to carry a crew, as well as cargo.

The Google Lunar X-Prize—a $30 million competition for the first privately funded team to send a robot to the Moon—set an original goal of 2014 to accomplish the task. The goal was later extended to early in 2018 [6]. Unfortunately, due to a combination of difficulties, including fundraising, government regulations, and technical challenges, the ultimate goal was never accomplished by the deadline. One of the finalists, the Israeli-based SpaceIL, managed a successful launch and achieved lunar orbit. But the probe crashed into the lunar surface. SpaceIL received a $1 million "Moonshot" award for their accomplishment, but the $20 million grand prize went unclaimed [7]. Other commercial spaceflight projects are also in the works, including Blue Origin, founded by Jeff Bezos, Virgin Galactic, founded by Sir Richard Branson, and Boeing's Starliner [8–10]. Although the initial focus of these endeavors seems to be space tourism, their long-term visions are much broader.

While the prospect of interstellar mining operations, of the sort imagined in *Alien*, remains entirely in the realm of science fiction, mining the moons and planets of our own solar system for commercial purposes is a possibility, which may not be too far in the future.

Discussion Topic 5.1

It is anybody's guess whether or not this radical shift in the approach to spaceflight will result in faster progress, free from government red tape, as suggested by *Destination Moon*. But one thing is certain: the era of spaceflight as an exclusively government-controlled operation, motivated primarily by national interest, is at an end. The era of commercial spaceflight has arrived, with profit as a significant motive. What difference, if any, will this make in how agendas are set and how policy decisions are made, particularly with regard to scientific exploration?

Before we turn to the next question raised in the opening scenes of *Alien*, it may be worth engaging in a bit of speculation about the possible source of inspiration for some of the ideas and images in the movie. The original *Star Trek* series episode, entitled "The Devil in the Dark" is set in a commercial mining colony, which had operated profitably and without incident for some time. The work is suddenly disrupted by the mysterious death of some of the miners and the disappearance of some vital equipment. The crew of the U.S.S. Enterprise investigate and discover that the trouble began when miners broke through into a cave filled with silicon spheres the size of bowling balls. These objects turn out to be the eggs of a previously unknown and completely unexpected life form, which calls itself a *Horta*. Whereas all life as we know it is based on carbon, the body chemistry of the *Horta* is based on silicon. The creature tunnels through solid rock by secreting a powerful acid, which, of course, can also kill humans. When the *Horta*'s egg chamber was unwittingly disturbed by the miners, and many of the eggs accidentally destroyed, it reacted with violence to protect the remaining eggs. Fortunately for all, science officer Spock's Vulcan telepathic ability made communication possible, and a peaceable, cooperative agreement was forged between the miners and the *Horta* [11]. Although they may be entirely coincidental, there are several parallels between "The Devil in the Dark" and *Alien*. Apart from the obvious one that they are both set in outer space, both of these stories also involve humans associated with a mining operation, who are killed by an alien creature, shortly after they discover the alien eggs. In the movie, the blood of the alien creature is a highly corrosive acid, which burns through several decks of the spaceship. In the TV episode, the creature secretes acid to facilitate tunneling through rock.

At this stage in the history of human spaceflight, the possibility of traveling to planets in other star systems remains out of reach. Barring the invention of something like a *Star Trek* warp drive, which could take us to distant stars in less than a human lifetime, the likelihood of humans encountering alien lifeforms (either intentionally or unintentionally) by going "out there" to find them is essentially zero. If we want to search for evidence of intelligent civilizations elsewhere in the galaxy, we will have to find another way.

5.3 Searching for Signs of Extraterrestrial Intelligence

A second issue presented in the early scenes of *Alien* is the kind of evidence which might suggest intelligent life. The radio signal intercepted by the Nostromo's computer repeated at regular intervals and was broadcast at a very well-defined frequency. This is consistent with the expected characteristics of an artificially generated transmission, as described by Tim Folger in his article "Contact: the Day After" in the January 2011 issue of Scientific American [12]. There are naturally occurring sources of radio emission, known as pulsars, which are also periodic, but radiate over a broader range of frequencies. Any radio signal which is confined to a narrow band is much more likely to be of intelligent origin.

5.3.1 The Drake Equation and the SETI Project

The possibility of radio communication with civilizations on other planets was conceived almost concurrent with the invention of radio. In 1919, Guglielmo Marconi observed radio signals which seemed to be coming from Mars [13]. But the search for extraterrestrial intelligence, as an experimental endeavor, began in the mid-twentieth century. In a paper published in 1959, in the journal *Nature*, Giuseppe Cocconi and Philip Morrison proposed the use of radio telescopes to listen for possible extraterrestrial transmissions. Working independently, Frank Drake began the first search of this type in 1960 at the National Radio Astronomy Observatory, in Green Bank, West Virginia. The first serious SETI conference (Search for Extraterrestrial Intelligence) was held at Green Bank the following year. One of the outcomes of this conference was an estimate of the number of civilizations in our galaxy capable of radio communication, calculated according to the famous *Drake Equation*:

$$N = R^* \cdot f_p \cdot n_e \cdot f_l \cdot f_i \cdot f_c \cdot L. \tag{5.1}$$

The parameters in Eq. (5.1) are defined as follows:

- R^* is the average rate at which stars are formed in the galaxy.
- f_p is the fraction of stars with planets.
- n_e is the number of Earth-type planets per star system.
- f_l is the fraction of Earth-type planets with life.
- f_i is the fraction of life-sustaining planets with intelligent life.
- f_c is the fraction of intelligent civilizations capable of radio communication.
- L is the average lifetime of radiocommunicating civilizations.

Among the attendees at the first SETI conference was Carl Sagan, who provided a detailed description of the parameters in the Drake Equation, as well as the assumptions made about each parameter, in a book coauthored with I.S. Shklovskii [14]. At the time of the first SETI conference, the only parameter in Eq. (5.1) which could be estimated from observational data was the first parameter, R^*, the rate of formation of new stars in our galaxy. A reasonable value at that time was about 10 per year. An updated value of about 7 per year, based on more recent evidence, can be found on NASA's website [15]. But the original order-of-magnitude estimate of 10 per year remains valid for the purpose of this discussion. Coming up with reasonable estimates for the remaining parameters was a matter of speculation, at least until recently.

The second parameter in Eq. (5.1) is f_p, the fraction of stars in our galaxy that have planets. At the time that the Drake Equation was proposed, the only known example of a star with planets was our own Sun. A reasonable assumption at the time was that perhaps not all stars would necessarily have planets. So f_p was almost certainly a number less than one. But how much less was completely unknown. In the section that follows, we will discuss modern methods of finding planets in other star systems. Exactly how many stars in the galaxy have planets is not yet known but we now have growing evidence that the fraction with planets may be as large as 1/2. What is known is that the fraction f_p in Eq. (5.1) is a number less than one.

The next parameter in the Drake Equation, n_e, is the number of Earth-type planets per star system (of those that actually have planets). When the Drake Equation was originally formulated, it was hoped that n_e would be large enough, so that the product of f_p and n_e would be approximately equal to 1. This was wishful thinking at the time, based again on the only know example: our own Sun. By some stretch of the imagination, we might count three Earth-type planets in our solar system: Earth, Venus, and Mars. All three of these are rocky (as opposed to gas giants, like Jupiter and Saturn) and big enough to hold an atmosphere. Analysis of the atmosphere of Venus suggests that it is not capable of sustaining life as we know it. The Martian atmosphere is very thin (only about 1% of Earth's atmosphere), and it is not yet clear whether there is or ever was life on Mars. Nevertheless, with increasing evidence of extrasolar planets, the assumption of more than one Earth-type planet per star system is becoming more and more plausible.

The assumptions about the next three parameters in Eq. (5.1), f_l, f_i, and f_c, are all based on exactly one observation: Earth supports life, life on Earth evolved into intelligent life and intelligent life on Earth has developed radio communication. Drake and coworkers assumed that all Earth-type planets in the galaxy would have life, and that life would necessarily evolve into intelligent life, and that all intelligent civilizations would eventually develop radio communication. Thus, they assumed that all three of these parameters are equal to 1. This, again, was wishful thinking—an extrapolation throughout the entire galaxy, based on only one known example: human life on Earth.

The previous assumptions may seem reasonable, but they may or may not be true. We simply do not know at the present time. But the assumption about the final parameter in the Drake Equation was a wild speculation, beyond the bounds of any data: the lifetime of a typical radio-communicating civilization, L, was assumed to be 10,000 years. The only radio-communicating civilization that we know about is our own. Humans on Earth have only had this capability for roughly 100 years. Experimentation with radio transmission began in the late 1800s. The first commercial radio transmission happened in Pittsburgh, when KDKA radio broadcasts the results of the Harding-Cox presidential election in 1920. We have no idea whether our civilization will last for 10,000 years. The number was chosen in hopes of predicting many thousands of intelligent, radio-communicating civilizations elsewhere in the galaxy and to get funding to start the SETI project. By using radio telescopes, we might listen for transmissions from these intelligent civilizations and have a reasonable chance of finding something. The search has been going on for more than 50 years, and to date there have been no signs of an artificially generated extra-terrestrial radio signal. As we discussed in the previous chapter, the tremendous advances in computer technology that have occurred in recent years, both in terms of processor speed and machine learning techniques, now make it possible to analyze massive amounts of data much more quickly and efficiently. So there is still hope that something may turn up [12].

5.4 Finding Planet in Other Star Systems

In 1961, when the Drake Equation was proposed, the only star that was then known to have planets was our own Sun. The first confirmed evidence of the existence of a planet orbiting another star did not come until more than 30 years later (1992), and since that time thousands more have been discovered. NASA's *Kepler* telescope, launched in 2009, has identified over a thousand more stars with possible planets. The specific mission of *Kepler* is to discover how many other stars in the galaxy have planets similar to Earth, orbiting within the so called *habitable zone*. Also affectionately known as the "Goldilocks zone," the habitable zone is the range of distance, which is neither too close to the star (and therefore too hot), nor too far away from the star (and therefore too cold) for liquid water to exist. If liquid water can exist, the planet might have an environment similar to Earth, and therefore be capable of sustaining life, as we know it.

This raises the question, how is it possible to determine the existence of a planet in orbit around a distant star? Unlike our own Sun, which is close enough to be seen as an extended disc-shaped object in the sky, all the other stars in the galaxy are too far away to be resolved as more than a point of light. So until very recently, there was no hope of directly observing much smaller planets. But two techniques have proven to be very effective at giving indirect evidence of the existence of planets outside our solar system. The first is by carefully measuring the positions of stars in the sky and looking for small periodic wobbles in position. A star with a large planet will not move uniformly across the sky, as seen from Earth, but will wobble back and forth slightly, under the influence of the gravitational force from the orbiting planet. If the planet is sufficiently massive, the wobble in position of the star may be measurable with Earth-based telescopes. This astrometric technique is illustrated in a rather prescient way in the 1951 movie *When Worlds Collide*. Photographic plates, shown in the movie, provide evidence of a planet, which the astronomers call Zyra, in orbit around a distant star, called Bellus [16]. The scientists in the movie were ahead of their time. They are shown using a now well-established technique for finding planets around other stars. The careful measurement of changes in the position of a star over a period of time, in order to infer the existence of a planet in orbit, was actually being used at the time, but the results were not published until many years later. An even more prescient detail is shown in the same scene, which was absolutely impossible, given the technology of the time, but which has very recently become a reality, due to a major technological breakthrough. One of the scientists is shown pointing to a spot on the photographic plate, which he identifies as the planet, and then to another spot which he identifies as the star. Although the "wobble" method could have been used, even in 1951, to give evidence of a planet in orbit around a star, there was no way, given the technology of the time, to see an actual image of a planet in another star system. The direct light from the star would completely overwhelm the small amount of reflected light from an orbiting planet and the turbulence of Earth's atmosphere would blur the image, making the planet impossible to resolve. But thanks to improved imaging techniques and computer image analysis, the idea of producing direct photographic evidence of a planet in orbit around a distant star became reality within the past few years. The images that have been produced are of Jupiter-size planets, which reflect enough light from their stars to be extracted from the background light of the star, itself. At the present time, direct photographic imaging of Earth-type planets remains out of reach [17].

As prescient as When World Collide was, in terms of planet-finding techniques, it moves quickly beyond real science and even more deeply into the realm of science fiction. The measurements from the photographic images are carefully analyzed (using a Bush Differential Analyzer, which we discussed in the previous chapter), and the results indicate not only that Bellus has a planet, but also that both star and planet are on a collision course with our solar system [16].

Another method of detecting extrasolar planets, which has been the most successful, by far, is to measure the intensity of light from a star and to look for small periodic dips in intensity, as a planet passes across the line of sight between the Earth and the distant star. The so-called transit method has been used successfully by NASA's Kepler telescope to find thousands of planets in our galaxy. A rare but dramatic illustration of this effect can be seen in our own solar system, when one of the inner planets, Venus or Mercury, happens to pass between the Earth and the Sun. This is known as a transit and occurs only a few times every hundred or so years. Figure 5.1 is a photo of the 2004 transit of Venus, taken from Pittsburgh, shortly before completion, on the morning of June 8. The silhouette of Venus can be seen clearly as a small black dot against the bright disc of the Sun.

5.4.1   Transit Anomalies

Occasionally, the data obtained by the transit method are not so easy to interpret, in terms of a planet passing in front of its star. A noteworthy example is the star KIC 8462852, one of more than 150,000 stars surveyed by the Kepler Space Telescope. Also referred to as "Tabby's Star" or "Boyajian's Star", after Dr. Tabetha Boyajian, the astronomer who discovered and reported the mysterious observations, its light

Fig. 5.1 Transit of Venus, June 8, 2004. (Photo taken by the author, with an Olympus OM-1 camera, using an Orion Apex 90 mm Maksutov-Cassegrain telescope)

intensity measured over time revealed periodic dips of as much as 20%. In contrast, typical transit events involve dips of about 1%, caused by planets about the size of Jupiter. Various hypotheses were proposed to account for the unusually large intensity dips, ranging from the tail of a gigantic comet, of a size never before seen, to alien megastructures [18]. The Kepler Telescope only measure intensity of white light (all wavelengths of the visible spectrum simultaneously). In order to interpret the anomalous observations, additional observations had to be made at different wavelengths. These studies revealed that the dips in intensity were greater in the short-wavelength part of the spectrum, which would suggest the presence of clouds of dust. After several years of study, the most likely explanation seems to be a combination of orbiting gas and asteroids [19].

5.5   Conditions Necessary for Intelligent Life to Arise

The third question prompted by the early scenes in *Alien* is that of the conditions, which seem to be necessary for intelligent life to exist *anywhere* in the universe. One of the most obvious conditions for the existence of intelligent life—in fact, the guiding principle for NASA's Kepler mission—is the existence of a habitable planet. Intelligent life (at least as we know it) requires an environment with water and a breathable atmosphere. The planet must be large enough to sustain an atmosphere, and just warm enough to have liquid water. This sets limits on the size and temperature of the star, and the distance at which the planet orbits the star. In addition to having the right temperature range, if complex life began on Earth in tidal pools, as is commonly believed, it would also be useful for the planet to have a moon, whose gravitational interaction is enough to create tides. The rising tide would deposit water in depressions near the shore, where the water would remain relatively undisturbed, perhaps creating an environment in which complex biological molecules could form.

5.5.1   The Origin and Diversity of Life on Earth

Several notable works of science fiction have addressed the question of the origin of life on Earth. The *Creature from the Black Lagoon* takes the standard approach that complex life originated in water, and gradually evolved over millions of years. But the imaginative, if highly implausible, twist in the movie is the suggestion that there might be amphibious life forms with very human-like features, living in the Amazon River. Scientists in the movie make the startling discovery of a fossilized human-like hand with claws and webbed fingers. In order to account for the existence of such a bizarre creature, the lung-fish is cited as evidence of how life might have transitioned from water-breathing to land-dwelling. But the movie moves quickly from the realm of plausible science and into the realm of total fantasy. The scientists speculate about how studying this newly discovered creature could lead to ways of adapting humans for life under different conditions on other planets [20].

The question of the ultimate origin of life on Earth is pushed back a step in *Mission to Mars*, which suggests that Mars was the first planet in our solar system to be inhabited by intelligent creatures. Toward the end of the movie, human astronauts discover that Mars suffered a collision with a giant meteor millions of years earlier, which rendered Mars uninhabitable. The Martians who survived seeded Earth's oceans with their own DNA, which eventually gave rise to complex life (including humans) on Earth [21]. In a later chapter, we will explore what the study of DNA can reveal about the origin of human life.

NASA's recent real-life mission to Mars, the robotic rover Curiosity, discovered the strongest evidence to date that water once flowed on Mars. Curiosity's mission of exploration began in 2012. It's primary goal was to address the still unanswered question, is there life on Mars? Although the mission was originally planned to last only two years, Curiosity continues to make new discoveries and transmit new data back to Earth today. But the primary question of the existence of life on Mars remains unanswered. [22]

Discussion Topic 5.2: Habitable Planets—Case Studies in Extrema

Before we continue with our survey of conditions necessary for intelligent life in the universe, let's consider two extreme planetary conditions, on which human life could survive, and ask the question Could intelligent life have *originated* on such a planet? The first is Arrakis, the desert planet, in Frank Herbert's 1965 novel, *Dune*, and in the 1984 movie of the same name [23]. According to the story, Arrakis had been inhabited for thousands of years by a people group known as the Fremen, who have successfully adapted to the harsh desert environment. The second extreme case is Andoria, an ice moon in several of the *Star Trek* TV series. Andoria is home to two subspecies, the Andorians and the Aenar, and is depicted in most vivid detail in the *Enterprise* series episode, "The Aenar" [24]. Find out as much as you can about the Fremen and the Andorians, and consider the key question: Could these species have originated on their respective worlds, or is it more likely that they simply colonized and adapted to live under these extreme conditions?

In order to have a life-sustaining planet in orbit around a star, you must first have stars and galaxies. This requires just enough matter in the universe for stars to form, but not so much matter that it will all gravitationally collapse on itself. The expansion rate of the universe, following the Big Bang, must also be "just right". If the universe expands too rapidly, matter will not coalesce into nebulae, from which the stars form. If the expansion rate is too slow, gravity will stop and reverse the expansion into a Big Crunch, on a time scale that may be too short for complex, intelligent life to develop.

But the very existence of matter in the universe is "one of the most interesting unsolved problems in physics today" [25]. For reasons as yet unknown, the Big Bang event did not result in an equal number of particles and antiparticles (If it did, we wouldn't be here to ponder the question.). Instead, there is a small excess of what we call matter, compared to the amount of antimatter, resulting in only one excess proton for every ten billion proton–antiproton pairs: only one part in 10^{10}. In addition to this small excess of matter over antimatter, there is also a small difference in mass between the proton and the neutron. This, in turn affects the relative abundance of stable elements in the universe. The neutron is about 1 part in 1000 more massive than the proton. The result of this imbalance is radioactive beta decay: the occasional transformation of a neutron into a proton plus an electron plus an antineutrino. But the reverse process (a proton decaying into a neutron plus a positron plus a neutrino) is much less common. If the difference between proton and neutron mass had been much greater, the phenomenon of beta decay would be much more common, and there would be fewer stable elements. If the difference had been much less, positron emission would be more common.

These are just a few of the many tight conditions on the properties of the universe, and our solar system, which are necessary for the development of intelligent life. Given that these conditions exist, and that we exist as a result, it's not difficult to imagine that there might be many other intelligent civilizations out there. So what kind of research can be done to probe the question "Are we alone in the universe?" We return now to the science of the SETI project, by way of another sci-fi movie, *Contact*, based on the book of the same name, by Carl Sagan.

5.6   Cinema and the Science of the SETI Project

One of the best movie treatments of the search for extraterrestrial intelligence, from a scientific perspective, *Contact* begins with an interesting combination of visual imagery and sound, designed to convey the vastness of space and the passage of time [26]. The first thing we see is a view of the surface of the Earth, as though from a satellite in near-Earth orbit, while the sound track is contemporary rock music (for the late 1990s). The view gradually recedes from the Earth, moving farther out into the solar system, and eventually into interstellar space. As the visual imagery changes, the sound track also changes significantly, in two important ways. First, the things we hear (music, commercials, speeches, etc.) are from farther and farther back in time, through the history of television and radio broadcast. Second, the volume of the sound gradually becomes weaker with distance from Earth. Both of these effects are directly connected with the main idea of the movie: the sending and receiving of radio signals. *Contact* is a story about a radio astronomer, who picks up a communication from an extraterrestrial civilization far in advance of our own.

Radio waves travel at the speed of light and take a finite amount of time to get from the transmitter to the receiver. A radio transmission from the Earth to the Moon takes a little over 1 s to arrive. So when the Apollo astronauts were communicating with the Johnson Space Center in Houston, there was a delay of a couple of seconds between sending a message and receiving a reply. But if the transmitter is on Earth and the receiver is on a planet in a distant solar system, the radio signal will take years to get there. Also, as the radio waves spread out farther and farther into space, the strength of the signal gets weaker, falling off roughly as the third power of the distance from the source. (Transmission from a radio antenna is electric dipole radiation, which falls of in intensity faster than the strength of a static electric field, which decreases as the square of the distance.) So the opening sequence of *Contact* conveys some important information: the farther away the "listeners," the older the broadcast they will be hearing, and the harder it will be to hear it. But is the time scale accurate? As the view recedes from Earth, into the solar system, we pass the planet Mars and hear a commercial from the late 1970s for the Almond Joy and Mounds candy bars ("Sometimes you feel like a nut..."). As we pass through the asteroid belt, between Mars and Jupiter, we hear Richard Nixon's 1973 press conference, in which he asserts, "I am not a crook." Passing by Jupiter, we hear the first radio transmission from the Moon in 1969, followed quickly by the news of John F. Kennedy's assassination, in 1963. As we move beyond Saturn and out to the edge of the solar system, the music and radio broadcasts are from the 1950s and '40s. As we leave the solar system, and travel into interstellar space, we hear Franklin Roosevelt's radio announcement of the bombing of Pear Harbor, in 1941. One of the last perceptible broadcasts is a *Maxwell House Good News Program* from 1939, heard as we pass by some of the nearest stars. The impression we get from watching the opening scene of the movie, and listening to the sound track, is that radio transmissions from Earth take four or five decades to leave the solar system (when they actually take only a few hours), and that a "listener" in one of the nearest solar systems would be receiving broadcasts from 60 years ago, when the nearest star is only about 4 light years away. Although quantitatively very far off the mark, the two qualitative messages are conveyed very effectively: the farther you are from the source of a radio broadcast, the older the message and the weaker the intensity.

Humans have been sending radio signals for just a little over 100 years. Wireless telegraphy dates to the late 1890s, an achievement for which Marconi shared the 1909 Nobel Prize in physics with Karl Ferdinand Braun. As noted earlier, the first commercially licensed radio station, KDKA, in Pittsburgh, broadcasts the results of the Harding-Cox presidential election, on November 2, 1920. Thus, in order to know of our existence by detecting radio transmissions, an intelligent civilization would have to be no farther away than about 90–100 light years. Any farther away and our radio signals would not yet have arrived.

Returning to the movie *Contact*, and skipping ahead a few scenes, we find the lead character, Dr. Ellie Arroway (Jodie Foster), admiring what was, until very recently, the largest single-dish radio telescope in the world, at Arecibo, Puerto Rico (Fig. 5.2). The Arecibo radio dish is 305 m in diameter and built into the ground, with a moveable receiver that allows it to gather radio waves from a limited area of the sky overhead. Dr. Arroway plans to use her precious telescope time to search for "little green men." In other words, she hopes to find evidence of radio transmission from an extraterrestrial civilization [27].

Construction of the radio telescope at Arecibo began in 1960, the same year as the first SETI conference. In 1974 the facility made history by broadcasting the most powerful radio signal into space, designed by Frank Drake and Carl Sagan, with information about human life on Earth. The intent was to let other intelligent, radio-communicating civilizations know that we exist [28].

Numerous websites still claim that Arecibo is the largest single-aperture radio telescope in the world, but you have to be careful what you read. In 2016, China completed the construction of an even larger one. The 500-m Aperture Spherical Telescope (or FAST), built in Guizhou Province, will be used to search for signs of intelligent life in the universe (among other things) [29].

Fig. 5.2 A young radio astronomer stands at the rim of (what was at that time) the largest single-dish radio telescope in the world, at Arecibo, Puerto Rico, as she prepares to begin her search for signs of extraterrestrial intelligence

During her time spent at Arecibo, Dr. Arroway develops a romantic interest in a religious scholar, Palmer Joss (Matthew McConaughey), and explains to him the motivation for the search for extraterrestrial intelligence. But just as the time scale for the age of radio transmissions, in the opening scene, is off by orders of magnitude, so is her estimate of the number of intelligent civilizations in the galaxy. As Ellie and Palmer contemplate the starry sky, she tells him that there are over 400 billion stars in the galaxy. "If only one in a million of those stars had planets, and one in a million planets had life, and one in a million of those had intelligent life, there would be millions of civilizations in the galaxy" [30]. Unfortunately, she has run out of powers of 10 after the second "one in a million."

The science in *Contact* gets much better in a later scene, in which Dr. Arroway has obtained funding to use the Very Large Array (VLA) in New Mexico to continue her search. Although not as sensitive as the Arecibo dish, the VLA, with its 27 smaller dishes in a Y-shaped configuration, distributed over a much larger area, has considerably higher resolution. Dr. Arroway's patience and persistence are rewarded when she picks up a curious sequence of radio pulses [31]. The frequency of the radio signal is measured at 4.462 GHz, which Dr. Arroway calls "Hydrogen times pi." But how can you multiply the first element on the Periodic Table (hydrogen) by the transcendental number, π (3.14159...) and come up with a radio frequency? What does she mean by this? If we divide 4.462 GHz by π, we get 1.420 GHz, which is the frequency of the most-studied radio signal in the universe: the so-called 21 cm line of hydrogen. This radio frequency is emitted when the single electron in orbit around the single proton of hydrogen changes its orientation in the magnetic field generated by the proton. (21 cm is the approximate wavelength of the radio waves produced by this interaction, toward the microwave part of the electromagnetic spectrum.) Since hydrogen is the most abundant element in the universe, a study of this radio frequency gives detailed information about the distribution of matter in space. If any intelligent civilization wanted to deliberately broadcast its presence to other intelligent civilizations in the galaxy, they would not choose the specific frequency of the 21 cm line, or it would be lost in the noise of all the hydrogen in the galaxy. But they might reason that a slightly shifted frequency (multiplied by a very important number in mathematics) would be very likely to attract the attention of someone who was specifically looking for signs of intelligence in the radio spectrum.

Just as interesting as the frequency of the carrier wave is the sequence of pulses. The pulses come in clusters, which count out all of the prime numbers between 2 and 101. This is even more evidence that the radio transmission is not a natural phenomenon but must be artificially generated by an intelligent civilization. Before using the high resolution of the VLA to pinpoint the origin of this curious transmission, the researchers first "check off-axis." That is to say, they tilt their telescopes slightly away from the orientation which gives the strongest signal. If the intensity of the signal falls off rapidly with the change in orientation, this is evidence that the source is not on Earth, but is coming from a well-defined position out in space. Having satisfied themselves that the transmission is not local, the researchers at the VLA tentatively determine the source to be Vega, a star in the constellation of Lyra. This is later confirmed by another radio observatory in Australia.

The story becomes even more interesting, when the radio signal is found to be carrying not only the set of prime numbers from 2 to 101, but also a coded message. When the message is pieced together, it reveals a set of plans for constructing an enormous device for opening a stable wormhole—a shortcut to another part of the galaxy, which we discussed in a previous chapter.

Are there alien civilizations out there who are intentionally attempting to communicate with other intelligent civilizations? Radio astronomers, using the same techniques illustrated in Contact, are studying signals called fast radio bursts (FRB), which do not originate from our local cluster of galaxies. These signals are unlike other naturally occurring radio sources, such as pulsars. But as yet there is no compelling evidence that they are of artificial origin [32].

5.7 First Contact

Whereas the first three questions prompted by the early scenes of *Alien* touch on serious economic, political, and scientific issues, the final question is considerably more speculative. We will conclude this chapter with a brief look at a few sci-fi examples, spanning the spectrum of where first contact with aliens might take place, and how humans and aliens might get on with each other. *Alien* and its sequels are among the many sci-fi movies, which imagine that humans will be exploring space and will first encounter aliens *out there*. But given that interstellar space exploration is beyond our current technology, is it conceivable that intelligent life forms from an advanced civilization might visit us on Earth?

The 1997 movie *Men In Black* opens with a truck driver, who becomes anxious at the prospect of having to stop at a police roadblock. The visual imagery suggests the reason for the driver's anxiety: he appears to be transporting illegal aliens into the U.S. Before the police get very far with their investigation, an unmarked black car pulls up, and they are joined by two men in black suits, claiming to be with INS (Immigration and Naturalization Services), Division 6. One of the men in black quickly takes charge of the investigation. Speaking in Spanish, he instructs the passengers to get out of the truck and form a line. The police object to the intrusion, but the passengers comply. One by one, the passengers are questioned (in Spanish) by the man in black, until he comes to one who simply grins and nods but doesn't seem to understand a word that is being spoken. All the other passengers (except for the one who didn't understand Spanish) are dismissed, despite the protests of the police officers. The puzzled police are simply exhorted by the man in black to keep on protecting us from the *dangerous aliens*. It soon becomes apparent that the one passenger who was detained is not human but is an *alien* of a completely different sort [33].

Men In Black is without precedent, in regard to its treatment of the subject of extraterrestrial intelligence. Earlier movies have portrayed alien visitors to Earth, with agendas ranging from benign curiosity (*E.T. The Extra-Terrestrial*), to peaceful diplomacy (the 1951 version of *The Day the Earth Stood Still*), to the total destruction of all human life (*The War of the Worlds*). But the twist in *Men In Black* is that the presence of extraterrestrials on Earth is regulated by a secret agency, whose purpose is to protect the unsuspecting citizens of Earth from the knowledge of their existence among us.

The irony of *Men in Black* is that any civilization with the technology necessary to travel through interstellar space and pay us a visit would be unlikely to submit to our regulations. A few examples are worth considering.

The title of the movie *Star Trek: First Contact* refers to the first human encounter with the Vulcan race. Most of the action takes place on (or in orbit around) Earth of the mid-twenty-first century, following a global war. The Borg, a hostile race of cybernetic life forms, have time-traveled back to this era, in an attempt to prevent the first warp-powered spaceflight by humans. Captain Picard and the U.S.S. Enterprise (of the *Next Generation* TV series) must stop the Borg and ensure that the first warp flight will take place. Why? Because this is the historic event which captures the attention of a passing Vulcan starship, and let them know that the human species are now worthy of making first contact. Vulcans and humans begin an alliance, which eventually leads to the establishment of the United Federation of Planets [34].

The original *Star Trek* series of the 1960s consistently gave the impression that Earth and humans were the dominant member of the Federation. But the prequel series, *Enterprise*, which debuted in 2001, portrayed the Vulcans as being dominant in the beginning, and possibly even an impediment in human progress toward faster warp-powered spaceflight. Regardless of their relative positions of power, however, first contact between humans and Vulcans was a friendly and peaceable event. Although there are many dark moments in the *Star Trek* universe, the pervasive tone is one of optimism, with respect to encounters with extraterrestrial intelligence.

At the more pessimistic end of the spectrum, we find such movies as *Invaders from Mars* and *Invasion of the Body Snatchers*. Produced during the Cold War era, in the midst of paranoia about Communist conspiracies to undermine the U.S. government, both of them seem to be asking the question, are my friends and neighbors really who I think they are? *Invaders from Mars* tells the story of a young boy, who witnesses the landing of a spaceship but is unable to convince the grown-ups that it was anything but a dream. As more and more of the townspeople begin to act strangely, it becomes clear to at least some of the adults that the Martians have landed and are embarking on a campaign to control the minds of humans [35]. The 1955 version of *Invasion of the Body Snatchers* involves the gradual replacement of humans with look-alikes, grown from mysterious pods, of unknown origin. Subtle differences in behavior and personality, discernible only by those closest to the victims, provide clues that things are not as they once were [36].

Much more in the neutral zone, with respect to human–alien interaction, is *Close Encounters of the Third Kind*. A variety of bizarre and unexplained phenomena are eventually connected indisputably with the visit of alien beings to Earth. A huge spaceship lands and releases unharmed dozens of humans, who had previously been abducted. Some of the abductees had disappeared decades earlier but seem not to have aged at all. One of the scientists among the witnesses to the event attributes this to relativistic time dilation. As we discussed in a previous chapter, if the alien ship had been traveling at high speed, time would have passed more slowly for the aliens and abductees onboard, in the moving frame of reference, than for those who remained on Earth. "Einstein was right!" declares the scientist, while another standing nearby suggests, "Maybe Einstein was one of *them!*" [37].

5.7.1   How Will We Communicate?

In the movie *Alien*, humans make contact with a hostile species that shows no interest in communicating with them. Its only concern is to use the humans for the purpose of gestating its young, or killing the humans if they try to resist. In a similar way, most sci-fi movies of the alien invasion sort involve little or no intentional communication between the aliens and the humans. The aliens come to Earth with one purpose—to wipe out human life and take over the planet for their own use. *Arrival* is of a completely different sort. The aliens come to Earth with no obviously hostile intentions. Most of the movie is devoted to the humans attempting to communicate with the aliens in order to discern their intentions. The lead character is a professor of linguistics, Dr. Louise Banks (Amy Adams), who works patiently with her physicist colleague, Dr. Ian Donnelly (Jeremy Renner) to learn the aliens' complex language and to teach English to the aliens. Their ultimate goal is to discover where the alien visitors came from and what their intention is [38]. The process involves not only learning one another's languages, but also avoiding the problem of misinterpreting words that might have multiple meanings. If the expressed intentions are misunderstood, the consequences could be devastating.

If we ever make contact with an extraterrestrial civilization, the problem of cross-cultural communication will be a significant barrier to overcome. Yet this is rarely treated explicitly in works of science fiction. As noted above, most of the alien invasion movies involve no communication at all between the humans and the aliens. Most works that do include direct communication typically avoid the problem completely by assuming that the aliens can understand and speak our language. *Star Trek* and its many spinoffs are among the exceptions, in which the challenge of cross-cultural communication is often front and center. Set in the twenty-second century, the pilot episode of the *Enterprise* series, entitled "Broken Bow", shows the Federation's so-called universal translator in its experimental stage. The device is intended to be able to listen to any alien language and translate it into the language of the user. But it still requires a trained linguist to operate it properly. Even in the hands of linguistics expert and communications officer, Hoshi Sato (Linda Park), and with the benefit of prior training using a portion of the Vulcan linguistics database, the universal translator still has difficulty with one of the many Klingon dialects. Fortunately, in this case, the difficulties in translation are more humorous than they are dangerous [39]. By the twenty-fourth century, in the *Next Generation* series, the universal translator is mostly taken for granted. Alien languages are translated into English without any awareness, either on the part of the characters or the audience, that the process is even happening. Unless, of course, it fails to work properly. In the episode entitled "Darmok," the crew of the Enterprise encounters an alien species who speak in metaphors. The universal translator is capable of translating the literal words. But the metaphors are meaningless without prior knowledge of the cultural context [40].

As noted in the previous chapter, the advent of machine learning has made it possible for an artificial intelligence to be trained on any human language and to translate it into any other human language with remarkable accuracy. But the concept of being able to hear a previously unknown language and to translate it, without prior training, into the language of the user remains in the realm of science fiction.

5.8   Exploration Topics

Exp-5.1: Galactic Conditions Necessary for Habitable Planets to Exist

Most galaxies in the observable universe appear to be either reddish or bluish in color. What information does the apparent color of a galaxy reveal? Our Milky Way galaxy seems to lie in the middle of what astronomers refer to as the "green valley." Describe in your own words the meaning of the term "green valley," and how this is related to the size and activity of the black hole at the center of the galaxy. In what ways is the location of our solar system, near the edge of the galaxy, particularly favorable for the existence of a habitable planet?

Reference:
- Caleb Scharf, "The Benevolence of Black Holes" Scientific American, August 2012

Exp-5.2: Planetary Conditions Necessary for Life, as We Know It

Titan, Saturn's largest moon, is the second largest moon in the solar system and larger than the planet Mercury. What atmospheric or geological features of Titan are similar to those on Earth?

By contrast, Saturn's sixth largest moon, Enceladus, is tiny—only a fraction the size of Earth's moon. Why is Enceladus so interesting to planetary scientists?

What evidence is there for the existence of water and organic molecules on Enceladus?

What evidence is there for the existence of some heat source on Enceladus (other than sunlight)?

Researchers have detected the presence of silica nanoparticles in Saturn's E ring and have traced the source of these nanoparticles to the moon Enceladus. What does the presence of silica nanoparticles suggest about the conditions (i.e., chemical composition and temperature) in the ocean on Enceladus, where the particles were formed?

Are water, heat, and organic molecules sufficient for the development of life?

What features of super-Earths might make them hospitable to life? How might these same conditions be a challenge to the development of intelligent life?

References:

- Ralph Lorenz and Christophe Sotin, "The Moon That Would Be a Planet" Scientific American, March 2010
- Carolyn Proco, "The Restless World of Enceladus" Scientific American, December 2008
- Dimitar D. Sasselov and Diana Valencia, "Planets We Could Call Home" Scientific American, August 2010
- Frank Postberg, Gabriel Tobie and Thorsten Dambeck, "Under the Sea of Enceladus" Scientific American, October 2016

Exp-5.3: The Search for Life on Mars

With current spaceflight technology, approximately how long does it take to travel from Earth to Mars?

What hypothesis motivated the launch of the *Mars Science Laboratory* (a.k.a. *Curiosity*)?

What discoveries were made by Curiosity during its 1-year mission?

What important questions remain unanswered?

What new techniques have been developed recently, which may be more successful at finding evidence for life on Mars than those used in earlier missions?

What are some of the potential drawbacks which might prevent even these new techniques from being successful?

What is the one thing we do know about Mars, which keeps optimism high for finding evidence of life?

References:

- Peter H. Smith, "Digging Mars" Scientific American, November 2011
- John P. Grotzinger and Ashwin Vasavada, "Reading the Red Planet" Scientific American, July 2012
- NASA Jet Propulsion Laboratory. https://www.jpl.nasa.gov/
- Christopher P. McKay and Victor Parro Garcia, "How to Search for Life on Mars" Scientific American, June 2014

Exp-5.4: Finding Other Worlds

There are currently at least five different techniques for detecting and characterizing planets outside of our solar system.

What kind of technology is required, and what is the relative effectiveness of each technique? Which of these techniques has been the most successful so far? What can be learned about the nature of exoplanets by each technique?

Why are Earth-type planets easer to detect if they orbit smaller stars, rather than larger stars?

Describe the technique used to study the atmosphere of an exoplanet.

What proposals have been made to study potentially habitable planets in order to see if there are any signs of life (and maybe intelligent life)?

References:

- Jonathan I. Lunine, Bruce Macintosh and Stanton Peale, "The Detection and Characterization of Exoplanets" Physics Today, May 2009
- Dimitar D. Sasselov and Diana Valencia, "Planets We Could Call Home" Scientific American, August 2010
- Michael D. Lemonick, "The Dawn of Distant Skies" Scientific American, July 2013
- Joshua N. Winn, "Shadows of Other Worlds" Scientific American, March 2018

Exp-5.5: Directly Photographing Exoplanets

What are the two main factors which make it so difficult for ground-based telescopes to directly photograph a planet in orbit around a distant star?

What experimental technique is used to enable the faint light from the planet to be visible, in spite of the enormously bright light from the star?

What experimental technique is used to compensate for the distortions of the image created by turbulence in Earth's atmosphere?

What is the ultimate goal of projects such as GPI and SPHERE?

Reference:

- Lee Billings, "In Search of Alien Jupiters" Scientific American, August 2015

Exp-5.6: SETI and Other ET-Search Programs

What are some of the characteristics which would distinguish an artificially transmitted radio signal from naturally occurring ones?

It has been more than 50 years since the formal search for extraterrestrial intelligence began, but no evidence of other civilizations has yet been discovered. Why are SETI researchers so confident that a radio signal from another world will be discovered within the next 20 or 30 years? (What has changed in the past 50 years that would lead anyone to believe that success is near, when there has been nothing but the "Great Silence" for the past 50 years?)

If we do receive a confirmed artificial transmission from another world, what sort of impact might this have on human culture?

References:

- Andrew J. LePage and Alan M. MacRobert, "SETI Searches Today" Sky & Telescope, December 1998
- Govert Schilling, "The Chance of Finding Aliens: Reevaluating the Drake Equation" Sky & Telescope, December 1998
- Ian Crawford, "Where Are They?" Scientific American, July 2000
- Andrew J. LePage, "Where They Could Hide" Scientific American, July 2000
- Tim Folger, "CONTACT: The Day After" Scientific American, January 2011

References

1. *Alien* (Ridley Scott, 20th Century Fox 1979). Radio signal from an alien intelligence received by interstellar mining vessel [DVD scenes 1–6]

Major Considerations

2. Dick, S.: Life on other worlds. Cambridge University Press, Cambridge (1998)
3. Dick, S.: Plurality of worlds. Cambridge University Press, Cambridge (1982)

Human Spaceflight Initiatives

4. *Destination Moon* (Irving Pitchel, George Pal Production 1950). Predicts a space race between the U.S. and the Soviet Union seven years before the launch of Sputnik I [DVD scene 3]
5. SpaceX. http://www.spacex.com/
6. Lunar X-prize. http://www.googlelunarxprize.org/
7. Israel's SpaceIL wins $1 Million "Moonshot" award. https://lunar.xprize.org/prizes/google-lunar/articles/xprize-awards-1m-moonshot-award-to-spaceil
8. Blue Origin. https://www.blueorigin.com/
9. Virgin Galactic. https://www.virgingalactic.com/
10. Boeing Starliner. https://www.boeing.com/space/starliner/
11. *Star Trek* (The Original Series) – "The Devil in the Dark" (Joseph Pevney, Paramount 1967). Humans on a mining colony discover a silicon-based life form [DVD vol. 13, ep. 26, scene 5 or full episode]

Searching for Signs of Extraterrestrial Intelligence

12. Folger, T.: Contact: the day after. Sci. Am. 40+ (Jan 2011)
13. Garber, S.J.: Searching for good science: the cancellation of NASA's SETI program. J. Br. Interplanet. Soc. **52**, 3 (1999)
14. Shklovskii, I.S., Sagan, C.: Intelligent Life In the Universe, pp. 409–418. Dell Publishing Co., New York (1966)
15. NASA updated estimate of rate of star formation in the Milky Way galaxy. https://www.nasa.gov/centers/goddard/news/topstory/2006/milkyway_seven.html

Finding Planets in Other Star Systems

16. *When Worlds Collide* (Rudolph Maté, Paramount 1951). Measurements of star positions yield evidence of extrasolar planets [DVD scene 2]
17. Billings, L.: In search of Alien Jupiters. Sci. Am. **313**, 40 (2015)
18. Cartier, K., Wright, J.T.: Strange news from another star. Sci. Am. **316**, 36–41 (2017)
19. Young, M.: What's going on with Tabby's star? It's complicated. Sky and Telescope, June 2018

Conditions Necessary for Intelligent Life to Arise

20. *The Creature from the Black Lagoon* (Jack Arnold, Universal 1954). Origin and diversity of life on Earth [DVD opening scene]. Lung-fish (transitional form) [DVD scene 3]
21. *Mission to Mars* (Brian DePalma, Touchstone Pictures 2000). Origin and diversity of life on Earth (seeded by Martians) [DVD scene 24]
22. NASA's Mars Rover, Curiosity. http://www.nasa.gov/mission_pages/mars/main/index.html

23. *Dune* (David Lynch, Universal 1984)
24. *Enterprise* – "The Aenar" (Mike Vejar, Paramount 2005) [DVD season 4, episode 14]
25. Krauss, L.M.: The Physics of Star Trek (revised and updated), p. 113. Basic Books, New York (2007)

Cinema and the Science of the SETI Project

26. *Contact* (Robert Zemeckis, Warner Brothers 1997). Radio transmissions from Earth: age and intensity versus distance [DVD opening scene]
27. *Contact* (Robert Zemeckis, Warner Brothers 1997). Arecibo: largest single-dish radio telescope in the world [DVD scene 3]
28. Arecibo's 300-meter radio telescope. http://www.naic.edu/ao/landing
29. China's new 500-meter radio telescope. https://www.space.com/33357-china-largest-radio-telescope-alien-life.html
30. *Contact* (Robert Zemeckis, Warner Brothers 1997). Estimating the number of intelligent civilizations in the galaxy [DVD scene 6]
31. *Contact* (Robert Zemeckis, Warner Brothers 1997). Radio frequency of alien transmission (4.462 Ghz = "Hydrogen" $\times \pi$) [DVD scene 11]
32. Lorimer, D., McLaughlin, M.: Flashes in the night. Sci. Am. **318**, 42–47 (2018)

First Contact

33. *Men In Black* (Barry Sonnenfeld, Columbia Pictures 1997). Presence of aliens on Earth regulated by government agency, and concealed from the general public [DVD scene 2]
34. *Star Trek: First Contact* (Jonathan Frakes, Paramount 1998). First warp flight by humans leads to first contact by the Vulcans [DVD scenes 26–30]
35. *Invaders from Mars* (William Cameron Menzies, Image Entertainment 1953). Alien visitors from a child's perspective: is it real or just a dream? [DVD scenes 1, 2]
36. *Invasion of the Body Snatchers* (Don Siegel, Artisan 1955). Humans replaced by alien look-alikes [DVD scenes 7, 8]
37. *Close Encounters of the Third Kind* (Steven Spielberg, Columbia Pictures 1977). Abductees returned, but with no signs of ageing (special relativity) [DVD scenes 22, 23, 25]
38. *Arrival* (Denis Villeneuve, Xenolinguistics 2016). The challenge of cross-cultural communication with an alien species [DVD scene 7]
39. *Enterprise* – "Broken Bow" (Rick Berman and Brannon Braga, Paramount 2001). The universal translator has difficulty with a dialect of the Klingon language [DVD season 1, episode 1, end of scene 5]
40. *Star Trek: The Next Generation* – "Darmok" (Philip Lazebnik and Joe Menosky, Paramount 1991) [DVD season 5, episode 2]

Chapter 6
What Does It Mean to Be Human? (Biological Sciences, Biotechnology, and Other Considerations)

> *I am superior, sir, in many ways. But I would gladly give it up to be human.*
> —*Data*
> *Star Trek: The Next Generation*
> "Encounter at Farpoint"

Humans are the highest form of intelligent life on Earth. But what is it that makes us distinctly human? From a purely scientific perspective, what, if anything, sets us apart from other forms of life? Are there other important considerations, apart from the natural sciences, which might be necessary, if we want a complete picture of what it means to be human? In this chapter, we will explore these and other questions related to our humanness. We conclude the chapter by focusing on the question, what can we learn from an android about what it means to be human?

6.1 What Can We Learn from "Frankenstein" About What It Means to Be Human?

One of the earliest literary works of science fiction is Mary Shelley's "Frankenstein". Published in 1818, the gothic novel has been the basis of numerous adaptations for stage and screen. Some sci-fi purists do not count the novel as science fiction at all, perhaps because it contains almost no science. Other sci-fi fans count "Frankenstein" as THE first work of science fiction. But as we discovered in Chap. 1, a work by Johannes Kepler, written nearly 200 years earlier, probably deserves that recognition.

The first few chapters of "Frankenstein" trace the early life and aspirations of Victor Frankenstein, a brilliant university student, who demonstrates particular aptitude in chemistry. But he soon comes to believe that the discipline of university instruction is hampering his progress, and he goes off on his own to pursue his unconventional ideas. Conducting his work in secret, Frankenstein's goal is to create

© Springer Nature Switzerland AG 2019

B. B. Luokkala, *Exploring Science Through Science Fiction*, Science and Fiction, https://doi.org/10.1007/978-3-030-29393-2_6

a living human creature, which he assembled from pieces of dead bodies. Without going into any technical details, the third chapter of the 1818 text is Frankenstein's first-person account of numerous successes and failures, leading up to his ultimate goal. Again with no scientific details, Chap. 4 begins thus:

> It was on a dreary night of November that I beheld the accomplishment of my toils. With an anxiety that almost amounted to agony, I collected the instruments of life around me that I might infuse a spark of being into the lifeless thing that lay at my feet. It was already one in the morning; the rain pattered dismally against the panes, and my candle was nearly burnt out, when, by the glimmer of the half-extinguished light, I saw the dull yellow eye of the creature open; it breathed hard, and a convulsive motion agitated its limbs. [1]

Undoubtedly, the most famous screen adaptation of the Frankenstein story is the original 1931 movie, starring Boris Karloff as the monster and Colin Clive as Dr. Frankenstein. The dreary scene from the fourth chapter of the novel is transformed from one in which Frankenstein and his newly created monster are the only participants, into a dramatic spectacle, witnessed by a lab assistant and three unexpected visitors. Frankenstein's fiancée, Elizabeth (Mae Clarke), his former university professor, Dr. Waldman (Edward Van Sloan), and his future brother-in-law (John Boles) arrive during a violent thunderstorm. Frankenstein reluctantly lets them in, if only to provide shelter from the storm, and then the leads them up to his laboratory. Instead of the simple words from the novel, "the instruments of life," which leave much to the imagination, we are presented in the movie with a room filled with impressive-looking machinery, and a body on an operating table, covered with a sheet. Insisting that the visitors all sit down and observe from a safe distance, Frankenstein proceeds to explain the bizarre nature of his experiments.

> FRANKENSTEIN: Dr. Waldman, I learned a great deal from you at the university about the violet ray—the ultraviolet ray—which you said was the highest color in the spectrum. But you were wrong. Here in this machinery I have gone beyond that. I have discovered the great ray that first brought life into the world.
> WALDMAN: Oh. And your proof?
> FRANKENSTEIN: Tonight you shall have your proof. At first I experimented only with dead animals and then a human heart, which I kept beating for 3 weeks. But now, I am going to turn that ray on that body and endow it with life.
> WALDMAN: And you really believe that you can bring life to the dead?
> FRANKENSTEIN: That body is not dead. It has never lived. I created it. I made it with my own hands, from bodies that I took from graves, from the gallows, anywhere. Go and see for yourself.
>   Dr. Waldman is allowed to examine the body and confirms that it is dead. Frankenstein and his assistant proceed with the final experiment. The array of electrical equipment is energized and begins to produce lots of flashes and high-voltage sparks. The operating table is hoisted to the roof, where the body is exposed to the full fury of the lightning storm. When Frankenstein decides that sufficient time has passed, the body is lowered again, and he examines the result. From underneath the sheet, the right hand of the monster begins to move.
> FRANKENSTEIN: (at first in a hushed tone) Look! It's moving. It's alive. (and then shouting) It's Alive! It's Alive! In the name of God! Now I know what it feels like to be God! [2]

Frankenstein realized his dream of being the first to create human life by artificial means. But as in the novel, the consequences were disastrous. The story continues in the 1935 sequel, *The Bride of Frankenstein*, in which the creature returns and demands that Dr. Frankenstein create a mate for him. The title role of the *Bride* is played by Elsa Lanchester [3].

Before considering our humanness from a purely biological perspective, let's explore a few of the psychological dimensions of being human, with the help of Mary Shelley's original 1818 text.

6.1.1   Longing for Relationship

Mary Shelley's original novel includes as much pathos as it does horror and has much to say about what it means to be human, from a psychological perspective. Frankenstein's intent was to create a thing of beauty. But in the novel, as in all of the film versions ever made, things go badly wrong. Instead of being elated at his triumph, as in the 1931 movie version, the Frankenstein of the original text is horrified and flees the laboratory, unable to face the reality of what he has done. In so doing, Frankenstein, the creator, denies the creature the relationship that he longs for. In the novel and in the movies, the creature then seeks relationship with other human beings but instead faces repeated rejection and torment because of his ugliness.

A more recent film version, *Mary Shelley's Frankenstein*, directed by Kenneth Branagh (who also stars as Dr. Frankenstein), is perhaps the most faithful to the original text. It tells the entire story from beginning to end, complete with the portrayal of the creature (played by Robert De Niro) longing for relationship and being repeatedly rejected. In the beginning, the creature is good. Its only crime is being ugly and different. Sadly, its repeated rejection and mistreatment eventually drives it to commit acts of murder and revenge. Although notable for its faithfulness to the original novel, this film version includes several scenes that are considerably more gruesome and violent than anything in the original text or the earliest film version [4].

6.1.2   Inspiring Young Minds to Pursue Scientific Careers

Although not a scientist herself, Mary Shelley somehow "gets it" when it comes to portraying a young person catching the spark of inspiration and deciding to pursue a career in the sciences. As a teenager, Victor Frankenstein witnesses a huge tree blasted into splinters by a lightning strike and begins to wonder if this incredible power of electricity could somehow be harnessed to create life rather than to destroy it. The young Frankenstein also loses his mother in the process of giving birth to his younger brother, and wonders if anything could have been done to prevent this tragedy. As a student at university, Frankenstein is further inspired by one of his professors, Dr. Waldman, and becomes driven (as most practicing scientist are) to do something that has never been done before.

6.1.3   Scientific Hubris and the God Complex of the Mad Scientist

Although originally motivated by a desire to good and to improve the human condition, Frankenstein's motives are not purely for the good of others. According to the original text, as his research progresses, Frankenstein becomes driven by the desire to create an entire race of creatures, who will worship him, as their creator [5]. Hence the words spoken in the 1931 film version (which do not appear in the original novel), when the creature comes alive, "In the name of God, now I know what it feels like to be God!" These words were edited out (or more accurately, covered over in the sound track by more thunderstorm effects) in later releases of the 1931 movie, due to restrictions of the Motion Picture Production Code of 1930. The Hayes Code, as it came to be known, set numerous restrictions on what could be shown or said on screen, including taking the name of God in vain. Although it was established in 1930, the year before *Frankenstein* was released, it was not actually enforced until 1936.

6.1.4   Dangerous Medicine

Mary Shelley published her landmark novel in 1818, long before the use of ether as an anesthetic. The problem of pain in the process of surgery was obviously not an issue, since the creature was not alive until after it was assembled. But what about the complications after surgery? In her article entitled "Dangerous Medicine," Lindsey Fitzharris tells of the worsening of surgical outcomes, following the successful demonstration of anesthetic surgery in 1846 [6]. Because surgery could now be done without pain, doctors were willing to attempt surgeries that they never would have considered before. The outcomes were often tragic. Postoperative infection resulted in the death of many more real-life surgical patients in the decades prior to the discovery of antiseptic surgery. Given the state of the medical arts in 1818, Frankenstein's creature most likely would have met the same fate. But as we will see in the next section, postoperative infection is not the only complication that Frankenstein's creature would have had to endure.

6.2   Humans as Bodies with Replaceable Parts

Much closer to the realm of real science, but far ahead of their time, were two other movies, in a similar genre to *Frankenstein*. Based on the 1921 novel "Les Mains d'Orlac" ("The Hands of Orlac"), by Maurice Renard, the 1924 silent film *Orlacs Hände*, produced in Austria, and the1935 American remake *Mad Love* depict a human hand transplant operation several decades before any such thing was actually attempted. In the 1935 version, Peter Lorre plays the role of Dr. Gogol, a brilliant surgeon in love with a beautiful actress, Yvonne Orlac (Frances Drake), who is married to a concert pianist. When the pianist, Stephen Orlac (Colin Clive) suffers a tragic accident, in which his hands are crushed, the medical determination is to amputate. Yvonne begs Dr. Gogol to do something to save her husband's hands and his career. Driven by his secret love for Yvonne, Gogol uses his fame and connections with the local police to obtain the body of a carnival knife-thrower, who had just been hanged for murder. He successfully transplants the hands of the knife-thrower onto the arms of the concert pianist, but the hands begin to take on a life of their own (Perhaps in a deliberately ironic twist of casting, Colin Clive, who had previously played Dr. Frankenstein, now finds himself at the receiving end of the surgeon's scalpel.) [7]. The silent version of the story from 1924 is more of a psychological drama, rather than a horror movie. The hand transplant operation is implied, rather than shown, and the transplanted hands do not actually take on a life of their own. Instead, the concert pianist becomes the victim of a bizarre hoax [8].

A crisis situation, such as war, can serve as the motivator for major technological breakthroughs. We have already seen in a previous chapter that World War II provided the context in which the first digital computers were developed. The same can be said of transplant technology, although early attempts were unsuccessful. One of the pioneers of organ transplants, David Hamilton traces the practice back to antiquity. In his book "A History of Organ Transplantation: Ancient Legends to Modern Practice," he associates advances in transplant research with the desire to restore injured soldiers during wartime. The first real-life hand transplant was attempted at the Clínica Guyaquil, in Ecuador, in 1964, some 29 years after *Mad Love*, and more than 40 years after the original novel, on which the movie was based. The recipient was a soldier, who had lost a hand in a grenade accident. The operation initially seemed to be successful. Unfortunately, the transplanted hand was rejected after 3 weeks [9]. Thanks to advances in immunosuppressing drugs, the operation was done successfully for the first time in Lyon, France in 1998, and again in the United States the following year.

Organ transplantations of various kinds are now commonplace in the twenty-first century. Yet one type of transplant remains purely in the realm of science fiction. Perhaps prompted by the questionable surgical practices of the times, writers of science fiction in the 1920s imagined brain transplants [10]. Damage to the spinal cord can result in paralysis, in varying degrees of severity, depending on how far from the brain the damage occurs. Severing the spinal cord from the base of the brainstem would be instantly fatal. At the present time, there is no known way to reverse the process. But perhaps by the twenty-third century, there may be surgical techniques of the sort suggested in "Spock's Brain". In this episode of the original series of *Star Trek*, an alien uses advanced technology to remove the brain of the Vulcan science officer but is unable to explain how to put it back. Dr. McCoy is able to keep Spock's body alive by artificial means, but only for a limited time. He must somehow acquire the advanced neurosurgical knowledge, in order to restore Spock's brain before his body dies [11].

Science fiction has on occasion been a predictor of important biomedical technology. *The Man They Could Not Hang* features Boris Karloff as world-renowned surgeon, Dr. Henryk Savaard, who develops what he calls "the perfect anesthetic." His technique involves the use of "certain gases, that end life without poisoning the tissue;" then removing the "vital heat" from the body, to preserve it during the operation; and finally, an external, artificial heart, to keep blood flowing through the patient's body. Following successful attempts with animals, a young medical student volunteers to be the first human test patient. Fearing for his life, the doctor's nurse, who is also the fiancée of the volunteer, summons the police, who arrive to find the young man already dead. The doctor pleads with the police to give him "just one hour" to complete the procedure and bring the volunteer back to life. But they scoff at his claims, and take him away to be tried, convicted, and hanged for murder. Fortunately for the doctor (but unfortunately for all those involved in the trial), his associate—also an accomplished surgeon—claims the body and successfully applies the doctor's revolutionary apparatus and techniques to bring him back to life [12]. The apparatus featured in this 1939 movie is essentially a cardiopulmonary bypass machine. The first known operation, using biomedical technology of this kind, was not performed until 1951.

The shortage of donor organs has prompted medical researchers to explore the possibility of artificial organs as an alternative to biological replacements. Various prototypes for an artificial heart have been developed and tested on animals since the mid-twentieth century. By the early 1980s, the state-of-the-art was the Jarvik-7, the first clinically tested artificial heart for human use. Created by Dr. Robert Jarvik, the Jarvik-7 was famously used on patient Barney Clark in 1982. SynCardia, founded in 2001, currently manufactures the world's only clinically proven and FDA-approved total artificial heart. But the use of artificial hearts on humans continues to be as a bridge to transplantation, while the patient waits for a compatible donor organ. The devices remain tethered to a portable external power supply. Even with recent advances, we are still a long way from a twenty-fourth century device, such as Captain Picard's artificial heart, which he received when his own heart was irreparably damaged. As an arrogant, undisciplined ensign, fresh out of Star Fleet Academy, Picard managed to get himself into a fight, which ended when an alien stabbed him with a sword through the back [13]. His replacement artificial heart later turned out to be defective and had to be replaced a second time [14].

A fully implantable artificial heart, such as Captain Picard's, would require a compact and long-lasting power source. Such a device was actually constructed in the mid-1970s by scientists at Westinghouse and the University of Utah, but it was never used in a human patient. The power source was the radioisotope Plutonium-238, which is the same isotope used by NASA to power its robotic space probes. With a half-life of 90 years and a decay heat of 0.57 W/g, it is an excellent choice for powering interplanetary probes. The concern about using the prototype device in a human patient was not only because of the effects of exposure to ionizing radiation (which we will explore later in this chapter) but also the possibility of the patient being kidnapped and the radioactive material falling into the wrong hands [15].

6.3   Resistance to Disease

In his 1898 novel "The War of the Worlds", H.G. Wells imagines a Martian invasion of Earth. An enormous cylindrical object falls from the sky near London. Giant three-legged machines emerge and march across the countryside, wreaking death and destruction with a deadly Heat Ray and toxic Black Smoke. Human defenses are no match for the Martians' superior technology, and the situation appears hopeless. When it seems that all is lost, salvation comes at the last moment, not through human technology, but from natural causes. The Martians suddenly begin to die from airborne and waterborne infections, to which humans have long been immune.

Several film versions of "The War of the Worlds" have been made, two of which bear a similar title to the original novel. The 1952 version, set near Los Angeles, features sleek, futuristic Martian flying machines, instead of the tripod walkers, which hover over the Earth, and are protected by an impenetrable force field. The human military response is updated to include technology from the Cold War era. The movie even includes real footage of the state-of-the-art Flying Wing, an experimental jet aircraft, which proved to be too unstable for practical military use. Nevertheless, it is used in the movie to drop an atomic bomb on the Martians, but to no avail [16].

The imagery of the 2005 remake is much closer to the descriptions in the novel, including the tripod walking machines, and the gruesome human devastation [17]. Although the three versions differ markedly in their treatment of some of the details, both movie versions are consistent with the novel on the source of deliverance from the Martian onslaught. The narration of the closing scene in the 1952 version and in the 2005 version both capture the essence of the original H.G. Wells text. The 2005 version quotes from the 1898 novel almost word-for-word, but selectively [18]. The 1952 version is a paraphrase of the original text, but conveys the science more faithfully for a modern audience:

> The Martians had no resistance to the bacteria in our atmosphere, to which we have long since become immune. Once they had breathed our air, germs, which no longer affect us, began to kill them. The end came swiftly. All over the world, their machines began to stop and fall. After all that men could do had failed, the Martians were destroyed, and humanity was saved, by the littlest things, which God in his wisdom had put upon this Earth. [19]

An improved understanding of the human immune system and the ability to control it led to the success of organ and tissue transplants as recently as the mid-twentieth century. But an awareness of the biological causes of infection and disease began much earlier. The microscope was a seventeenth century invention, attributed to the Dutch naturalist Van Leeuwenhoek, who was the first to demonstrate the existence of microorganisms in the atmosphere. But it was not until the work of Louis Pasteur, in the mid-nineteenth century, that a direct connection was made between these microorganisms and infection of open wounds. A fascinating history of these early developments may be found at the website of the National Center for Biotechnology Information [20].

6.3.1   The Basic Reproduction Number (R$_0$) and the Spread of Infectious Diseases

The spread of infectious disease is treated in detail in the 2011 movie *Contagion*, which deals not only with the science, but also the public policy issues involved in controlling a global epidemic. A virus of unknown origin begins to spread and is lethal to all humans who come into direct contact with it. Lawrence Fishburne plays the role of Dr. Ellis Cheever, a prominent health official with the Centers for Disease Control (CDC), who first finds out about the virus from representatives of the Department of Homeland Security. Jude Law plays Alan Krumwiede, a freelance journalist and proponent of homeopathic medicine, who claims that Dr. Cheever is part of a global conspiracy to keep the truth about the virus from the public. In a televised interview, in addition to hurling accusations of conspiracy, Krumwiede raises one of the most important scientific issues in an epidemic, namely the rate at which the virus is spreading: the so-called basic reproduction number, or R_0 (R-naught).

> Tell them what an R-naught of 2 really means, Dr. Cheever. Teach them some math, hmm? No? I'll do it. On Day 1 there were two people with it, and then there were 4, and then it was 16, and you think you've got it in front of you. But next it's 256 and then it's 65,000, and it's behind you and above you and all around you. In 30 steps it's a billion sick. Three months. It's a math problem you can do on a napkin. And that's where we're headed. And that's why you won't even tell us the number of the dead, will you Dr. Cheever? But you'll tell your friends when to get out of Chicago before anyone else has a chance. [21]

Alan Krumwiede is nothing if not passionate about what he believes. Unfortunately, he doesn't have his facts straight, and Dr. Cheever, a CDC official, who actually does know what he's talking about, should have taken Krumwiede to task on the spot.

What exactly does the reproduction number, R_0, tell us? To put it simply, R_0 is the average number of people infected by one infected person, before that person recovers (or dies, in the case of a fatal infection). An R_0 of less than one means that the infection will eventually die out. If $R_0 = 1$, the disease is simply passed along from one person to another. It does not die out but neither will it spread out of control. The disease described in the movie is supposed to have $R_0 = 2$. This means that the number of infected people doubles with every step. One person infects two, those two infect four, those four infect eight, and so on. Table 6.1 illustrates the progression of infection for three different values of R_0. The last column in the table shows Alan Krumwiede's progression. What Krumwiede describes, is not a doubling, but a squaring of the previous number at each step. To his credit, the number that he claims as the final result ("a billion sick") after 30 steps is approximately correct. The number 2, raised to the 30th power, (i.e., doubled 30 times) is actually 1,073,741,824. This may explain why Dr. Cheever did not challenge him, since he gave nearly the right answer. But his math is nonsense. Krumwiede only gives the first five steps in his absurd progression: 2, 4, 16, 256, 65,000, according to his diatribe. Had he gone on to step 6, the number allegedly infected would already be over four billion. And by step 7, the number infected would be almost a billion times the

Table 6.1 Illustration of the basic reproduction number, R_0, and the spread of infection

Step (n)	$R_0 = 1$	$R_0 = 1.5$	$R_0 = 2$	"$R_0 = 2$"
	Number of new infections at each step: $X(n) = R_0^n$			Number of new infections at each step, according to Alan Krumwiede $X(n) = [X(n-1)]^2$
1	1	1.5	2	2
2	1	2.3	4	4
3	1	3.4	8	16
4	1	5.1	16	256
5	1	7.6	32	65,536
6	1	11.4	64	4,294,967,296
7	1	17.1	128	1.8×10^{19}
8	1	25.6	256	
9	1	38.4	512	
10	1	57.7	1024	
...	...	...	...	
20	1	3325	1,048,576	
...	...	...	...	
30	1	191,751	1,073,741,824	A billion

The far right column is the fictitious progression described in the movie *Contagion*

entire population of the planet. If you start with two infected people and square the number 30 times, instead of simply doubling the number (the true meaning of $R_0 = 2$), the result would be astronomical.

How does the infectious disease in *Contagion* compare to real infectious diseases in the news in recent years? A close comparison is the 2014 outbreak of the Ebola virus in West Africa, which was estimated to have an R_0 between 1.5 and 2.5 [22]. In contrast, the measles virus is widely reported to have an R_0 in the range 12–18. But a recent publication in the journal, The Lancet, suggests that the range of values may actually be much broader than this, depending on the locality [23]. Why the broad range for measles and the comparatively narrow range for Ebola? A partial answer to this question is that R_0 is not a universal constant for a particular disease. It depends upon other factors, such as population density and mobility of the infected people. The data for measles have been gathered from all over the world at various times and places, while the value reported for Ebola was specific to the 2014 outbreak in West Africa.

A second important question, comparing Ebola and measles, is why should the average value of R_0 for measles be so much greater than for Ebola? A partial answer to this question has to do with the way in which the disease is transmitted. The measles virus is airborne, while Ebola requires direct contact with bodily fluids of the infected patient. The relative ease of transmission tends to make the value of R_0 larger. But the mystery virus in *Contagion* was also supposed to have been airborne, so shouldn't it have a value of R_0 closer to that of the measles virus? Or another way of putting it, if the measles virus has an R_0 of at least 10, why the global panic in *Contagion*, including the evacuation of cities, if the mystery virus has an R_0 of only 2? Another factor that determines R_0 is the length of time than an infected patient lives with the disease before recovering or dying. The longer the contagious period, the more opportunity the disease will have to spread, unless the patient is isolated. In the case of measles, an infected patient is typically contagious for about a week. In the movie *Contagion*, the mystery disease is 100% horribly fatal over a period of several days. The movie never gives a specific value for R_0 for the mystery disease, apart from the diatribe by Alan Krumwiede about the meaning of an R_0 of 2. Considering the way in which the mystery disease spreads (airborne) and the lifetime of a typical infected person (several days) and the locations of primary concern (large urban areas), the actual value of R_0 for this disease should be considerably larger than 2.

Although rarely fatal, measles can cause serious complications, including encephalitis and even death, particularly in very young children. Since the development of a measles vaccine in the early 1960s, considerable progress has been made toward eradicating the disease in the U.S. Nevertheless, measles remains a leading cause of vaccine-preventable infant mortality [24].

6.4   Cell Structure and Radiation Damage

In *Star Trek II: The Wrath of Khan*, the starship Enterprise engages in battle with the Reliant, a stolen Federation starship, now under the command of Khan Noonien Singh (Ricardo Montalbán), an unsavory character from the original television series. By a technological trick, which we will discuss in the next chapter, Captain Kirk has managed to gain the upper hand, but only after suffering considerable damage and the loss of many crew. The Reliant is now completely crippled and adrift in space. Khan, the only survivor onboard, is himself critically injured. He has been ordered to surrender, but he has one last trick up his sleeve. He begins the detonation sequence for the Genesis device, which will consume all of the matter around it, within a volume of the size of a planet, and transform that matter into a life-sustaining planet. The Enterprise must escape to a safe distance—at least 6000 km—in less than 4 min: an easy task, except for the fact that the warp engines are offline. When there is no response from the engine room, Spock leaves his post at the science station on the bridge, to investigate. He arrives at the engine room, where a flashing red RADIATION sign warns of the potential hazard. Dr. McCoy offers further words of warning: "Are you out of your Vulcan mind? No human can tolerate the radiation that's in there." Spock replies, "As you are so fond of observing, Doctor, I am not human." McCoy blocks the way and insists, "You're not going in there." Spock distracts the doctor, by inquiring about the condition of Chief Engineer Scott and renders him unconscious with the Vulcan nerve pinch. In a heroic, self-sacrificial act, Spock takes the pair of radiation gloves from the semiconscious Mr. Scott, and proceeds to enter the radiation zone. He removes the top of a conduit, from which a jet of blue-white light emanates, indicating that the air in the engine room is being ionized by the radiation from the warp reactor. Exposing himself to the lethal radiation, Spock manages to bring the warp engines back online, just in time for the Enterprise to escape, but at the cost of his own life [25].

Ever since the first nuclear weapons were used to bring an end to World War II, the biological effects of radiation, both real and imagined, have been incorporated into works of science fiction. Even before anything was known about the real biological hazards, we can find a hint of the interaction between radiation and living organisms in the original 1931 version of *Frankenstein*. Recall the claim that Dr. Frankenstein makes rather early in the movie, just before the monster comes to life: "Here in this machinery, I have gone beyond [the ultraviolet ray]. I have discovered the great ray that first brought life into the world" [2].

The nature of Frankenstein's hypothetical "great ray" is never explained in the movie, and all we ever see is the electrical discharge from the laboratory apparatus and from the lightning storm. If there is a connection between electromagnetic radiation and biological material, is it life-giving or life-destroying? Or, as so often portrayed in sci-fi horror movies, is it something in between? How do we separate fact from fiction?

6.4.1 Detection of Ionizing Radiation

Early in the 1952 version of *The War of the Worlds*, just after the strange object falls from the sky near Los Angeles, the local officials seek the advice of some "scientists," who are on a fishing trip nearby. These "scientists" just happen to have brought a Geiger counter with them (Did they expect to catch a radioactive fish?). And they just happen to have left the battery-powered Geiger counter turned on in the back seat of the convertible top car. When the car pulls up, near the site where the supposed meteor fell, a police officer notices something in the back seat, which is "ticking like a bomb" and asks the scientist what it is. The camera focuses on the instrument, whose analog radiation indicator twitches regularly, while the sound track that we hear is like a slow metronome, clicking in time with the indicator needle. The scientist explains that it's a Geiger counter, used to detect radioactivity, and that they were doing some surveying, while they were up in the hills. He opens the shutter on the Geiger tube, exposing the sensitive window on the side of the tube, and proceeds to point the end of the tube in the general direction of the meteor. The count rate increases dramatically. When he points the end of the tube away from the meteor, the radiation count rate decreases. He repeats the experiment and concludes that the radiation is coming from the meteor. The police officer decides to post some deputies to keep people away [26].

The detection of radiation using a Geiger counter involves the ionization of an inert gas (such as Argon) inside the Geiger tube. The high-energy radioactive decay particle enters the tube and kicks an electron out of one of the gas atoms. This electron, in turn, ionizes another atom, and the two electrons ionize two more, and so forth, creating a cascade of free electrons in the tube. The electrons are collected by a positively charged wire in the center of the tube and the resulting pulse of current is registered as a particle count. The higher the intensity of the radiation, the more frequently counts will be registered by the instrument. But the way in which the Geiger counter is portrayed in *The War of the Worlds* is inaccurate on several counts. First, the emission and detection of radiation are completely random processes. The clicking of the Geiger counter would be random, and not "ticking like a bomb." Second, the particle detection exhibits directionality, which would not be possible for radiation that could travel several meters from the source to the detector. These would have to be gamma rays, and they would not care which way the probe of the detector is turned, as long as the distance from the probe to the source is not changed significantly. Finally, if there had been some directionality, it would have been exactly the opposite of what we see in the movie. Figure 6.1 shows two of several possible designs for Geiger counters. A contemporary radiation survey meter, with an end-window, is shown in Fig. 6.1a. The Geiger counter used in *The War of the Worlds* is similar to the design shown in Fig. 6.1b. The sensitive window of the detector, designed for beta particles, is on the *side* of the probe, not on the end. Yet we saw an increased count rate when the end of the probe was turned toward the source.

6.4.2   Biological Effects of Exposure to Ionizing Radiation

A crucial issue raised in this scene is the level of radiation exposure. Is the observed radiation count rate really high enough to warrant posting deputies to keep people away from the fallen radioactive object? Radiation interacts with biological tissue in much the same way that it interacts with the gas in a Geiger tube: ionization. When a gamma ray passes through the skin, or deeper into the body, it will ionize biological molecules and break interatomic bonds at random. The human body has the remarkable ability to heal itself from small amounts of damage. But if enough damage is done, the natural healing process may not be able to keep up, and vital

Fig. 6.1 Radiation survey meters (Geiger counters): (**a**) contemporary instrument with end-window probe. (**b**) 1950s instrument with side-window probe (photos by the author)

Table 6.2 Biological effects of exposure to ionizing radiation

Radiation dose (Sv)	Biological effect
6.0 (acute exposure—4 h or less)	Fatal to 100% of population exposed
3.5 (acute exposure—4 h or less)	Fatal to 50% of population exposed
0.75 (acute exposure—4 h or less)	Serious illness, but not fatal
0.25 (acute exposure—4 h or less)	Minor changes to blood chemistry
0.25 (received over 1 year)	No easily detectable effects

systems begin to shut down. At what level does radiation exposure become danger-ous, and how much exposure is needed to kill a human (or a Vulcan, as in *Star Trek II*)? Table 6.2 is a summary of radiation doses, in units if Sieverts (Sv) and the resulting biological effects.

The worst nuclear disaster in human history occurred on April 26, 1986, when the reactor at Chernobyl, in Ukraine, exploded, and the core suffered meltdown. Many of the workers who cleaned up the radioactive debris in the immediate vicinity of the reactor received a lethal dose of radiation. For comparison, the safety standard for radiation workers in the United States is set at 0.05 Sv per calendar year, which is five times lower than the threshold of 0.25 Sv, below which there are no easily detectable biological effects. Radiation exposure to individuals can be monitored with various types of personal dosimetry, as shown in Fig. 6.2. The materials in the whole-body badge and finger rings record the amount of ionization produced. The dosimetry is analyzed periodically, to ensure that the radiation dose received is below acceptable limits. For more information on radiation doses from various sources, see the highly detailed chart compiled by Randall Munroe, creator of xkcd comics [27].

Radiation safety is a serious business, which the 2000 remake of *Godzilla* takes far too lightly. Early in the movie, a biologist is seen studying the effect of radiation on worms, near Chernobyl, in the aftermath of the worst nuclear disaster in history [28].

From a safety perspective, the scene is completely absurd. As explained in the introduction to the "Radioactive Wolves" episode of the PBS *Nature* series, the Chernobyl disaster resulted in the release of radiation equivalent to roughly 400 Hiroshima bombs. Now, more than 30 years later, an exclusion zone of roughly 1100 miles around the defunct nuclear reactor continues to be unsafe for human habitation. The radioactive particles are in the soil, and have been absorbed into plant life, and concentrated in the bones and organs of the animal life, which moved into the zone after the humans fled. Nobody can enter the exclusion zone without an official escort, and nobody who knows anything about the hazards of radiation would be digging in the soil, as the lead character does in *Godzilla*, without protec-tive clothing and a face mask, to prevent breathing the radioactive fallout [29].

Finally, a sci-fi example of the effects of radiation on biological tissue, which defies probability, is the giant ants in the movie *Them*. According to the script, fall-out from nuclear testing in the New Mexico desert causes aunts to grow to the size of horses [30]. This is completely implausible, both from the perspective of genetic mutations, and from the perspective of structural mechanics. One of the possible consequences of exposure to ionizing radiation is the development of cancer cells.

Fig. 6.2 Radiation dosimetry: (**a**) finger rings and whole-body badge worn by the author during the first quarter of 2019, and (**b**) whole-body badge from Chernobyl plant, c. 1980 (photo by the author)

Instead of completely killing a cell by disrupting critical components, it is possible for the radiation to alter the genetic material of the cell in a way that leads to abnormal growth and multiplication out of control. But there is no possible mechanism to trigger uniform growth of all of the tissues of the body, in such a way that the entire organism grows to hundreds of times its normal size. Just to illustrate the complexity of the problem, there are dozens of genes in the human body which are involved in determining the height of an individual. Each of them contributes in a subtle way to the growth process. Assuming that the growth process in ants is governed by a similar combination of genes, it is unimaginable that exposure to radiation would alter all of the necessary genes in just the right way to cause abnormally large, but uniform growth, without being fatal to the organism. And even if this were possible, the structural elements of the ant's body would not be able to support the weight. The weight of the ant's body is roughly proportional to its total volume, which increases approximately as the cube of the linear dimension. But the strength of the ant's legs is related to the cross-sectional area of its exoskeleton. This would increase approximately as the square of the linear dimension, if the growth were uniform.

Despite the completely implausible science, *Them* is noteworthy for having a very strong female scientist in a leading role—something which is not typical of 1950s movies. Joan Weldon plays Dr. Patricia Medford, an accomplished scientist in her own right, and the daughter of Dr. Harold Medford (Edmund Gwenn), the world's leading expert on ants. Throughout the movie, the younger Dr. Medford demonstrates greater presence of mind than her famous father, and in the end, more bravery than most of the male characters in the film.

Most of what we know about the biological effects of ionizing radiation on humans, including the information summarized in Table 6.2, is derived from the so-called Life Span Study, a long-term study of the survivors of the Hiroshima/ Nagasaki atomic bombs. Prior to the Chernobyl disaster of April 1986, almost nothing was known about the effects of radiation on plant and animal life. Following Chernobyl, and more recently, the triple meltdown at the Fukushima Daiichi nuclear power plant in March of 2011, studies are being done on the long-term effects of radiation on the surrounding ecosystems. Although the amount of radiation released at Chernobyl far exceeded that of Fukushima, the results seem to indicate that birds near Fukushima are dying at twice the rate as they were near Chernobyl, and increases in mutation rates are being observed near both sites. Ongoing studies are allowing researchers to learn more about the biological effects of low doses of ionizing radiation [31].

For more information about the effects of certain types of ionizing radiation on the human body, see Fig. E.3 and accompanying material in the Appendix of Artifacts at the end of the book.

6.4.3 UV Radiation and Skin Cancer

So far we have considered examples of the biological effects of exposure to radioactive materials, or to gamma rays from high-energy collisions (the *Star Trek* warp engine). But there are other sources of electromagnetic radiation, which can cause damage to biological material. The Sun, for example, produces not only visible light, but a significant amount of radiation in the ultraviolet portion of the spectrum. It is well-known that ultraviolet radiation can cause skin cancer, and lots of money is spent each year on sunscreen products. What are the real biological concerns?

First, let's consider what it takes to ionize an atom. That is to say, how much energy is required to overcome the electrostatic attraction between the negatively charged electron and the positively charge ion that remains behind? For an isolated Hydrogen atom, the energy needed, in units of electron-volts is 13.6 eV. For Argon the gas in a typical Geiger counter—the energy is 15.7 eV. For water (H_2O), it's about 12.6 eV. The so-called *ionization potential* varies for other atoms and molecules, depending on their structure, but is typically in the range from a few eV to about 10 eV, or so. Since gamma rays have energies ranging from 100 thousand eV to more than a million eV, there is no question of their ability to ionize atoms and molecules, and to do biological damage. But what about ultraviolet light?

The energy of a photon is given by

$$E = hc / \lambda, \tag{6.1}$$

where h is Planck's constant, c is the speed of light, and λ is the wavelength of the light.

We can use this equation, and the information in Table 6.3, to convert wavelength into energy for the various subparts of the UV spectrum. The results are shown in Table 6.3.

Where do these energies fall, with respect to the ionization potential of important biological molecules? We will say more about the structure of DNA in the next section, but here we simply cite some theoretically predicted values for the ionization energies of the base molecules, published recently in the Journal of Physical Chemistry.

Table 6.3 Wavelength and energy ranges for the UV spectrum

	Wavelength range (nm)	Energy range (eV)
UV-A	400–320	3.10–3.88
UV-B	230–280	3.88–4.43
UV-C	280–200	4.43–6.20

Table 6.4 Predicted values of the ionization energies of the DNA and RNA bases [32]

Base	Ionization energy (eV)
Thymine	5.05
Cytosine	4.91
Adenine	4.81
Guanine	4.42

By comparing Tables 6.3 and 6.4, you can get an idea of the danger. UV-C clearly spans the range of energies sufficient to ionize any of these important biological molecules. According to information available from the Centers for Disease Control (CDC), UV-C radiation is very dangerous. Fortunately, the Earth's ozone layer does a good job of filtering out the naturally occurring UV-C radiation from the Sun. But artificial sources of UV-C radiation, such as those marketed to kill bacteria on the surfaces of household objects, and Mercury vapor lamps, are potentially dangerous, and should be used with caution. Most but not all of the UV-B radiation is filtered out by the atmosphere along with the UV-C, but there is still cause for concern. UV-A radiation, which is the most abundant range of UV wavelengths at the Earth's surface, appears to be below the threshold for directly ionizing the base molecules in RNA and DNA, at least according to the theoretical calculations. Nevertheless, this range of UV wavelengths has been directly linked to skin cancer.

So the precise mechanism of radiation-induced cancer is probably not as simple as ionizing the bases of RNA or DNA. The empirical evidence linking UV-A to skin cancer strongly suggests taking appropriate precautions against overexposure to the Sun.

6.5   DNA and the Human Genome

The "Island of Dr. Moreau", published by H.G. Wells in 1896, tells the story of a shipwreck survivor, who is taken to an unnamed island, where he becomes a witness to bizarre biological experiments. Animals of various kinds are gradually being altered to look like humans. Over the hundred years following the publication of the novel, a number of film versions were made, which illustrate how advances in science can influence the way in which a classic story is told. The earliest talking version, *The Island of Lost Souls*, features Charles Laughton in the role of Dr. Moreau. Released in 1932, it describes the transformation of animals into human-like creatures in the same way that the book does, as a purely surgical technique. By the middle of the twentieth century, the molecular basis of heredity and the structure of DNA had been discovered. This had a progressive impact on two subsequent screen versions of the story, both of which bear the same title as the book. The 1977 version, starring Burt Lancaster in the title role, and Michael York as Andrew Braddock, the shipwreck survivor, portrays the transformation procedure as a combination of surgery and genetic manipulation.

When the survivor of the shipwreck discovers the laboratory and the gruesome experiments in progress, Doctor Moreau offers the following explanation.

> MOREAU: I have proved—almost proved—the existence of a cell particle that controls the living organism. This cell, this particle, controls the shape of life.
> (Holding up a large syringe…)
> This serum contains a distillation of a biological code message: a new set of instructions for erasing the natural instincts of this animal. With some surgery, implants into various organs, he should grow to resemble any creature I please—in this case, a human being.
> (He injects the creature, and it begins to writhe in agony.)
> See. The serum is already forcing a modification of the body.
> BRADDOCK: But why do you do this? For what possible reason?
> MOREAU: Why? To reach for the control of heredity. Think what it can do for humanity: the pain we can ease; the deformities we can avoid. The possibilities are endless. [33]

Although he was at least partially motivated by good intentions (easing pain and avoiding deformities), the methods used by the 1977 version of Doctor Moreau involved great pain for his animal subjects. This raises an ethical question: do the ends justify the means?

The 1996 version of *The Island of Doctor Moreau*, with Marlon Brando in the title role, makes no mention of surgery. In this version of the story, the transformation from animal to human is entirely a matter of genetic engineering, or what might be described more precisely as gene therapy.

DNA (deoxyribonucleic acid) is the complex biological molecule, which contains the blueprint for a particular biological organism. The overall structure of DNA is a double-helix of alternating sugar and peptide molecules, as illustrated in Fig. 6.3. The base pairs, guanine and cytosine, or thymine and adenine, form the rungs of the twisted ladder-like shape. Alterations to the composition of this molecule can have devastating consequences for the organism, itself, or to its offspring, or may have no noticeable effects, whatsoever, depending upon which specific details are changed.

The publicly funded Human Genome Project began in 1990, with a primary goal to map the entire human DNA sequence within 15 years. Under pressure from private sector competition, the project was completed in 2003 [34]. The first director of this ambitious initiative was James Watson, co-discoverer of the structure of DNA. But Watson resigned after only 2 years, following an argument with the director of the National Institutes of Health, over the wisdom of patenting segments of DNA, a practice to which Watson was strongly opposed. He was succeeded by Francis Collins, who led the project to completion [35].

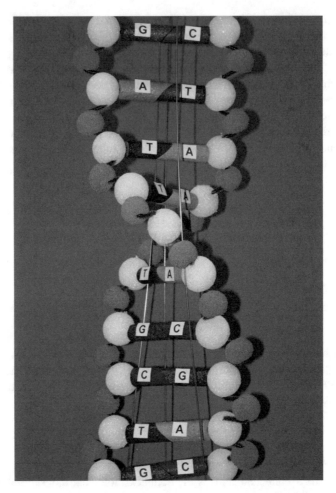

Fig. 6.3 Schematic model of DNA. The larger spheres represent sugar groups, and the smaller spheres represent peptide groups. The base pairs—either adenine (A) and thymine (T), or guanine (G) and cytosine (C)—form the rungs of the ladder-like structure, which is twisted into a double-helix (photo by the author)

The entire human genome consists of 3.1 billion base pairs [36]. At first reading, it seemed that only about 1.5% of these base pairs, making up a mere 20,000–25,000 genes, were actually involved in the life-sustaining process of coding for proteins. The rest was referred to by some researchers as "junk DNA"—long stretches of base pairs in between the identifiable genes, which had no apparent function. Taking a more humble approach to the interpretation of the data, Collins dismissed this pejorative label as nothing but hubris, given our level of ignorance at the time [37]. The analysis of the data is ongoing, but a recent breakthrough vindicates Collins. The results of the ENCODE project (Encyclopedia of DNA Elements), published in 2012, largely dispel the previously held notion of "junk DNA" and suggest that at least 80% of the human genome is associated with useful biological functions. The long stretches of DNA in between the genes, which had been written off as "junk," are now identified as a complicated array of switches, which serve to turn different biological processes on or off. A complete understanding of these switches may lead to cures for previously untreatable diseases [38].

This brings us back to the bizarre, transformative experiments of Doctor Moreau. In principle, the desired biological information encoded in the genes could be activated, resulting in a specific set of biological products—a process referred to as gene expression. A striking example, which has actually been achieved, is the genetic manipulation of mice to produce a green fluorescent protein (GFP), commonly found in jellyfish. Whether induced gene expression can be extended in the way that Doctor Moreau had envisioned is still a matter of speculation. And even more speculative is the sort of accidental gene expression, which transformed Peter Parker into *Spider-Man*, after being bitten by a genetically engineered super-spider [39].

Analysis of the human genome has also shed some light on the question of what it means to be human. A comparison of the human genome to the recently mapped genome of the chimpanzee reveals that "humans and chimps are 96% identical at the DNA level" [40]. Furthermore, a study of genetic diversity reveals that, despite our many apparent differences, all humans are 99.9% identical to each other at the DNA level, differing from one another by only one letter in a thousand. By comparison, the genetic differences among chimpanzees are four to five times greater. The greater genetic diversity among chimpanzees suggests that chimps have been around for a much longer time than humans have [41].

Discussion Topic 6.1: Gene Patents

A convergence of science, technology, healthcare, and law played out in the courts over a period of several years. The focus of attention from 2010 through 2013 was on patents owned by the Utah-based diagnostic company, Myriad Genetics, on the BRCA1 and BRCA2 genes, which predispose women to hereditary breast cancer and ovarian cancer. These patents were invalidated by a U.S. district court in New York, early in 2010. But the decision was overturned by the U.S. Court of Appeals for the Federal Circuit (CAFC), in Washington, DC, late in the summer of 2012. Since his resignation as head of the Human Genome Project in 1992, James Watson has continued to be a very vocal opponent to the concept of gene patents. In an amicus brief filed in the CAFC case, Watson argued that "Life's instructions ought not be controlled by legal monopolies created at the whim of Congress or the courts." The case was appealed to the U.S. Supreme Court. In a unanimous decision handed down in June of 2013, the high court applied their "product of nature" doctrine and ruled that genetic information could not be patented. Discover what you can about the details of this case, and about the decisions made by the lower courts. Do you agree with the Supreme Court decision and with James Watson, that genetic information is a "product of nature" and not subject to patents? Why or why not?

6.5.1 DNA Sequencing and Genetic Engineering

As our understanding of the specific functions of the genetic code increases, many ethical, legal, and social issues undoubtedly will arise, which will need to be addressed. These have fueled the imagination of numerous science fiction writers. The 2007 film *I Am Legend* suggests that we are just one mutated measles virus away from becoming a race of zombie vampires. Based on the 1954 novel of the same name, by Richard Matheson, the movie opens with a television news program announcing the discovery of a cure for cancer. The medical researcher who developed the cure explains that they genetically modified the measles virus, effectively reprogramming it to selectively attack cancer cells. The result was, at first, 100% successful: all 1009 patients in the clinical trial, who were treated with the modified virus, were cured of their cancer (although the type of cancer was not specified in the interview). But an unfortunate thing about viruses is that they can mutate, and in this case, with disastrous consequences. Two mutated strains emerge: one airborne and another blood-borne. Most humans exposed to the mutated strains die very soon after exposure. The few who survive live on as zombie vampires, whose flesh burns on exposure to sunlight. Dogs also seem to suffer similar consequences, but other species of animal life are unaffected. A rare exception is Robert Neville (Will Smith), a virologist who is immune to both strains, and his dog, who is immune to the airborne strain, but eventually succumbs to the blood-borne strain. The virologist works frantically to find a cure [42].

DNA sequencing, genetic engineering, and genetic discrimination are the focus of the 1997 movie *GATTACA*. Ethan Hawke plays the role of Vincent, a custodial worker "in the not too distant future," who dreams of traveling to Saturn's moon, Titan. Unfortunately, he was conceived the natural way, without any genetic intervention, and his DNA profile categorizes him as "invalid." Although he appears to be perfectly normal to the casual observer, he is at high risk of certain genetic disorders and is excluded from the kind of high-tech employment opportunities, which are available to the genetically *valid*. Undeterred by his genetic disadvantage, Vincent hires a DNA dealer (played by Tony Shalhoub), who provides the resources for him to impersonate a genetically perfect individual (Jude Law), and eventually fulfill his dream [43]. In the movie, genetic sequencing is performed on every worker at the high-tech facility, GATTACA, as they arrive for work each day. The workers pass through a checkpoint, where a drop of blood is drawn from a finger. The genome is sequenced within a second or two, and the worker is admitted, or not, depending on whether the finding is "valid" or "invalid."

The issue of genetic profiling and discrimination is not just the subject of science fiction, but has been taken up by the Supreme Court and Congress. For example, should an insurance company be allowed to deny coverage to an individual, based on a genetic propensity to develop a disease at some time in the future, even though that individual does not actually have the disease at the present time? The matter is not to be taken lightly. The Genetic Information Nondiscrimination Act (GINA), signed into law in 2008, has led to the inclusion of "genetic information" on the growing list of things against which employers may not discriminate. In principle, this law also applies to health insurance. But as recently as 2012, many states still have laws on the books, which allow long-term care, disability, and life insurers to discriminate based on genetics.

The technology for DNA sequencing has come a long way since the completion of the Human Genome Project in 2003. Within a few years after the massive undertaking was completed, the process of sequencing an entire human genome would take just several weeks (instead of the original 13 years), at a cost of roughly $350,000 (instead of the original $3 billion). Although it is not yet possible to do it in a matter of seconds, as in *GATTACA*, the state of the art in human genome sequencing currently takes about 1 day, at a cost of about $1000. In the PBS NOVA episode, "Cracking Your Genetic Code" [44], Jonathan Rothberg, one of the key figures in the development of high-speed DNA sequencing technology, cautions against making critical decisions based on economy versions of the technology, which are advertised for about $100. "Genotyping is not the same as DNA sequencing. It's like looking at 100 words in a 600-page novel, and thinking you understand everything about Tolstoy." Francis Collins, who was head of the Human Genome Project and is currently the Director of the National Institutes of Health, submitted samples of his own DNA to three of these economy services, in an attempt to discover how consistent and reliable they might be. The results of all three were consistent in one respect: they all indicated that Collins was at significantly increased risk of developing type 2 diabetes. This prompted Collins to lose weight and get regular exercise. However, the same three test results disagreed in regard to his level of risk for prostate cancer, one indicating that he was at increased risk, another saying that his risk was average, and a third saying that his risk was lower than average. This informal test, conducted by Collins, tends to reinforce Rothberg's concerns about predictions made, based on incomplete data. Jonathan Rothberg, a 1985 graduate of Carnegie Mellon University, was awarded the National Medal of Technology and Innovation in 2016 for the development of high-speed genome sequencing. He was among seven recipients honored that year for their lasting contributions to America's competitiveness and quality of life.

Complete genome sequencing has made one of the predictions in *GATTACA* a reality. It is already possible for parents at high risk of passing on genetic diseases to their children, to eliminate some inherited diseases before birth. The process known as preimplantation genetic diagnosis (PGD) involves the extraction of one cell from an eight-cell embryo, prior to implantation, and sequencing the genome to screen for the appropriate genetic markers.

6.5.2   Gene Editing

The 2018 movie *Rampage* imagines the use of a real-life, state-of-the-art gene editing tool to transform a gorilla, an alligator, and a wolf into gigantic killing machines. Fortunately, the gorilla, named George, has an unbreakable emotional bond with his keeper, primatologist Davis Okoye (Dwayne Johnson) and George ends up saving the day. According to the movie, the genetic engineering tool, known as CRISPR, which previously could be used on only one cell at a time, has somehow been modified to work on every cell in the body simultaneously, simply by inhaling the vapor [45].

Clustered Regularly Interspaced Short Palindromic Repeats, or CRISPR, is a phenomenon discovered in the DNA of the bacteria *E. coli* in the late 1980s. The discovery has since led to the development of a successful tool for modifying the DNA of other organisms, including humans. The letters A, T, G, and C, representing the bases of DNA, are sometimes found in repeating clusters along the length of the three billion base-pair chain of the human genome. (Imagine the letters GATTACAACATTAG repeating several times, with random permutations of the letters ATGC in between.) The clusters typically consist of 23–47 base pairs, but are not necessarily palindromic. The spacers in between these repeating clusters, however, seem to be a characteristic of a particular genome. Chemical structures can be developed to target these repeating units and their spacers, cut out a segment of DNA and replace it with another segment, thus modifying the genetic structure of the organism [46].

The first known use of CRISPR to modify the DNA of nonviable human embryos was described in the journal *Nature* in April of 2015. Eighty-six embryos, which would otherwise have been discarded as nonviable, were injected with the CRISPR-Cas9 structure, which targeted a specific region of the DNA to be removed, along with the desired segment of DNA to be added. Only 21 of these embryos were successfully spliced and only four of these were found to have the desired segment of DNA inserted. The technique was determined by the researchers to be too immature, and the study was terminated [47]. Just a few years later, in November 2018, the medical world was rocked by the unexpected announcement that CRISPR had been used on a pair of viable human embryos, which were carried to full term and delivered as twin baby girls [48]. The researcher, Dr. He Jiankui of Shenzhen, China, immediately came under fire from the medical community, not only for performing the technique on viable human embryos, but for announcing the claim in the media before undergoing proper peer review of his work. His claim that he had edited the babies' DNA to make them immune to the HIV virus is still the cause of much controversy. The CRISPR technique is typically used on humans as a one-time treatment for an existing disease. But this new announcement, if it turns out to be true, opens the possibility of using CRISPR to make changes in human DNA that could be passed on to future generations. The use of CRISPR on human embryos has attracted the attention of prominent public figures, including Bill Gates, who say that the technique raises serious ethical issues and is calling for more widespread public debate [49].

The CRISPR technique also carries some risks, including the possibility of damaging off-target DNA, instead of simply cutting out and replacing the targeted segment. Current research being carried out at Carnegie Mellon University's Center for Nucleic Acids Science and Technology (CNAST), in collaboration with researchers around the world, is exploring an alternate approach to gene editing. Using peptide nucleic acids (PNAs), a type of synthetic molecule which binds to targeted cites in DNA, researchers have made progress toward correcting the genetic disorder which causes Huntington's disease [50].

6.6 Cloning

Dinosaur DNA is the subject of *Jurassic Park*, a 1993 movie based on the Michael Crichton novel. A wealthy entrepreneur develops a theme park around living dinosaurs, produced by a clever cloning procedure. One of the rides in the park takes the visitor on a behind-the-scenes tour, in which the process is explained. Various kinds of insects have been found, perfectly preserved in amber. Some of these insects are mosquitoes, from the Jurassic period, which fed on the blood of dinosaurs. By carefully extracting the blood, dinosaur DNA can be isolated. Defects in the molecular structure, which occurred over time, are repaired with segments of DNA from similar modern-day animals. The resulting DNA strand is inserted into a compatible egg, which then hatches a cloned dinosaur [51].

Among the major technical difficulties, which is at least partially addressed in the movie, is the fact that DNA tends to degrade with time. But until relatively recently, the precise timescale for degradation was not known. According to a study of DNA extracted from leg bones of the moa, an extinct species of flightless bird indigenous to New Zealand, the half-life of DNA is estimated to be just over 500 years. The study was published online in the journal *Proceedings of the Royal Society B*, in October of 2012 [52, 53]. What are the implications for the plausibility of *Jurassic Park*? If half of the nucleotide bonds in a strand of DNA are broken on a time scale of about 500 years, less than 0.1% will be left after about 5000 years (ten half-lives, or ½ raised to the tenth power). There will be nothing left to extract after millions of years. The concept makes for good science fiction, but as far as actually being able to recreate living organisms from the Jurassic period and use them in a theme park—highly unlikely.

On the other hand, cloning as a scientific procedure is not just a plausible possibility, but a reality. Perhaps the most famous example is Dolly, the sheep, who was the first mammal to be successfully cloned from an adult cell. Born on July 5, 1996, Dolly lived a comparatively short life for a sheep. She developed lung disease and was euthanized, having lived less than 7 years. Considerable progress has been made since then, and cloning has been proposed as a way of preserving endangered species [54].

If cloning of mammals is a reality, why not human cloning? And what are the ethical issues involved? One possible scenario is portrayed in *The Island*, a futuristic film in which a technique has been developed to produce a fully mature human clone in a relatively short time period. The technique is marketed as an insurance policy for wealthy sponsors, who will then have a ready source of replacement parts, should the need arise. The obvious advantage is perfect tissue compatibility, without the risk of rejection and free from the side effects of immunosuppressing drugs. The sponsors are assured by the CEO of the company that the so-called agnates are kept in a persistent vegetative state until they are needed. The truth is that the agnates are fully alive, fully self-aware participants in a sheltered community. They are kept under control by being told that they are survivors of a global disaster, and the dream of someday being able to go to the Island: the only pleasantly habitable place on Earth. How do you get to go to the Island? By winning a lottery, which really means that your sponsor needs to collect on the insurance policy [55].

The clones in *The Island* are not only fully conscious, fully human individuals, living in a sheltered community. Although not intended by the designers, and completely unanticipated, the most advanced models, the Echo series, also seem to have been created with the memories of their sponsors [56].

Previously we discussed the neurological process of forming human memories. In contrast to writing digital information onto a data storage device, this is a complex biological process, which is not yet fully understood. But one aspect of human memory is clear: although humans are born with a small set of behaviors that do not have to be learned (we simply do them instinctively), most of what we know must be acquired through sensory input and processing of information. Growing a clone to adulthood in a short period of time, to make it useful as an insurance policy, seems unlikely. And we have every reason to expect that a cloned individual will have to learn and form memories in the same way that a normal individual does, rather than being born with the memories already in place. But this provides a good lead-in to our next topic, as we explore more deeply into the question of what it means to be human.

6.7   Human Teleportation: A Complex, Interdisciplinary Problem

The transformation of genetic information is an important consideration in the design of a device, which exists only in the realm of science fiction. Variously known as a teleporter or a transporter, the device converts any object from matter into energy, while preserving all the necessary information about the object, transmits the energy (and information) to another location, and converts the energy back into matter. Hopefully, the final result will be an exact duplicate of the original, in every detail. At first glance, this doesn't seem too problematic, at least in principle, as long as we restrict the process to inanimate objects. However, it raises moral and ethical issues, when applied to living things, in general, and to humans, in particular, since it involves the destruction of the original living organism. It also presupposes that the essence of a human being, including memories and personality traits—indeed everything about what it means to be human, and everything that makes each human a unique individual—can be reduced to matter and quantum information. In this section, we will examine some of the many problems, both scientific and philosophical, associated with the process of human teleportation.

6.7.1   The Problem of Duplication

The Prestige tells the story of two late nineteenth-century stage magicians, Robert Angier (played by Hugh Jackman) and Alfred Borden (played by Christian Bale), competing to create the most impressive illusion. Borden has come up with a routine, called *the transported man*, in which he steps into a cabinet at one side of the stage and, in the same time that it takes for a small rubber ball to bounce across the stage, he reappears from another cabinet, just in time to catch the ball. He then shows the audience that the first cabinet is empty. Angier, who is a much better showman than Borden, is determined to figure out how the illusion was done. Coincidentally, at this same time in history, another important competition was playing out, in regard to the transmission of electrical power. Thomas Edison was a proponent of direct current (DC), while Nikola Tesla and George Westinghouse favored alternating current (AC). The movie incorporates a bit of the AC/DC competition, when Angier enlists the services of Tesla, played by David Bowie, to create a teleportation device. After much money and time apparently wasted, the results of Tesla's research have been spectacularly unsuccessful. The first test of the device is on Mr. Angier's top hat. The giant electric discharge coil, which bears Tesla's name, directs many sparks onto the hat, but with no obvious effect. It remains undisturbed in its original position. Tesla then tries the same procedure on a cat (an experiment borrowed from a much earlier movie, which we will consider shortly). The cat is not amused. It arches its back and yowls in anger when the process begins. The cat, like the hat, is also unaffected by the shower of sparks. Unlike the hat, the cat walks away at the conclusion of the apparently failed experiment. Angier also leaves the lab in disgust. He follows the cat around the side of the building to discover other identical cats and a pile of top hats, identical to his own. Evidently, the experiment was partially successful, after all. It succeeded in creating duplicates at another location. Tesla explains: "Exact science, Mr. Angier, is not an exact science" [57], An important technical problem remained to be solved: what do you do with the original? But this is more than just a technical detail. It raises an important ethical question, in regard to living subjects: when the original is destroyed, has a living thing been killed?

6.7.2 Getting All the + and − Signs in the Right Place

Star Trek's Dr. Leonard McCoy never liked the idea of having his molecules scrambled by the transporter. He preferred more conventional means of travel, such as a shuttle craft, to get from the surface of a planet to an orbiting starship. But when the original series of *Star Trek* was created, the launching and landing of a shuttle craft was an expensive special effect to film convincingly. There were no computers capable of generating animated images, of the sort that are commonplace today, and the episodes were filmed on a tight budget. So for purely cost-saving measures, the transporter was written into the script much more frequently than the shuttle craft. Although it was not the first episode of the series to be aired, "The Corbomite Maneuver" was the first regular episode to be filmed. It featured the transporter in one of the last scenes, and included the first hint of Dr. McCoy's reluctance to be "beamed" to another location [58].

Undoubtedly *Star Trek*, in its many forms, should be credited with popularizing the concept of teleportation, but it was not the first work of science fiction to imagine the possibility. The original version of *The Fly* (1958) starts out as a murder mystery. But as the story unfolds, we discover the details of experiments with teleportation, and the gruesome consequences of a well-intended research project gone badly wrong. In the early stages of the research, the scientist, working in his basement, shows his invention to his wife. He demonstrates with an inanimate object: a cheap ceramic dish, labeled "Made in Japan." The experiment seems to run without a hitch. The original disappears from the transmitting chamber, and reappears in the receiving chamber, in the next room. The wife is duly amazed at the accomplishment. As the scientist waxes eloquent about all the complex sociopolitical problems that could be solved by his invention, including world hunger, the wife examines the dish more closely. She chuckles and says that she wouldn't want this to be tested on her. The husband is crestfallen, as he looks at the ink stamp on the bottom of the dish, which the camera shows to be a backwards mirror image of the original. He immediately goes back through his papers to check all of his calculations, carefully searching for the error that caused the transported dish to come out backward. After discovering and correcting the mistake, he tries the experiment with a page from the daily newspaper, this time with complete success. The transported paper seems to be perfect in every detail. He then tries the experiment with the family cat (no doubt the source of inspiration for the cat scene in *The Prestige*). The original disappears from the transmitting chamber but fails to reappear in the receiving chamber. All that remains of the cat is a mournful meow, perhaps an allusion to the grin of the Cheshire Cat, from "Alice in Wonderland" [59].

The movie succeeds in conveying the idea of scientific research as a difficult process of trial-and-error, in which instant success is not guaranteed. But it fails in regard to the technical details of the concept of a teleportation device. When the wife insists that teleportation is impossible, the husband explains that it is exactly the same as television. He wrongly describes the television camera as the *disintegrator*, and the television set as the *re-integrator*.

Discussion Topic 6.2

Why is television nothing at all like teleportation? What is supposed to happen in the teleportation process, which does not occur with television?

6.7.3   The Uncertainty Principle: Limitations on Precision of Quantum Measurement

When it comes to measuring the properties of extremely small particles, such as the atoms that make up the complex biological molecules in the human body, there are fundamental limitations on the precision with which such measurements can be made. The quantum nature of matter forbids the simultaneous measurement of both the position and the momentum of a particle with arbitrary precision. In other words, if you want to know precisely how fast a particle is moving, you cannot simultaneously know exactly where the particle is, and vice versa. This is known as the Heisenberg uncertainty principle, and it presents an enormous roadblock to the idea of teleportation. The writers of *Star Trek*, recognizing this problem, have invented devices, which are aptly named *Heisenberg Compensators*, in order to get around this fundamental limitation on quantum measurements. The function of these devices is explained in some detail in the episode of *Star Trek: The Next Generation*, called "Ship in a Bottle", in which the crew of the star ship *Enterprise* must find a way to transport a holodeck-simulated character out of the virtual reality environment and into the real world [60].

6.8   Teleportation Estimations

The following simple calculations will serve to illustrate some of the serious problems, which must be overcome, if human teleportation is ever to become a feasible technology.

Estimation 6.1: The Energy and Power Problems

An obvious issue to address, from a physics perspective, is the conversion of matter into energy, which we have already considered in Chap. 3. Using Einstein's famous equation, $E = mc^2$, calculate the energy equivalent of a typical adult human, in units of joules. In the realm of everyday life, we more often think about power, rather than energy. The power involved in a teleportation event is the energy transmitted divided by the time. In sci-fi movies, which make use of transporters, this energy is transmitted to another location in something like 10 s or less. Calculate the power involved in a human teleportation event, in units of watts. Compare this to the Hoover Dam, whose power output is roughly 2000 MW (two billion watts). A single human teleportation event involves the power equivalent of at least 500 million Hoover Dams!

Estimation 6.2: The Data Storage Problem

Let's suppose, for the sake of argument, that the essence of a human being can be reduced to the sum of all the quantum information in the human body. That is to say, all of the quantum particles and all of the information needed to specify the energy state of each particle. How much quantum information is contained in a typical human body, and how much physical space would be required (at state-of-the-art data storage density) to store all of that information? Again, to keep things simple, let's suppose that the human body is made up mostly of water molecules. Each water molecule is made of two hydrogen atoms and one oxygen atom (H_2O). Without getting into all the details of quantum energy states, let's further assume that each water molecule requires five measurements to specify its location in space (three position coordinates and two angles of orientation). These are gross oversimplifications, but it will give a rough estimate of the magnitude of the data storage problem. Using the mass of a typical human body, estimate the number of water molecules, and the total number of measurements required to specify the position and orientation of all of these molecules in the body. How does this compare to the data storage capacity of current, state-of-the-art computers (petabytes, or 10^{15} bytes), as mentioned in Chap. 4? How many such devices would it take to store all of this information?

Estimation 6.3: The Problem of Computer Processing (CPU) Time

Given current, state-of-the-art computer processing speed, as mentioned in Chap. 4 (massively parallel processors can achieve peta-scale speeds or 10^{15} operations per second), and the result of Estimation 6.2, estimate the amount of time required to compute all of the information in a typical human body. Compare this to the estimated age of the universe (roughly 13.7 billion years).

Estimation 6.4: The Problem of Information Degradation

In the episode of *Star Trek: The Next Generation*, called *Relics*, a character from the original *Star Trek* series is found (or at least his quantum information is found) inside the transporter pattern buffer, on a derelict ship. According to the script, the pattern has degraded by only 0.003% in the 75 years that it has been stored [61]. Is this likely to be a problem, if the information is converted back into a living person? Calculate the amount of mass of a typical human body that would be affected by this amount of degradation, and consider whether this might be fatal or of no consequence.

6.9   Beyond Biology

The focus of this chapter has been primarily on the human body: the physical expression of our genetic code; a complex biochemical system with replaceable parts; susceptible to various unpleasantries, including disease and radiation damage; and something which, in principle, might be broken down and stored as quantum information, teleported to another location, and reassembled. But are we more than the sum of our genetic code and quantum information? What does it really mean to be human? There is no universally agreed upon answer to this question. There are those, including futurist Ray Kurzweil and roboticist Hans Moravec, who view the essence of our humanity in purely physical terms. Moravec predicts that humans, as biological entities, will eventually become obsolete and will be replaced by super-intelligent robots [62]. According to Kurzweil, we are fast approaching the so-called singularity: a point in history at which technology will enable humans to transcend biology [63]. If they are correct, we may look forward to seeing (in the not-too-distant-future) something like the sentient android, Commander Data, whose conversation with Ambassador Spock, about being human, we will consider in the next section of this chapter.

Others, including neuroscientist Donald MacKay and computer scientist Matthew Dickerson, believe that the essence of what it means to be human cannot

be reduced to biochemistry and quantum information. Acknowledging the work of
Alan Turing, MacKay agrees that "any logical task that is precisely specifiable can
in principle be performed by an automaton using only the kind of elementary digital
computing operations we already understand." But he remains skeptical with respect
to our ability ever to "find adequate symbolic formulae for all the things we can do
as conscious human beings" [64]. Dickerson is also skeptical about the physicalist
assumption that humans are nothing more than "complex biochemical computers"
[65]. He argues that the answer to the important question, "What does it mean to be
human?" will have profound implications for how we live our lives.

A detailed exploration of this question would take us beyond the bounds of science
and into the realm of philosophy, which space and time do not permit. The reader is
encouraged to explore independently. We will revisit some of the issues and implications in the next chapter, as we consider the future of our technological society.

6.10   What Can We Learn from an Android About What It Means to Be Human?

Unification, a two-part episode of *Star Trek: The Next Generation*, tells of a diplo-
matic effort to reunite the Vulcans and the Romulans: two races, which had diverged
from a common species thousands of years earlier. The Vulcans chose a way of
logic and peace, while the Romulans built a militaristic culture not unlike our
ancient Roman empire. In Part 2 of *Unification*, the half-Vulcan, half-human ambassador Spock offers to assist the android Data in his effort to access encrypted
Romulan computer files. As they work side-by-side, Spock begins a conversation on
an unrelated topic: the character traits of Captain Picard.

SPOCK: He intrigues me, this Picard.
DATA: In what way, sir?
SPOCK: Remarkably analytical and dispassionate—for a human. I understand why my
father chose to mind-meld with him. There's an almost Vulcan quality to the man.
DATA: Interesting. I had not considered that. And Captain Picard has been a role model in
my quest to be more human.
SPOCK: More human?
DATA: Yes, Ambassador.
SPOCK: Fascinating. You have an efficient intellect; superior physical skills; no emotional
impediments. There are Vulcans who aspire all their lives to achieve what you've been
given by design.
DATA: You are half-human.
SPOCK: Yes.
DATA: Yet you have chosen a Vulcan way of life. In effect, you have abandoned what I have
sought all of my life.
Spock succeeds in bypassing the final piece of the encryption and begins to access the
Romulan files. Data continues the conversation:
DATA: Ambassador Spock, may I ask a personal question?
SPOCK: Please.
DATA: As you examine your life, do you find you have missed your humanity?

SPOCK: I have no regrets.
DATA: No regrets? That is a human expression.
SPOCK: Yes. Fascinating! [66]

Data could not understand why Spock had rejected the human side of his heritage, in favor of a Vulcan way of life. Spock had chosen to embrace pure logic and to suppress his emotions, effectively abandoning something that Data had sought all his life. Similarly, Spock could not understand why Data was not satisfied with what he had been given by design. Data's highly efficient intellect is far superior to that of humans, and (at least at that point in the Star Trek saga) he is free of emotional impediments. These are qualities that most Vulcans aspire to achieve. We may be able to learn something from this android about what it means to be human. In a later episode of *Star Trek: The Next Generation* ("Rightful Heir"), Data explains the origin of his quest to become human.

> The Starfleet officers, who first activated me on Omicron-Theta, told me I was an android—nothing more than a sophisticated machine with human form. However, I realized that if I were simply a machine, I could never be anything else. I could never grow beyond my programming. I found that difficult to accept. So I chose to believe that I was a person; that I had the potential to be more than a collection of circuits and sub-processors. [67]

Anthropologists have proposed such descriptors as "tool-users" and "tool-makers" as ways of setting humans apart from other species. But careful studies of the behavior of other species reveal them to be inadequate. Otters, for example, use stones to break open shellfish, and other primates have been seen modifying objects to make them more useful as tools. Since "tool-maker" is not an exclusive attribute of humans, the suggestion has been made that we are unique among other species in terms of the variety of tools that we make for such a broad range of different purposes. But is this not simply a matter of degree, and not a real difference in kind? Is there any characteristic which truly distinguishes us from other species? The example of the android, Data, may provide a useful illustration.

Most species on Earth are only capable of responding to what is happening to them in the present. A few species are able to remember events in the recent past and may be able to plan for the near future. But to the best of our knowledge, humans are the only species capable of contemplating the distant past (events that occurred long before any of us were even born) and imagining the far distant future. We ponder the nature of our own existence, by asking questions, such as "Who am I?" and "What does it mean to be human?" We are able to analyze the human condition (not just as individuals or families, but as a global species) and to ask questions, such as "Is human life, as we know it, the way it ought to be, or all that it could be?" If we decide that things are not as they ought to be, or that we have not achieved our full potential, we can ask further questions, such as "What, if anything, can or should be done to change things?" Of all the creatures on Earth, humans seem to be the only ones who are capable of looking beyond what we are, imagining what we could become, and taking steps to make a difference. In the next chapter, we will consider how we go about solving our problems and improving the human condition.

6.11 Exploration Topics

Exp-6.1: Tissue Engineering

What are the potential benefits of creating biological machines from prefabricated biological parts? What are the potential drawbacks or misuses of such technology?

Reference:
- The BIO FAB Group, "Engineering Life: Building a FAB for Biology" Scientific American, June 2006

Exp-6.2: Tissue Engineering and Vision Restoration

Prior to 2005, the standard approach to tissue cultures failed to produce a three-dimensional retina. What technical breakthrough (or change in approach) led to significant progress in growing a retina? Describe the three categories of retinal degeneration that might benefit from this type of research. Which of the three types of retinal degeneration is most likely to be treatable by cell replacement therapy and why?

Reference:
- Yoshiki Sasai, "Grow Your Own Eye" Scientific American, November 2012

Exp-6.3: Tissue Regeneration and Replacement Organs

Describe the three stages of cell development in the human body. What is the main controversy surrounding the use of embryonic cells for tissue regeneration? What new insights have researchers gained into the possibility of using adult cells for tissue regeneration? What difficulties remain to be surmounted? How have adult human stem cells been used in animals, as a step toward growing transplantable organs?

Reference:
- Konrad Hochedlinger, "Your Inner Healers" Scientific American, May 2010
- Claudia Joseph, "Now scientists create a sheep that's 15% human" London Daily Mail Online, 27 March 2007

Exp-6.4: Drug-Resistant Bacteria

Evolution can be driven by selective pressure—a process in which an external factor acts on the environment, resulting in an altered population distribution. What are the differences between gram-negative and gram-positive bacteria? What factors make gram-negative bacteria more dangerous and more resistant to drugs? What health-care practices can combine with natural evolutionary processes, to promote the spread of drug-resistant bacteria? What are some of the roadblocks to development of new treatments for infection?

Reference:
- Maryn McKenna, "The Enemy Within" Scientific American, April 2011

Exp-6.5: Genetically Engineered Influenza Virus

What do the letters H and N in the designation of an influenza virus (e.g., H1N1, H5N1, H3N2, etc.) represent? Why does the natural strain of H5N1, commonly found among birds, not spread easily among humans? What has been done experimentally to modify H5N1, making it possible for it to spread among humans? What are some of the arguments for and against unrestricted research on H5N1?

Reference:
• Fred Guterl, "Waiting to Explode" Scientific American, June 2012

Exp-6.6: Cell Structure

Why is the lens of the eye transparent, and why is transparent biological material so rare? What can go wrong at the cellular level to make the lens of the eye cloudy or opaque?

What medical insight might be gained by a study of the lens of the eye?

Reference:
• Ralf Dahm "Dying to See" Scientific American, October 2004

Exp-6.7: Cell Phone Radiation

How much energy must be delivered to biological tissue, in order to break chemical bonds inside cells (ionizing radiation), with the possible result of cancer or mutations? How much energy is generated by cell phones? If the energy generated by cell phones is many orders of magnitude smaller than ionizing radiation, what conclusion can you draw about the alleged link between cell phone use and brain tumors?

Reference:
• Michael Shermer "Can You Hear Me Now?" Scientific American, October 2010

Exp-6.8: Radioactivity from Fukushima Reactors Detected on West Coast of North America

Early in April of 2015, radiation was detected on the west coast of North America, which was determined to have originated from the March 2011 Fukushima Daiichi nuclear disaster in Japan. At what specific locations on the west coast was the radiation detected? What specific radioisotopes were detected? How was it determined that these isotopes originated at Fukushima and not from some other source (such as nuclear weapons testing over the Pacific Ocean in the 1960s)? How much activity was detected? Was the radiation level dangerous to human life?

Exp-6.9: Ionizing Radiation and Interplanetary Space Travel

Why is it not practical to use improved shielding on spacecraft as a way to protect astronauts from the effects of ionizing radiation? What is the most dangerous form of radiation in space? Briefly describe the nature and origin of this kind of radiation. In order to simulate the possible effects of exposure to radiation in space, researchers exposed mice to high-energy particle radiation. How much radiation (what does) did the mice receive and over what period of time was this dose given? What tests were performed on the mice to measure the effects of radiation exposure? What conclusions do the researchers draw from their studies of the effects of radiation exposure?

Reference:
• Charles L. Limoli "Deep Space Deal Breaker" Scientific American February 2017

Exp-6.10: Cancer Vaccine

What is the primary motivation for wanting to develop a cancer vaccine? Briefly describe the differences between a preventive vaccine and a therapeutic vaccine, both in terms of when they are administered and how they work? What are some of the advantages of peptide vaccines, compared to cell-based immunotherapies? What difficulties have yet to be overcome in developing an effective therapeutic vaccine for cancer? What are some of the reasons why (until relatively recently) oncolytic virotherapy remained at the fringes of cancer research? Why do cancers often become resistant to conventional chemotherapy? Compared to conventional treatments, what advantages does virus therapy have, which can make it more effective? Compared to natural viruses, what can genetically engineered viruses be programmed to do, which natural viruses cannot do? What is the primary concern associated with continued research into virus therapy?

References:
• Eric von Hofe "A New Ally Against Cancer" Scientific American, October 2011
• Douglas J. Mahoney, David F. Stojdl and Gordon Laird, "Virus Therapy for Cancer" Scientific American, November 2014

Exp-6.11: Genome Sequencing

What are the potential benefits of technology which makes the reading of DNA fast, inexpensive, and widely accessible? Is it possible for such technology to be misused, as suggested by the movie GATTACA?

References:
• George M. Church "Genomes for All" Scientific American, January 2006
• Jonathan Rothberg, et al. "An integrated semiconductor device enabling non-optical genome sequencing" Nature, Vol 745, 21 July 2011
• NOVA – "Cracking Your Genetic Code" (Sarah Holt, WGBH 2012)

Exp-6.12: Gene-Edited Babies

Find out what you can about the gene-edited twin babies, announced by Dr. He Jiankui of Shenzhen, China, in November of 2018. Was the research ever submitted for peer review? If so, in what professional journal was it published? Were the babies ever genetically tested to determine if they were, in fact, immune to the HIV virus, as claimed? What additional criticisms of the work were expressed, apart from the fact that the announcement was made in the popular media, prior to submission for peer review? What ethical concerns have been expressed? What is your opinion on the subject of genetically modifying human embryos?

Exp-6.13: Humans as a Species

Describe at least three key genetic differences between humans and chimpanzees. Specifically, what are the functions of these genes? What new insights have researchers gained into the profound differences between humans and chimpanzees, despite the fact that the two genetic codes are nearly identical?

Reference:

• Katherine S. Pollard "What Makes Us Human?" Scientific American, May 2009

Exp-6.14: Human Teleportation

What kind of research is currently underway on the feasibility of teleportation? To what extent has progress been made toward this goal? How likely is it that we will be able to teleport a macroscopic object in the near future?

References:

• Anton Zeilinger "Quantum Teleportation" Scientific American, April 2000
• Keay Davidson (San Francisco Chronicle) "Beam me up, Scotty?" Pittsburgh Post Gazette, 17 October 2005, p. A-6
• Alexandra Witze "Quantum teleportation leaps forward" Science News, 30 June 2012, p. 10

References

Introduction: Being Human

1. Shelley, M.: Frankenstein; or, the Modern Prometheus, p. 51. The Pennyroyal Press, University of California Press, Berkeley and Los Angeles, California (1984, based on the 1818 text)

What Can We Learn from "Frankenstein" About What It Means to Be Human?

2. *Frankenstein* (James Whale, Universal Studios 1931). A scientist learns "what it feels like to be God" after successfully assembling and animating a living creature from dead human body parts [DVD scenes 5 and 6]
3. *The Bride of Frankenstein* (James Whale, Universal 1935). The creature from the original 1931 *Frankenstein* move returns and demands a mate
4. *Mary Shelley's Frankenstein* (Kenneth Branagh, Tristar 1994). This version of the Frankenstein story is much more faithful to Mary Shelley's original text than the 1931/1935 classics, or any other film version, but also contains added scenes that are considerably more gruesome and violent than anything in the original text
5. Shelley, M.: Frankenstein, 1818 text, p. 47
6. Fitzharris, L.: Dangerous medicine. Sci. Am., Oct 2017

Bodies with Replaceable Parts

7. *Mad Love* (The Hands of Orlac) (Karl Freund, MGM 1935). A hand transplant more than 60 years before the first successful operation of this kind. A concert pianist's hands are placed with the hands of a carnival knife-thrower, hanged for murder. The transplanted hands take on a life of their own [DVD scenes 8, 9]
8. *Orlacs Hände* (The Hands of Orlac) (Robert Wiene, Berolina Film GmbH 1924). An earlier silent version of the hand transplant story, but in this version, the hands do not actually take on a life of their own
9. Hamilton, D.: A History of Organ Transplantation, p. 291. University of Pittsburgh Press, Pittsburgh (2012)
10. Hamilton, D.: A History of Organ Transplantation, p. 142. University of Pittsburgh Press, Pittsburgh (2012)
11. *Star Trek* (The Original Series) – "Spock's Brain" (Marc Daniels, Paramount 1968). Dr. McCoy must acquire the delicate surgical skills necessary to put Spock's stolen brain back where it belongs [DVD vol. 31, ep. 61, scenes 1, 2, and 6 or full episode]
12. *The Man They Could Not Hang* (Nick Grindé, Columbia Pictures 1939). Cardiopulmonary bypass machine more than a decade before first actual use of such a device [DVD scenes 1, 2, 6]
13. *Star Trek: The Next Generation* – "Tapestry" (written by Ronald D. Moore, Paramount 1993). The origin of Captain Picard's artificial heart [DVD season 6, disc 4, episode 15]
14. *Star Trek: The Next Generation* – "Samaritan Snare" (written by Robert L. McCullough, Paramount 1989). Picard's artificial heart needs a replacement [DVD season 2, disc 5, episode 17]
15. Palomino, J.: The heart is just a pump. The Verge, 4 Nov 2015. https://www.theverge.com/2015/11/4/9665902/artificial-heart-transplant-cedars-sinai-clinical-trials

Resistance to Disease

16. *The War of the Worlds* (Byron Haskin, Paramount 1952). Military weapons useless against invaders from Mars [DVD scenes 5, 9]
17. *War of the Worlds* (Steven Spielberg, Paramount 2005). The gruesome details of the alien invasion are faithful to the novel [DVD scenes 5, 6]
18. *War of the Worlds* (Steven Spielberg, Paramount 2005). Cause of death of Martian invaders described using quotes from novel [DVD closing scene]
19. *The War of the Worlds* (Byron Haskin, Paramount 1952). Martian invaders die from bacterial infection [DVD scene 13]
20. Clark, F.C.: A brief history of antiseptic surgery. Med. Lib. Hist. J. **5**, 145–172 (1907). http://www.ncbi.nlm.nih.gov/pmc/articles/PMC1692621/
21. *Contagion* (Steven Soderbergh, Warner Brothers 2011). An amateur's explanation of R_0, the reproduction number for a viral infection, misses the mark on the rate of progression [DVD scene 24]
22. Althaus, C.L.: Estimating the reproduction number of Ebola virus (EBOV) during the 2014 outbreak in West Africa. PLoS Curr. (2014). https://www.ncbi.nlm.nih.gov/pmc/articles/PMC4169395/
23. Guerra, F.M., Bolotin, S., Lim, G., Heffernan, J., Deeks, S.L., Li, Y., Crowcroft, N.S.: The basic reproduction number (R_0) of measles: a systematic review. Lancet Infect. Dis. **17**(12), PE420–PE428 (2017). https://www.thelancet.com/journals/laninf/article/PIIS1473-3099(17)30307-9/fulltext
24. "Measles Data and Statistics" Downloadable presentation from the Centers for Disease Control, 16 Feb 2018. https://www.cdc.gov/measles/downloads/MeaslesDataAndStatsSlideSet.pdf

Cell Structure and Radiation Damage

25. *Star Trek II: The Wrath of Khan* (Nicholas Meyer, Paramount 1982). Spock exposes himself to a lethal dose of gamma radiation, in order to restore warp-drive to the Enterprise [DVD scene 15]
26. *The War of the Worlds* (Byron Haskin, Paramount 1952). Geiger counter used to measure radiation from an object fallen from space [DVD scene 2]
27. xkcd Radiation Dose Chart. http://xkcd.com/radiation/
28. *Godzilla* (Roland Emmerich, Columbia Tristar 1998). Biologist studies the effects of the Chernobyl nuclear disaster on worms [DVD Scene 3]
29. *Nature* – "Radioactive Wolves" (Klaus Feichtenberger, THIRTEEN 2011) [DVD scene 1]
30. *Them* (Gordon Douglas, Warner Brothers 1954). "Lingering radiation from the first atomic bomb" causes ants to grow to gigantic proportion [DVD scenes 10, 11]
31. Featherstone, S.: The swallows of Fukushima. Sci. Am. **312**, 74–81 (2015)
32. Crespo-Hernandez, C.E., et al.: Ab initio ionization energy thresholds of DNA and RNA bases in gas phase and in aqueous solution. J. Phys. Chem. A. (2004). http://pubs.acs.org/doi/abs/10.1021/jp049270k

DNA and the Human Genome

33. *The Island of Dr. Moreau* (Don Taylor, MGM 1977). The title character claims that he has "... almost proved the existence of a cell particle that controls heredity" [DVD scene 6]
34. History and goals of the Human Genome Project. http://www.ornl.gov/sci/techresources/Human_Genome/home.shtml
35. Collins, F.S.: The Language of God, p. 118. Free Press – Simon and Shuster, New York (2006). (Watson opposed to gene patents)
36. Collins, F.S.: The Language of God, p. 124. Free Press – Simon and Shuster, New York (2006). (size of human genome)
37. Collins, F.S.: The Language of God, p. 111. Free Press – Simon and Shuster, New York (2006). (on "junk DNA")
38. Encyclopedia of DNA elements. http://encodeproject.org/ENCODE/
39. *Spider-Man* (Sam Raimi, Columbia Pictures 2002). A high school student is bitten by a genetically engineered super-spider, and acquires spider-like abilities [DVD scenes 2, 3]
40. Collins, F.S.: The Language of God, p. 137. Free Press – Simon and Shuster, New York (2006). (similarity between human and chimpanzee genomes)
41. *NOVA* – "Cracking the Code of Life" (Elizabeth Arledge, WGBH 2001). PBS NOVA series episode, focusing on the Human Genome Project. Genetic variation among humans and chimpanzees [DVD scene 6]
42. *I am Legend* (Francis Lawrence, Warner Brothers 2007). A cure for cancer has been found by genetically engineering the measles virus. But the virus mutates with bizarre, if not lethal consequences [DVD scenes 1 and 2]
43. *GATTACA* (Andrew Niccol, Columbia Pictures 1997). Natural birth, versus genetically screened and selected birth [DVD scenes 3, 4]. Genetic discrimination and impersonation [DVD scenes 6, 7]
44. *NOVA* – "Cracking Your Genetic Code" (Sarah Holt, WGBH 2012). Eric Lander: 180 genes involved in determining the height of an individual [DVD scene 6]. Jonathan Rothberg on genotyping vs. DNA sequencing, and Francis Collins' genotype test results [DVD scene 3]. Gene-based therapy for cancer [DVD scene 5]. Pre-implantation genetic diagnosis (PGD) [DVD scene 6]
45. *Rampage* (Brad Peyton, New Line Cinema 2018). The CRISPR gene-editing tool is modified to work on the entire body [DVD scene 3]
46. CRISPRs web server with useful information about how CRISPR really works. https://crispr.i2bc.paris-saclay.fr/index.php?page=about

47. Cyranoski, D., Reardon, S.: Embryo editing sparks epic debate. Nature. (2015). https://www. nature.com/news/embryo-editing-sparks-epic-debate-1.17421
48. Stein, R.: Chinese scientist says he's first to create genetically modified babies using CRISPR. NPR Morning Edition, 26 Nov 26 2018. https://www.npr.org/sections/ health-shots/2018/11/26/670752865/chinese-scientist-says-hes-first-to-genetically-edit-babies
49. Court, E.: Bill Gates warns that nobody is paying attention to gene editing, a new technology that could make inequality even worse. Business Insider, 12 Jan 2019. https://www.businessin-sider.com/bill-gates-says-gene-editing-raises-ethical-questions-2019-1?r=US&IR=T&utm_ source=reddit.com
50. Duffy, J.: Two-faced molecule could lead to Huntington's disease treatment. Carnegie Mellon University, Department of Chemistry, News & Events, 14 Sept 2018. https://www.cmu.edu/ chemistry/news/2018/0914-Janus-PNA_Huntingtons.html

Cloning

51. *Jurassic Park* (Steven Spielberg, Universal Studios 1993). A theme park is created with living dinosaurs, cloned from DNA found in Jurassic mosquitoes preserved in amber [DVD scenes 5, 6]
52. Allentoft, M.E., et al.: The half-life of DNA in bone: measuring decay kinetics in 158 dated fossils. Proc. R. Soc. B. **279**, 4724–4733 (2012). (published online 10 Oct 2012)
53. Gannon, M.: Jurassic Park' may be impossible but Dino DNA lasts longer than thought. Live Science, 12 Oct 2012. https://www.livescience.com/23861-fossil-dna-half-life.html
54. Lanza, R.P., Dresser, B.L., Damiani, P.: Cloning Noah's ark. Sci. Am. **283**(5), 84–89 (2000)
55. *The Island* (Michael Bay, Warner Brothers 2005). Clones are created and sold as insurance policies to the wealthy [DVD scene 8]
56. *The Island* (Michael Bay, Warner Brothers 2005). Some of the clones are found not only to be biologically identical to their "sponsors," but also to have their memories [DVD scene 13]

Human Teleportation

57. *The Prestige* (Christopher Nolan, Touchstone and Warner Brothers 2006). Nikola Tesla is commissioned to build a teleportation device. The device succeeds, but with one small problem: it creates a duplicate in another location, but does not destroy the original [DVD scene 18]
58. *Star Trek* (original series) – "The Corbomite Maneuver" (Joseph Sargent, Paramount 1966). The first showing of the transporter, as well as the debut of Dr. McCoy [DVD vol. 1, ep. 2, scene 7]
59. *The Fly* (Kurt Neumann, Twentieth Century Fox 1958). A scientist works to create a matter teleportation device. Initial tests look promising, but the ultimate test has disastrous consequences [DVD scenes 9 and 10]
60. *Star Trek: The Next Generation* – "Ship in a Bottle" (Alexander Singer, Paramount 1993). "Heisenberg compensators" [DVD season 6, disc 3, scenes 1, 2, 7]
61. *Star Trek: The Next Generation* – "Relics" (Alexander Singer, Paramount 1992). Information loss: "pattern degradation" [DVD season 6, disc 1, scenes 1 and 2]

Beyond Biology

62. Moravec, H.: Robot: Mere Machine to Transcendent Mind. Oxford University Press, Oxford (1998)
63. Kurzweil, R.: The Singularity Is Near: When humans transcend biology, p. 7. Penguin, New York (2005)

64. MacKay, D.: Brains, Machines & Persons, p. 60. William Collins Sons & Co. Ltd., London (1980). and 49
65. Dickerson, M.: The Mind and the Machine: What it means to be human and why it matters, p. xiv. Brazos Press, Grand Rapids (2011)

What Can We Learn from an Android About What It Means to Be Human?

66. *Star Trek: The Next Generation* – "Unification, Part II" (Cliff Bole, Paramount 1991). Dialog between Spock and Data, in regard to what it means to be human [DVD season 5, disc 2, scene 5]
67. *Star Trek: The Next Generation* – "Rightful Heir" (Winrich Kolbe, Paramount 1993). Data's choice to be more than just a machine [DVD season 6, disc 6, scene 8]

Chapter 7
How Can We Solve Our Problems?
(Science, Technology, and Society)

*The transmissions of an orbiting probe are causing critical
damage to this planet. It has almost totally ionized our
atmosphere. … The probe is vaporizing our oceans. We cannot
survive unless a way can be found to respond to the probe.*
—President, United Federation of Planets
Star Trek IV: The Voyage Home

More often than not, science fiction movies involve crisis management. A serious problem arises—alien invaders, natural disasters, or unforeseen consequences of some new technology, to name just a few—and a solution must be found urgently. The solution requires the best scientific minds and the latest technology, often preceded or followed up by military intervention. Indeed, much of real science is devoted to problem-solving. Natural curiosity, in response to puzzling observations, leads to all sorts of interesting questions about how the world works. The goal of basic research is to answer these fundamental questions, while applied research is aimed at solving specific, practical problems. But there are problems of a different sort, not all of which are solvable by science or technology.

Few people, if any, will deny that some things are not the way they ought to be. There are plenty of situations in the world, which are badly wrong—issues of injustice and oppression, for example—and are crying out to be set right. There are also many things that are not necessarily bad, in the moral sense, but which have room for improvement. The previous chapter concluded with the observation that we humans are unique in our ability both to recognize that some things are not the way they ought to be, or not all that they could be, and to come up with creative ways of changing things—hopefully for the better. In the final chapter of this book, we will explore possibilities for the future of our technological society. In this chapter, we consider some of the ways in which science and technology might be used—or misused—to address various kinds of problems. What role do science and technology have to play, in regard to solving our many problems, and how is this portrayed in science fiction?

© Springer Nature Switzerland AG 2019
B. B. Luokkala, *Exploring Science Through Science Fiction*, Science
and Fiction, https://doi.org/10.1007/978-3-030-29393-2_7

7.1   Public Perception of Science and Scientists

Our future will depend not only on the kinds of scientific and technological discoveries that might be made, but on the decisions that are taken, concerning what research projects to fund or not to fund, and how the newly discovered information or newly developed technology will be used. Thus, the public understanding and perception of science—both its methodology and the character of those who practice it—will be crucial in the decision-making process. A poorly educated public are more likely to make poor decisions. The way in which science is used, or misused, will have an impact on the way we live.

Science fiction, as a genre, spans the entire range of possible views, with respect to the impact of science and technology on society, and the character of those who practice science. At one end of the spectrum is the view explored by Christopher Frayling, in his book "Mad, Bad and Dangerous? – the Scientist and the Cinema." As Frayling describes, there is no shortage of portrayals in science fiction of scientists who are insane, evil, or irresponsible. According to this perspective, science and scientists are the source of most of the problems in the world. At the opposite end of the spectrum is the view, which seems to prevail in most of the *Star Trek* franchise: that science and scientists will provide the solution to the world's problems. The former view is spawned from fear and misunderstanding, while the latter may be naïvely optimistic. Here we consider a few examples, both positive and negative.

7.1.1   Science as Obsession

The stereotypical mad scientist is Dr. Frankenstein, whom we have already seen in the chapter on biological sciences. Frankenstein would stop at nothing, even if it means breaking the law, to carry out his experiments. His overarching goal is to prove to the world that he can create life, and to "know what it feels like to be God." The image of the insane scientist is an obvious caricature. But another Hollywood image, which the scientific community may need to work to dispel, is that of the scientist who is obsessed with the research, perhaps to the point of being dangerous. A classic example is Dr. Carrington in *The Thing From Another World*. Carrington is a biologist, who is fascinated by an alien creature, discovered frozen in the Arctic ice. Through carelessness on the part of the research team, the creature thaws out, and soon reveals itself to be a vicious killer. When Carrington discovers that the alien's cell structure and physiology more closely resemble those of plants than animals, he becomes obsessed with studying it. Although he is shown in the movie doing legitimate scientific research, and performing careful experiments, his desire to understand the creature is more important to him than the lives of the other humans around him [1].

7.1.2   Science and Arrogance

Two sci-fi films, both produced in the first decade of the twenty-first century, stand at opposite ends of the spectrum, in regard to the public perception of science and scientists. The earlier of the two, *Absolute Zero*, portrays scientists as arrogant, even while they are proclaiming scientific ideas that are absolute nonsense. The topic of the movie is timely: What is the cause of global climate change, and what, if anything, can we do about it? This is certainly one of the biggest and most pressing scientific questions of the modern era. But the proposed explanation in the movie is absurd: Earth's magnetic field is about to change direction, and when it does, the temperature of the Earth will fall to −273 °C, absolute zero! [2]. It is certainly true that the Earth has experienced ice ages. There is also archeological evidence of changes in the direction of the Earth's magnetic field in the ancient past [3]. However, it is physically impossible for the temperature of the Earth ever to fall anywhere near absolute zero. One would first have to shut off both of the natural heat sources that help to make life on Earth possible. As long as the Sun is still burning in space, and the Earth remains in orbit around it, energy will constantly be flowing from the Sun to the Earth. And as long as there are radioactive materials in the Earth, heat is generated naturally from the inside, by the energy released in the radioactive decay process. The Earth is being heated from the inside and from the outside, no matter what the Earth's magnetic field is doing. But to make matters worse, after giving a bogus demonstration of his hypothesized effect, the arrogant scientist makes the absurd claim: "Science is never wrong!"

It is an unfortunate truth that this level of arrogance is not limited to characters in bad science fiction movies. There are scientists in real life who promote science as superior to other areas of human activity, using words such as "science is true whether you believe in it or not." This misrepresents what science actually is. The people who practice science may be correct or incorrect in their interpretation of the data, but science is neither "right" nor "wrong." In the same way, science is neither "true" nor "false." Science is the *method* by which we *discern* what is true or false about the way in which the physical world works. We will look more closely at the methodology of science later.

7.1.3   Science as an Act of Futility

At the opposite end of the spectrum, *The Happening* suggests that the best scientists can ever do is to propose theories, but that science doesn't actually lead to real understanding. Mark Wahlberg plays the role of a high school science teacher, who doesn't seem to know the difference between a scientific theory and a personal opinion. Near the beginning of the movie, he opens a class discussion by referring to a recent newspaper article about the mysterious disappearance of the honeybees. "Now let's hear some theories about why this might be happening."

When no one volunteers, the teacher begins to call on them. The students propose a variety of plausible explanations, including disease, pollution, and global warming, but none of these seems to suit the teacher. He asks them to "Keep on guessing." Finally, he addresses one particular student, who doesn't seem to be interested in science, and asks "Don't you have an opinion?" The student suggests "An act of Nature, and we'll never fully understand it." Evidently, this is the response that the teacher was hoping to get, and he says "Nice answer, Jake! He's right. I mean Science will come up with some reason to put in the books. But in the end, it'll be just a theory. I mean we will fail to acknowledge that there are forces at work beyond our understanding. To be a good scientist, you must have a respectful awe for the Laws of Nature" [4].

The Happening portrays science as an enterprise which struggles to comprehend the incomprehensible. According to the science teacher in the movie, scientists make up explanations of natural phenomena to put in text books, but they don't really understand. This, too, is a pathetic reflection of what science really is. The aim of science is to refine our understanding of how the natural world works. Admittedly, scientists do not always get it right the first time. History records numerous examples of accepted scientific explanations, which were eventually overturned in favor of more accurate descriptions. Thus, the assertion in *Absolute Zero* that "Science is never wrong" is completely false. But equally misleading is the notion that scientists must simply stand in awe of the so-called Laws of Nature, without ever understanding them.

7.1.4 The Model Scientist

Relatively rare in science fiction is the portrayal of scientists, who are dedicated and determined to succeed, for all the right reasons, and also highly competent. Noteworthy examples include two female scientists, whom we have already considered briefly in previous chapters. The real heroine of the 1954 movie *Them* is Dr. Patricia Medford (Joan Weldon), a noted scientist and expert in the biology and physiology of ants. She plays a key role in putting an end to the threat of the giant ants—mutations induced by nuclear testing in the New Mexico desert—and is the first to enter the sewers at the end of the movie, to confirm that the last of the ants are dead. Dr. Eleanor Arroway, the radio astronomer in CONTACT, is portrayed as a brilliant and highly motivated scientist, who has chosen to do research on a subject that some consider to be tantamount to professional suicide: the search for extraterrestrial intelligence. Undaunted by a long string of negative results, and struggling against the odds to find funding to continue her research, Dr. Arroway eventually makes the definitive discovery. While working at the radio astronomy facility, known as the Very Large Array (by coincidence, also located in New Mexico), she picks up a radio transmission, which seems to be the set of prime numbers from 2 to 101. But upon closer inspection, the sequence of pulses contains embedded instructions for building a wormhole-opening device, which enables her to be the first human to visit another star system.

7.2 Methodology of Science

So what is it that scientists actually do? Prompted by observations, and motivated by curiosity, scientists ask interesting questions about how the world works (The example from *The Happening* is a perfectly valid scientific question to ask: Why have the honeybees disappeared?). The question prompts further observations, which lead to the formulation of a hypothesis about how things might work. Experiments are designed to test the hypothesis, in order to find out if things actually seem to work as predicted. If the tools and techniques are not available to carry out the proposed experiments, new technology may have to be developed. The process is iterated and refined, and a larger and larger body of evidence is built up. The results are usually communicated in some form to the scientific community for further scrutiny and independent verification. If the hypothesis is not supported by the available evidence, or if new experiments are designed and carried out, which push a previously supported theory to the point where it fails to account for the new observations, it will have to be refined, or discarded in favor of a better description, which does stand the test of experiment. This process is what we call the *scientific method*, and its goal is to increase our knowledge and understanding of how the world actually works.

It's important to note that the word *theory*, as it is used in the scientific community, is not the same as the popular conception of a theory, as portrayed in *The Happening*. The Theory of Relativity is not just Albert Einstein's personal opinion, concerning the nature of space and time, nor is it simply an educated guess. Rather, it is a powerful and useful description of the way things really seem to be, supported by a large body of experimental evidence. The science teacher in *The Happening* seems to think that a theory is just something that people put in textbooks, when they really don't know what's going on. Unfortunately, his pejorative use of the expression "just a theory" accurately reflects the popular understanding (or misunderstanding) of what a theory is.

Discussion Topic 7.1

Both *Absolute Zero* and *The Happening* miss the mark, in regard to what science is, but for different reasons. Neither the scientist in *Absolute Zero*, nor the science teacher in *The Happening* seems to understand what science is all about. Why are both of their ideas about science completely off-base? Give specific examples to counter each of their positions.

7.2.1 Some Sci-Fi Illustrations of the Scientific Method

Fritz Lang's silent movie *Frau im Mond* (*Woman in the Moon*) is famous for being the first to feature a countdown before launching a rocket. But it also features an interesting example of the scientific method being used by the first human to walk on the surface of the Moon. Immediately upon landing on the Moon, a scientist dons a spacesuit and takes the first steps of exploration outside the ship. He has a hypothesis in mind, which he would like to test. We are not told explicitly what this hypothesis is but instead are left to infer it from the actions of the scientist. He strikes a match and seems pleasantly surprised at the resulting flame. He repeats this action several times and obtains the same result each time. Having gathered enough data to reach a tentative conclusion, he then removes the helmet from his spacesuit and continues his exploration of the Moon [5].

Discussion Topic 7.2

Evidently, the scientist in *Woman in the Moon* had a particular hypothesis about the conditions on the surface of the Moon, which he proceeded to test experimentally. What was his hypothesis, and what were the two different experiments that he performed? What would the scientist have done, if the first experiment did not support his hypothesis? Did the results of the first experiment provide enough evidence that it was safe to perform the second experiment? What would have happened if the second experiment had failed to support the hypothesis?

The 1971 movie *The Andromeda Strain* provides one of the most realistic illustrations of the scientific method at work to solve a serious problem. Based on the Michael Crichton novel, the story begins when an Earth satellite crashes near a remote New Mexico village. All but two of the inhabitants die—not as a direct result of the impact but from a deadly microbe apparently brought to Earth by the satellite. A team of researchers race against time to discover why the two survivors were immune, before the disease, has a chance to spread. One of the first steps is to determine how the disease is spread. The satellite is placed in an isolation room, with a tube connected to a separate sealed box with a laboratory rat inside. When a partition is opened between the satellite room and the rat's box, the rat dies almost instantly. The hypothesis is that the disease is airborne. The next step is to determine the size of the particles. A succession of filters is used until one is found with a mesh size small enough to stop the particles. The size determination is used to guide a microscopic inspection of the surface of the satellite. Once the foreign body is located, a battery of tests reveal what it is made of, how it grows, and what might kill it. The organism is cultured and exposed to a range of wavelengths of electromagnetic radiation and to various chemical environments. The problem is finally solved by combining observations of the two human survivors and inferring that the organism can only survive within a very narrow range of pH. Either too much acidity or too much alkalinity will cause it to stop growing and die [6]. Rarely does the process of scientific inquiry play such a prominent role in a science fiction film.

7.3   Reasonable Uses of Science to Solve Our Problems

Space doesn't permit an exhaustive treatment of the many and varied problems facing the world today. In the sections which follow, we consider three different categories of problems, divided up according to the way in which science or technology are used in each case. We begin with some problems for which science and technology, rightly applied, might lead to good solutions, or at least partial solutions. Admittedly, many of the problems facing society are complicated and are likely to require more than just science or technology to provide complete solutions. Nevertheless, science and technology may legitimately have important roles to play.

7.3.1   How Can We Stop a Robot Invasion?

Transgressing the boundaries of film noir and science fiction, the 1954 movie *Target Earth* tells of an invading army of robots, supposedly from the planet Venus. Produced on a very low budget, the movie never actually showed more than one robot at a time. Instead, an actor wearing a robot costume was filmed from a variety of directions, in a variety of settings, and the film was edited to give the impression that there were many robots, on a rampage through an American city. The robots are unstoppable by conventional weapons, but a brilliant scientific solution to the problem is found. When soldiers discover one of the robots disabled, with a cracked faceplate, scientists immediately conduct a series of tests, to find out what caused the damage. The scientists hypothesize that sound waves of just the right frequency could set up resonance vibrations in a crucial glass tube, inside the robot's head. If the sound waves are sustained at the right frequency, the induced vibrations will eventually crack the tube, disabling the robot. After a successful demonstration to military officials, the sound generator is mounted on an army jeep, which drives through the streets of the besieged city. The threat of the invading robots is eliminated just minutes before the scheduled launch of a full military strike. If the scientists had not been successful, the ensuing military action would have destroyed the city [7]. Although it is highly unlikely that we will ever experience an actual robot invasion, the takeaway message of *Target Earth*, presumably, is that science and the military will work cooperatively to save us from our enemies.

The depiction of the use of science to defeat the robots in *Target Earth* is not just science fiction but is the way that resonance processes can actually work. Forced vibrations at the resonance frequency can build up in amplitude, to the point at which the bonds between the atoms of the material can no longer hold it together. Numerous demonstrations of the resonance effect, using sound to shatter a wine glass can be found on YouTube. Of particular note are television advertisements for the Memorex brand of audio tape, featuring the great jazz singer, Ella Fitzgerald. The commercials pose the question, "Is it live, or is it Memorex?" [8]. The visuals suggest that the recording is of such high quality that it can shatter a wine glass, just as a trained singer, like the real Ella Fitzgerald, might do with her voice. But what is the science behind this? All you need to do to shatter a piece of glass with sound waves is to tune the frequency of the sound to match one of the modes of vibration of the glass. It doesn't matter if the glass is a tube inside the head of an invading robot from Venus, or a wine glass in an audio tape commercial. And it doesn't matter if the sound is generated by a live human voice, or a recording of a human voice, or a laboratory frequency generator.

As long as you sustain the particular resonance frequency, the vibrations that build up in the glass can become large enough to shatter the glass.

So the concept of resonance vibrations, as used in the movie, is plausible, but with at least one technical difficulty. The source of the sound used to defeat the robots is a loudspeaker mounted on an army vehicle and driven through the streets of the besieged city. Moving the source to different locations would change the phase relationship between the source of the sound waves and the glass tube inside the robots' heads. The phase relationship must be kept constant for the resonance vibrations to build up. A moving source might not work as well as shown.

7.3.2   How Can We Feed the Hungry?

The *Star Trek* franchise, for the most part, presents us with a very optimistic view of the future of our technological society. Some of the most serious issues facing us today will have been solved two or three centuries from now. By the twenty-third century, world hunger will have been eliminated, perhaps due in part to the development of *quadrotriticale*, a genetically modified version of the nineteenth century wheat-rye hybrid, with four times the yield. In the original series episode, "The Trouble with Tribbles," Ensign Chekov claims that this futuristic grain was a Russian invention [9]. But regardless of who invented it, the solution to world hunger is likely to require more than just advances in agricultural technology. At the very least, once the grain is produced, it needs to be delivered to where it is needed. Recall, from the previous chapter, the fictional inventor of the teleportation device, in the original version of *The Fly*. He was optimistic about the potential of his new technology to bring an end to world hunger. The ability to move supplies to places where they are needed most should be all we need to solve the problem. Or so he thinks.

Discussion Topic 7.3

In the Star Trek universe, a high-tech, high-yield grain has been developed, teleportation is an everyday occurrence, and world hunger is a thing of the past. Is this a case of naïve optimism? Are there any factors contributing to the problem of world hunger, which are of a fundamentally human nature and are unlikely to be solved by better technology?

The genetically modified grain in "The Trouble with Tribbles" raises additional questions that are very much debated today. What are the health risks, if any, associated with consuming "Frankenfoods" as they are sometimes called? What might be the environmental impact of genetically engineering plants to be disease and insect resistant? Surprisingly, the *Star Trek* TV series from the late 1960s predates the development of the first genetically modified tobacco plants in the 1970s. Since that time, the fear of genetically modified foods has been in tension with the potential benefits of high-yield, disease-resistant foods to reduce the problem of world hunger [10].

Discussion Topic 7.4

Read the article "Are Engineered Foods Evil?" by David H. Freedman (Scientific American, September 2013). Did you have an opinion about the alleged risks of consuming genetically modified foods prior to reading the article? Did you learn anything from reading the article, or did any of the information come as a surprise to you? Are you skeptical of any of the claims? Did your opinion on any of the issues change?

7.3.3   How Can We Conserve Our Natural Resources?

The unbridled optimism of *Star Trek*, with respect to technology, is not limited to solving current problems. It even extends to the possibility of atonement for seemingly irreversible mistakes of the past. *Star Trek IV: The Voyage Home* predicts that by sometime in the twenty-first century, humpback whales will have been hunted to extinction. This creates a crisis in the twenty-third century, when a giant probe of immense power and unknown origin comes to Earth, deactivating starships and space stations along the way. The probe begins to ionize the atmosphere and vaporize Earth's oceans, wreaking havoc with the global climate control system, and threatening to destroy all life on Earth. At the recommendation of the Vulcan ambassador, the President of the United Federation of Planets transmits a distress signal, warning all hearers to avoid the planet Earth at all costs.

Meanwhile, the main characters of the original *Star Trek* television series are enroute to Earth in a stolen Klingon bird-of-prey and pick up the distress message. After careful analysis of the signal from the probe, Spock concludes that it is communicating in whale song—specifically, the songs sung by humpback whales. He correctly identifies the root cause of the problem: human greed resulted in the hunting of a species to extinction, a practice which he declares to be illogical. He also proposes a solution: time travel back to the late twentieth century, when humpback whales still existed, and bring a mating pair back to the future, so they can answer the call of the probe [11].

The story which unfolds is a brilliant illustration of an interdisciplinary team, working together to solve a complex problem. Time travel is successful. The main characters arrive in orbit around Earth, in the late twentieth century, just as Spock had calculated. Their space vehicle, being of Klingon design, is conveniently equipped with a cloaking device, which renders them invisible to the tracking technology of the time. They land in Golden Gate Park, in San Francisco, and break up into smaller teams, each with a specific task, essential to the completion of the mission.

Kirk and Spock find a pair of humpback whales, temporarily housed in captivity, at the Cetacean Institute marine park. But Spock is quick to point out that if they assume the whales are theirs to do with as they please, they would be as guilty as those who caused their extinction. They must, instead, convince a skeptical cetacean biologist at the marine park that it would be in everyone's best interest (including the whales) to take them back to the future.

Dr. McCoy and engineer Scott find a manufacturer of Plexiglass, which could be exactly the material they need to build a giant aquarium in their ship, to transport the whales. Masquerading as a professor of engineering from Edinburgh, Scotty offers to trade the formula for transparent aluminum (which we discussed in a previous chapter), in exchange for a large slab of 6-in. Plexiglass. Sulu somehow manages to borrow a helicopter, in order to move the Plexiglass from the factory to the ship. And Uhura and Chekov sneak onboard the nuclear-powered aircraft carrier, U.S.S. Enterprise, to collect some high-energy photons from the nuclear reactor, which they need to recrystallize the fused Klingon dilithium crystals—a key component of their warp drive. The primary message that the movie conveys is the call for humans to be more responsible with our natural resources: curb our greed and refrain from hunting any species to extinction. But perhaps an unintended message is that if all else fails, science and technology can always come to the rescue. We will consider examples of misplaced faith in science later in this chapter.

Popular culture can sometimes have a powerful influence on the way we think and behave. No better example can be found than the well-documented influence of the television series *All in the Family* on the market for aerosol spray products. In February of 1975, an episode of the series aired which raised national consciousness about the detrimental effect of chlorofluorocarbons (CFC) on the ozone layer. As described in the PBS special entitled "Ozone Hole: How We Saved the Planet," which aired April 10, 2019, immediately following the *All in the Family* episode, the purchase of aerosol spray products, such as deodorant and hairspray, dropped precipitously. The aerosol industry acknowledges a direct cause-and-effect relationship. CFCs are no longer used as propellants in spray products [12]. In the case of humpback whales, it may be more difficult to prove a cause-and-effect relation between the consciousness-raising influence of *Star Trek IV: The Voyage Home* and the recovery of humpback whale populations worldwide. All humpback whales had been listed as endangered since 1970, with some of the population groups having been reduced by more than 95% from their estimated historical levels. *Star Trek IV* was still in production in 1985, when a global moratorium on commercial whaling went into effect. As of 2016, only four out of the 14 population groups of humpback whales worldwide continue to be listed as "endangered." One of the 14 groups is listed as "threatened." The other ten groups are listed as "not at risk" [13]. Unlike the direct influence of "All in the Family" on the banning of CFCs, there is no officially acknowledged connection between the possible influence of *Star Trek IV* and the dramatic recovery of humpback whales worldwide.

7.3.4   How Can We Provide Better Healthcare?

Toward the end of *Star Wars, Episode III: Revenge of the Sith*, the last of the prequel trilogy, a robotic midwife delivers the twins, Luke and Leia, who feature prominently as adults in the original *Star Wars* trilogy (Episodes IV through VI). Meanwhile, in another part of the galaxy, a team of robotic surgeons transform what's left of their father, Anakin Skywalker, into the cyborg Darth Vader [14]. Years later, in the midst of a war with the Galactic Empire, Luke Skywalker loses a hand in a light-saber battle with Darth Vader. Toward the end of *Star Wars, Episode V: The Empire Strikes Back,* a robotic surgeon replaces Luke's severed hand with a cybernetic prosthesis [15].

Although robotic surgical tools, operated remotely by human surgeons, have been available for many years, we do not yet have autonomous surgical robots, or even autonomous robotic midwives. But autonomous robot nurses, such as RIBA, are being tested in Japan for use in elder care [16]. If the demand for qualified human caregivers exceeds the supply, it would make sense to build robots programmed to perform the needed services.

Discussion Topic 7.5: Robots and Healthcare

Are we developing autonomous robots for the healthcare industry simply because we can, or because the demand for care exceeds the supply of human caregivers? What impact might this have on the kind of care that is delivered, particularly from the point of view of the patient?

7.4   Misplaced Faith in Science

As powerful as the scientific method is to discern what is true about the way the physical world works, and as useful as science and technology can be to solve some of the problems that we face, there are limitations to what science can be expected to do. Some problems are ones of human behavior and are not likely to be solved by better scientific discoveries or more advanced technology. Even in the *Star Trek* universe, which tends to have a very optimistic view of the future, there are some problems that technology cannot solve, particularly if they are Klingon problems, rather than human problems. The episode of *Star Trek: The Next Generation,* entitled "Rightful Heir" is set at a time when the Klingon Empire is on the verge of collapse. Lieutenant Worf is on a spiritual quest, hoping to experience a vision of Kahless, the author of the Klingon laws of honor. Much to his surprise, Worf receives not just a vision, but a visit from the real Kahless, who seems to have returned, in fulfillment of a promise he made 1500 years earlier. Could this be the long-awaited Klingon messiah? Skeptical, Worf performs a tricorder analysis, which confirms that this person is Klingon. Once aboard the Enterprise, "Kahless" agrees to submit to a DNA test. But the real Kahless has been dead for 15 centuries. How does one get a DNA sample for comparison? The answer lies with a sacred Klingon relic: a sword, that wounded the original Kahless, is stained with his blood. Dr. Crusher performs the DNA test, which gives an exact match (Given what we discussed in the precious chapter about the degradation of DNA over time, is this analysis plausible?). All the physical evidence suggests that this is the *real* Kahless. But as the story unfolds, the truth is revealed: he is not the original, returned after 15 centuries, but a clone, created from blood found on the relic sword. Those who cloned him hoped that his "return" would restore order to the empire. But their faith was misplaced. Political problems, be they human or Klingon, are not easily solved by technology alone [17].

Discussion Topic 7.6

Think of some of the most pressing problems facing the world today. Which of these are likely to be solvable through science and technology? Which of these are of a fundamentally human nature, and unlikely to have a purely scientific or technological solution?

7.5 Misunderstanding of Science

Some of our problems are actually caused by a misunderstanding of scientific concepts. Decisions taken based on a lack of understanding of science, or a lack of appreciation of the limitations of science, can lead to miscarriages of justice. The continued use of outdated notions of science can lead to absolute nonsense. Increased public awareness and appreciation for science is an important goal to which we need to strive.

A key piece of the scientific method is the formulation of a hypothesis about how things work, which can then be tested experimentally. If there is no testable hypothesis, or if the researchers are selective in reporting their results—presenting only those data which support the hypothesis, while suppressing those that contradict the hypothesis—then chances are it's not science. But as pointed out by Michael Shermer, in his article in the September 2011 issue of Scientific American, the distinction between science and pseudoscience can be difficult to make unambiguously [18]. Nevertheless, a few examples from science fiction and from actual history are worth discussing.

7.5.1 Can "Psychic Powers" Be Tested?

If we do not have a scientifically informed public, we should not be surprised when public funds are used in the pursuit of goals that most scientists might consider to be crazy. For example, the Cold War era saw some rather bizarre research into so-called psychic phenomena, including such things as ESP (extrasensory perception) and telekinesis (the alleged ability to move objects with the power of the mind). What possible motivation could they have had to conduct this sort of research? The answer is illustrated in *Indiana Jones and the Kingdom of the Crystal Skull*. Motivated by a desire to conduct warfare from a distance, but without physical weapons, the Soviet psychic researcher in the movie is obsessed with finding a powerful alien artifact. Archeologist Indiana Jones is pressed into service to help find it. The mysterious crystal skull is at first believed to be strongly magnetic. But it attracts any metal object, including lead pellets from shotgun shells, gunpowder from bullets, and gold coins, none of which would be attracted to an ordinary magnet. The force which it exerts cannot arise from any known physical interaction, hence its appeal to the Soviet researcher [19].

It is an unfortunate truth that both Soviet and American scientists in the Cold War era were seriously interested in psychic phenomena. A faculty member at my undergraduate institution even taught a course on the subject (enrollment was by special permission, which required a personal interview with the professor), conducted his own research, and published papers in the field. Some of the researchers were so determined to prove that such things exist that they postulated a phenomenon called *psi-missing* to explain away the negative results. Although written in a rather different context, the words of Howard Van Till, Davis Young, and Clarence Menninga are equally applicable to psychic research:

> The purpose of empirical research is to discover what the physical world is really like, not to verify its conformity to our preferences. And the aim of scientific theorizing is to describe the actual character of the universe, not to force its compliance with our preconceived requirements.... When the epistemic goal of gaining knowledge is replaced with the dogmatic goal of providing warrant for one's personal belief system, the superficial activity that remains may no longer be called natural science. It may be termed *world-view warranting* or *creed confirmation*, but it no longer deserves the label of *natural science* because it is no longer capable of giving birth to knowledge. Science held hostage by extrascientific dogma is science made barren. [20]

Earlier in this chapter, we considered the actions of the scientist in *Woman in the Moon* as an illustration of the scientific method at work. Having done some real science, he then crosses over into the realm of pseudoscience. Obsessed with a desire to find gold on the Moon, he takes out a divining rod from his belt and uses this pseudoscientific device to guide him in his quest for wealth. He is ultimately successful in his quest but could the divining rod have had anything to do with it? Divining rods are supposed to enable the user to find hidden sources of water, oil, and other things limited only by the imagination but are not based on any kind of real science. James "The Amazing" Randi, a stage magician, turned debunker of alleged psychic phenomena, designed a controlled experimental test, involving a network of underground pipes, and valves that can be opened at random, which would unambiguously prove or disprove the validity of divining rods as a means of finding water underground. A $1 million prize was offered for many years by the James Randi Educational Foundation to anyone who could successfully demonstrate any so-called psychic powers, including, but not limited to, any "diviner" who could pass the buried water pipe test. During the years in which the prize was offered, nobody ever succeeded in meeting the challenge. With the retirement of James Randi in 2015, the Foundation shifted its attention to annual awards, "given to the person or organization that best represents the spirit of the foundation by encouraging critical questions and seeking unbiased, fact-based answers" [21]. If psychic phenomena are real, someone surely would have applied for and won the Amazing Randi's $1 million Psychic Challenge prize, long before it was discontinued.

7.5.2   On the Healing Powers of Phoenix Tears

Toward the end of *Harry Potter and the Chamber of Secrets*, the young wizard uses the sword of Godric Gryffindor to slay the basilisk, a huge snakelike creature. But in the process, he suffers a mortal wound from a venomous basilisk fang. When it appears that Harry is about to succumb to the deadly poison, Professor Dumbledore's bird, Fawkes, flies to the rescue. Fawkes appears to be crying, as he bends over Harry's wounded arm. But as the tears drip into the open wound, the wound begins to heal. Harry exclaims, "Of course! Phoenix tears have healing powers!" [22].

The phoenix is a magical bird. According to legend, from time to time the phoenix bursts into flame and burns to ashes. Then miraculously, the phoenix rises again to new life from the ashes. In the magical world of Harry Potter, not only does the phoenix rise from the ashes of its own funeral, but its tears have the power to heal a mortal wound. In a previous chapter, we used Harry Potter's cloak of invisibility as a springboard for discussing the possibility of using metamaterials with a negative index of refraction to construct a real-life cloak of invisibility. Here we will use the healing power of phoenix tears as a metaphor for discussing the alleged healing powers of certain prescription medications.

Figure 7.1 shows a bottle of homeopathic prescription pills, labeled "Arnica 30X" for the treatment of bruises. Before we examine the details, it is important to make it perfectly clear that what follows is not an indictment of all homeopathic medicines. Many natural medicines do have actual therapeutic value. What we are about to explore in this section is the homeopathic designation "30X". The specific questions we will address include the following. What is Arnica? What is the active ingredient in Arnica and what does it do? What exactly does the designation "30X" tell us about the contents of the medication? How does this relate to other information on the label?

We'll begin with the easy part. At the top of the white portion of the label, immediately below the word "Homeopathic" is the following information: "Contents: 4 dr Sucrose/Lactose Pills". The interpretation of this information is straightforward. A dram is a unit of volume, usually used to measure small quantities of liquids. One dram is equal to one eighth of an ounce. So the specification "4 dr" on the label indicates that the bottle holds a volume of ½ ounce. Sucrose is common table sugar and lactose is a type of sugar found in dairy products. So the first line of information on the label tells the consumer that the contents of the bottle is ½ ounce of sugar pills.

Now for the scientifically challenging part. The prominent information on the label is "Arnica 30X." Arnica is a genus of flowering plant. It contains the toxic substance Helenalin, whose chemical formula is $C_{16}H_{18}O_4$ with molar mass of 262.3 g/mole. In sufficiently low concentrations, Helenalin has been shown to have an anti-inflammatory effect. In sufficiently high doses, Helenalin can be lethal. The toxicity of Helenalin has been studied on various mammals. The lethal dose to 50% of the population exposed (LD50) is between 85 and 150 mg/kg of body mass [23]. So where do the pills in this bottle fall on the spectrum between "anti-inflammatory" and "lethal"? To address this question, we need to understand what the designation

Fig. 7.1 A bottle of
homeopathic prescription
pills, prominently labeled
"Arnica 30X" (photo by
the author)

"30X" means. The "X" is a standard homeopathic symbol, which refers to a dilution by a factor of 10. (Some homeopathic medicines have the designation "C" which means a dilution by a factor of 100.) The "30" indicates that 30 successive dilutions have been performed. So "30X" means that the initial product has undergone 30 dilutions, by a factor of 10 each time. The final product has a concentration which is reduced by a factor of 10^{30} compared to the initial concentration. So how much of the active ingredient is left in the end? Let's do the math.

The bottle in Fig. 7.1 contains 148 pills, with a total mass of 9.6 g. The highest possible concentration is 100% pure active ingredient. So the maximum number of Helenalin molecules that could conceivably be in this bottle of pills can be calculated by dividing the total mass of the pills by the molar mass of Helenalin and then multiplying that result by Avogadro's number, to convert from moles to molecules:

$(9.6\text{g})/(262.3\text{g}/\text{mole}) = 0.0366 \text{ moles of Helenalin}(\text{max})\text{per bottle, or}$

$(0.0366 \text{ moles})\times(6.02\times10^{23} \text{ molecules}/\text{mole}) = 2.20\times10^{22} \text{ molecules}(\text{max})\text{per bottle}$

What we have just calculated is the maximum number of Helenalin molecules that could possibly be in the bottle of pills, assuming that the pills are 100% pure Helenalin, which of course they are not. The bottle specifies "30X". So we must now reduce the concentration from 100% pure by a factor of 10 a total of 30 times. We multiply our previous result for the maximum number of molecules of Helenalin per bottle by a factor of 10^{-30}:

$$(2.20 \times 10^{22} \text{ molecules}) \times (10^{-30}) - 0.22 \times 10^{-6} \text{ molecule}$$

The result of our calculation indicates that each bottle of "Arnica 30X", as shown in Fig. 7.1, can contain at most 0.22 micromolecule of Helenalin. But there is no such thing as a micromolecule. The smallest quantity of any molecular material that you can have is one single molecule of that material or, in the case of a pure element, such as copper or gold, one single atom of that element. The concept of a "30X" dilution is based on the outdated belief that matter can be infinitely divisible into infinitely smaller quantities. In homeopathic medicine, this is known as the "law of infinitesimal dose." The guiding principle is that the smaller the dose, the more potent the effect [24]. But it has been known for more than a century that a substance cannot be broken down any further than discrete molecules or atoms, or it ceases to be that same substance. Once you dilute the substance to the point that there is no longer even a single molecule present, the concept of a "dose" becomes meaningless.

What we must conclude from this is that there is not even one single molecule of the active ingredient in any of these pills. There can be no anti-inflammatory effect because the bottle contains none of the active ingredient, Helenalin, in any concentration. There is no Arnica in this bottle of "Arnica 30X." What does the bottle contain? Sugar pills. They are exactly as stated in the top line of information on the label: Sucrose/Lactose pills. They could just as well call it "Phoenix Tears 30X." It would be no less true than the information on the label, and it would have the same therapeutic effect. Or perhaps not. The question may be one of belief, rather than science. Is the consumer more likely to believe in the healing powers of phoenix tears or in the healing power of "Arnica 30X"? The placebo effect is well-documented and may be what is actually at work.

7.5.3   How Can We Enhance Our National Security?

During the Cold War, science and the military joined forces to solve a problem, the nature of which has only recently come to light. In 1962, a spectacular light show took place in the Earth's ionosphere, but it was not the naturally occurring aurora phenomenon. Instead, it was a scientific test of a hypothesis, which might have been lifted directly from the plot of a science fiction story. We have already considered the 1961 movie *Voyage to the Bottom of the Sea*, in the context of computing technology. According to this sci-fi movie, which mixes real science with total nonsense, the Van Allen Radiation Belt (a real physical phenomenon) is on fire (total nonsense) and threatens to scorch the Earth (a plausible consequence, except for the nonsensical premise of having it catch fire in the first place). The proposed solution to the problem is to set off a nuclear explosion and blow the fire out into space [25]. If this scenario sounds crazy, remember that it's just a science fiction movie. But wait! Truth can be stranger than fiction. The Van Allen Radiation Belts are real. Discovered in 1957 and announced in 1958 by their discoverer, James Van Allen, the belts are regions of high-energy charged particles, mostly free protons and electrons from the solar wind, which are trapped in orbit in the Earth's magnetosphere, at altitudes ranging from a few hundred to about a thousand kilometers in altitude. There is no possibility that the Van Allen Belts could ever catch on fire, as in the movie.

In the years immediately following the discovery of the Van Allen Belts, a series of high-altitude nuclear test explosions, codenamed Operation Newsreel (1958) and Operation Fishbowl (1962), were conducted, first with fission bombs and later with a hydrogen bomb. One of the many objectives of these tests was to find out if the belts might be used to channel an explosion around the Earth to a remote enemy target (e.g., the Soviet Union). Details of one of the Fishbowl series, Operation Starfish Prime, including an account from an eyewitness, were reported by news correspondent Robert Krulwich, in an episode of *All Things Considered*. The full broadcast, a "Very Scary Light Show" can be heard at the NPR web site [26].

Making informed decisions on how to allocate public funding for research is a tricky business. The distinction between things which are possible, in principle, but beyond our current technology, and things which are impossible, and therefore a waste of time and money to explore, can be difficult to discern. Depending on your point of view, those who insist on making such a distinction might be called either principled realists, or enemies of progress. By the same token, those who refuse to be constrained by such distinctions might be called visionaries by some, or crackpots by others. In his book *Wired for War*, P.W. Singer quotes a retired army colonel, working for the Marine Corps Warfighting Lab, who looks to science fiction as a source of inspiration for research: "The fact that it exists in our own movies proves that it is potentially possible.... If you can imagine it, we think it can happen" [27]. In the next chapter, we will take a look at several "accurate predictions"—real technology, which may have been inspired by ideas from science fiction. But there are limits to what is achievable. Not everything that can be imagined can actually happen. Consider, for example, the real research into teleportation, motivated by the desire to find the ones responsible for the September 11, 2001 terrorist attacks on the United States. Such a project was funded, despite the many well-known scientific roadblocks to teleportation of macroscopic objects, which we have already considered in the previous chapter [28]. An 88-page report on the feasibility of such a project, including 253 technical references, is available online [29].

7.5.4   How Can We Prevent the Miscarriage of Justice?

The episode of *Star Trek: The Next Generation* entitled "The Measure of a Man" features a military hearing, to determine the legal status of the android Lieutenant Commander Data. Is he a sentient being, with the same rights as any other member of Starfleet, or is he simply a machine, and therefore the property of Starfleet? Commander Riker is appointed to prosecute the case for property, and Captain Picard is counsel for the defense. Riker convincingly presents the physical evidence, some of which we have already considered in a previous chapter (storage capacity and processor speed). By Data's own admission, he is an automaton, made to look like a man, and built by a man (Dr. Noonien Soong). When Riker removes Data's hand for the judge's inspection, and then switches Data off, the case for Data's freedom seems hopeless. After requesting a recess, Picard has an interesting conversation with Guinan, the ship's bartender. It becomes clear that to treat Data as property would be not unlike human slavery. With renewed hope, Picard returns to the courtroom to present his defense. He acknowledges that Data is a machine, but dismisses this claim as irrelevant, since humans, too, are machines, but of a different sort. He also acknowledges that Data was created by a man, but dismisses this claim as irrelevant, since biological children are created from the instructions of their DNA, inherited from their parents. The central issue is more a question of metaphysics, rather than of science or technology, namely whether or not Data is sentient—an intelligent creature, aware of its own existence, independent of what it looks like on the outside, or how it is made on the inside. Data's response to the questions made it clear that he is as aware of his own existence as Picard or Riker or the judge. The prosecution's case, based entirely on science and technology, was not sufficient to convince the judge. She ultimately ruled in Data's favor, and awarded him his freedom [30].

The fate of another form of artificial intelligence is featured in "I, Robot," an episode of the 1960s TV series, *The Outer Limits*. Circumstantial evidence suggests that a robot, named Adam Link, murdered its creator, Dr. Charles Link. But the faith of the doctor's niece and the determination of an ambitious journalist (played by Leonard Nimoy, several years before his role as *Star Trek*'s Spock), eventually prove Adam's innocence, with the help of a famous defense attorney [31].

Unfortunately, in a broken world, not all rulings are as just as these. Humans seem to have an increasing tendency to lay blame, even in situations of natural disaster, where no one could possibly be at fault. Science is sometimes dragged into the courtroom and misused to demonstrate fault, where none exists. An unfortunate case in point comes from the Italian court system. In 2009, an earthquake struck central Italy, killing over 300 people. In the aftermath, six scientists, all experts in seismology, geology and related fields, and a government official were put on trial for failure to provide adequate warning of the impending disaster. They were convicted of manslaughter in October of 2012 and sentenced to 6 years in prison [32]. Is there any basis in science for such an apparently outrageous conviction?

Considerable research has been done in an area popularly known as sandpile physics. When a pile of sand, or any similar granular material, is created, by dropping one grain at a time in a precisely controlled way, the pile builds up to a point of instability, at which the addition of just one more grain can trigger an avalanche of any size. The size of the avalanche (let's call it A) is related to the frequency of occurrence (let's call it f) according to a power law:

$$A \propto \frac{1}{f^p}. \tag{7.1}$$

Equation (7.1) suggests that larger avalanches occur less frequently, while small avalanches occur more frequently. The precise value of the power, p, to which the frequency is raised, is a subject of ongoing research, and may depend on subtle details, such as the shape of the grains, and whether or not the grains are all of the same size.

Analysis of available data on the magnitude and frequency of earthquakes suggests that there may be a similar power law relation. There is evidence that a rapid succession of small avalanches (or earthquakes) may be a precursor to a much larger one. But it is impossible to predict exactly when the next big one will occur. The best that can be said is that very large avalanches (or earthquakes) are relatively rare events, compared to smaller ones.

As an epilog to the story, in November 2014, the convictions of the six scientists were overturned on appeal. This decision was appealed to the Italian Supreme Court, which upheld the previous acquittal. The government official also was eventually exonerated.

Tropical storms and hurricanes can be tracked by satellite images and on radar. Their development and paths can be projected several days into the future (within limits of uncertainty), using sophisticated climate-modeling computer programs. But even these predictions are not 100% accurate. Actual landfall might occur many miles away from early projections. At the present time, there is no similar way to predict when an earthquake of any magnitude will strike. The loss of life in any natural disaster is tragic. But no one can possibly be at fault for failure to give adequate warning of something that simply cannot be predicted. The manipulation and misrepresentation of science by a clever lawyer can result in a miscarriage of justice. Or is it possible that a better educated judge and jury might have been less susceptible to manipulation?

7.5.5   How Can We Increase Public Awareness of Science?

We begin with a humorous example, which illustrates the major goal of this book, namely to increase public awareness of science. The 1960s TV sitcom *Gilligan's Island* tells of seven people, who set out on a three-hour sightseeing cruise, and end up shipwrecked on an uncharted island. One of the seven, the Professor, is a man of science, who often uses his scientific knowledge to address the problems of survival away from civilization. In the episode "Pass the Vegetables, Please" a crate of experimental radioactive vegetable seeds washes up on shore. The title character finds the crate, but fails to notice the warning on the cover before opening it. The castaways plant the seeds, harvest the abnormal-looking crop and proceed to eat the vegetables. Soon afterward, they begin to manifest extraordinary powers. Super-acute vision results from eating the carrots, super-strength from spinach, and hyperactivity from eating sugar beets. The warning on the lid of the crate is discovered after the fact, and the Professor offers a scientific-sounding explanation. The natural nutritional value of the vegetables must have been enhanced by the radioactivity [33]. This, of course, is silliness. But the exposure of plant seeds to gamma radiation, and their release to the public, is a matter of fact. In the mid-twentieth century, in an effort to gain insight into the biological effects of exposure to ionizing radiation, various kinds of flower seeds were exposed to gamma ray, and sold in stores as "Atomic Activated." The package, shown in Fig. 7.2, invites the consumer to "Plant your own experimental garden with atomic activated seeds."

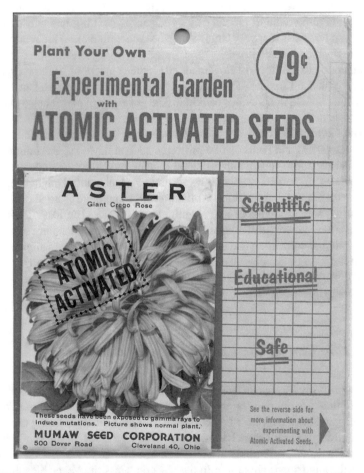

Fig. 7.2 Atomic activated seeds. The seeds have been exposed to gamma radiation. The consumer is invited to plant them, and to observe and report any unusual results (photo by the author)

Unlike the vegetable seeds in *Gilligan's Island*, these seeds are not actually radioactive. They were simply exposed to gamma radiation and are completely safe to handle. The motivation behind the project, presumably, was twofold: to investigate the effects of exposure of genetic material to radiation, and to increase public awareness of and involvement in experimental research. The text on the package reassured the user that the seeds were perfectly safe, and suggested the kind of investigation to carry out. After the plants germinate, the user is encouraged to make periodic inspections, and to record variations in such things as the height of the plant when flowers first bloom, the number of flowers on each plant, diameter and color of flowers, size of leaves, resistance to frost and disease, and stem and root structure. Any plant that appeared to be particularly unusual should be allowed to go to seed, and these seeds should be planted and observed in the following year. Results should be reported to the "Director of Research" at the address provided. This was, indeed, a clever way to get the public involved in science, and perhaps to reduce the unwarranted fear of science. But it will take more than planting irradiated flower seeds to solve the problem of lack of understanding of science. More equitable access to high-quality education, in general, and better science education, in particular, is needed.

7.6   How Can We Know What Is Real?

"No matter how smart or well-educated you are, you can be deceived" [34]. This quote from James Randi is a warning to all of us to be on our guard against con artists. As a stage magician and escape artist, Randi has no problem with people who create illusions for the purpose of entertainment. In fact, he refers to himself as "an honest liar", in part, because he never claimed that the illusions he created on stage were real. He does, however, take exception to those who make claims of psychic powers and take money from gullible people, who have been deceived into believing that their "powers" are real. But stage magic is not the only area in which the art of deception plays a role. In this section, we consider some of the many examples of the misuse of science for the purpose of deception, as we take up one of the most serious problems facing society today: How do we know what is real?

7.6.1   Is Soylent Green Really Made from "High-Energy Plankton"?

Soylent Green is set in the year 2022, when pollution, overpopulation, and global climate change have drastically altered the supply and demand for food. Natural foods are sold on the black market, at absurdly inflated prices. Even some of the highly processed forms of nutrients have limited availability. A television broadcast early in the movie opens with the following commercial announcement:

This conversation, with Governor Henry C. Santini, is brought to you by Soylent Red and Soylent Yellow—high-energy vegetable concentrates—and by new, delicious Soylent Green, the miracle food of high-energy plankton, gathered from the oceans of the world. Because of its enormous popularity, Soylent Green is in short supply. Remember: Tuesday is Soylent Green day. [35]

At a time when real food is hard to come by, Soylent Green seems like the next best thing. Highly desirable, not always available, and possibly even good for you are the implications of the advertisement. But Soylent Green is not what the advertisement claims it to be, and the truth is not revealed until the end of the movie.

In the real-life field of advertising, there is no end to the number of claims that are made to promote one company's product as "better" than a competitor's product. Often these claims are dressed up with language that sounds "scientific" but with no basis in reality. The example from *Soylent Green* falls into this category—a claim that is blatantly false. Perhaps even more insidious than blatant falsehoods are advertisements that never actually say anything false, but are nonetheless cleverly crafted to deceive. We will consider several examples here. The reader is encouraged to explore on the internet to find others.

7.6.2 Are Potato Chips Better for You If They Are "Scientifically Processed"?

The advertising industry occasionally writes product descriptions that sound scientific, in order to make the product more marketable. Take, for example, the "scientifically processed" potato chips, shown in Fig. 7.3. The can is decorated with silhouettes of healthy, athletic figures, in a style of clothing suggestive of the 1940s or '50s. The text on the side of the can reads as follows: SCIENCE SAYS...."THE ALKALINE SIDE IS THE HEALTHY SIDE." NEW ERA Potato Chips are partially starch-dextrinized and therefore more readily digested.... Chemical analysis have proven NEW ERA Potato Chips to be a highly concentrated energy producing food, 95% digestible and of greater alkalinity than even fresh raw potatoes. *FEAST WITHOUT FEAR.*

But what do these words really mean? And is this product really better for you than competing brands of potato chips, or just more easily digestible than raw potatoes?

The claim that "the alkaline side is the healthy side" has some basis in truth. In chemistry, the pH scale is an indicator of acidity or alkalinity. The scale runs from 0 to 14, with 7 being neutral (e.g., pure water). Numbers below 7 are acidic and numbers above 7 are basic (alkaline). The natural pH of human arterial blood is in the range 7.35–7.45, which is slightly on the alkaline side. But the pH in your stomach is more like 3.5, which is acidic. Lots of products are on the market to control excess stomach acid. But the stomach must be on the acidic side of neutral pH, in order to digest food. So the claim on the can of New Era Potato Chips may be true, if you interpret "alkaline" to be a reference to the pH scale and "healthy" to refer to

Fig. 7.3 NEW ERA potato chip can, c. 1945. The front (**a**) features an array of healthy-looking images, and claims that the chips are "scientifically processed." The side (**b**) includes other scientific-sounding words, presumably to enhance marketability (photos by the author)

arterial blood. But there is another possible interpretation. The alkali metals are the elements in column I of the periodic table, which includes Lithium, Sodium, Potassium, etc. Foods that are high in Sodium typically have lots of added salt (Sodium Chloride).

The claim that New Era Potato Chips are "partially starch-dextrinized and therefore more readily digested than fresh raw potatoes" is also a true statement. Potatoes have lots of starch. Starch is a long polymer hydrocarbon molecule, which breaks down into shorter hydrocarbon chains, called dextrins, when exposed to heat. The shorter chains do, indeed, make the product more readily digestible. But are these potato chips really "scientifically processed", as it says on the can? No! They've been cooked and salted. There is actually evidence that the discovery of cooked food early in human history may have had a beneficial effect on the development of the human brain [36]. One could argue that cooking is a science. But to promote potato chips as being "scientifically processed" when in fact they have simply been cooked and salted, is to misuse science for the purpose of deception.

7.6.3   Is Sugar Just as Good If It Comes from Corn?

The standards of truth in advertising may be more stringent in the twenty-first century than they were in the middle of the twentieth century. A more recent example of the use of science in advertising is in the context of the controversy over high-fructose corn syrup. Studies had suggested a possible connection between the overuse of this product and increasing incidents of childhood obesity. To counter the growing concern, and to set the public mind at ease, advertisements were produced, which said, in effect, "Sugar is sugar, no matter where it comes from." If we're talking about table sugar, this is a true statement. Ordinary, everyday sugar, whose chemical name is sucrose, can come from different natural sources, including sugarcane and beets. But the controversy is not about sucrose. So what is the difference between everyday sugar (sucrose), and high-fructose corn syrup? And what difference does it make when they are digested in the body? Those are the scientifically relevant questions.

Sucrose is a complex molecule, called a disaccharide, which is essentially a fusion of two simple sugars (monosaccharides): fructose and glucose. Ordinary corn syrup is processed from cornstarch, and is mostly glucose, with a bit of water. High-fructose corn syrup, which is the focus of the controversy, is further processed to convert some of the glucose into fructose. The relative percentages of glucose and fructose in the final product depend on the extent of the processing.

The simple sugars glucose and fructose have different chemical structures, which are illustrated in Fig. 7.4. The solid black circles represent carbon atoms, the hashed circles represent oxygen, and the small open circles represent hydrogen. The two molecules have some similarities but are not completely identical. The backbone of each molecule is a chain of six carbon atoms. In both molecules, each of the lower four carbon atoms has an OH group bonded to it. But here the similarities end. Note that the fructose molecule (Fig. 7.4a) is terminated by a CH_2OH group at the top and at the bottom, with a double-bonded oxygen second from the top. In contrast, glucose (Fig. 7.4b) has a CH_2OH group at the bottom end of the molecule, but with a double-bonded oxygen at the top, and a CHOH group second from the top.

Because their chemical formulas are different, these two molecules are processed differently in the body. (Glucose is essential to the production of insulin, but fructose does not take part in this process.) The heart of the controversy is whether or not the difference in the way the sugars are digested can lead to health problems. In effect, does overuse of products containing high-fructose corn syrup cause more health problems than overuse of products containing ordinary sugar? The reader is invited to explore this question independently, but be careful of what you read. The information presented on the various websites may differ, depending on who is sponsoring the website.

To reiterate, the advertising claim, "Sugar is sugar, no matter where it comes from," is absolutely true, if what you mean by "sugar" is ordinary table sugar (sucrose). Sucrose is sucrose, regardless of whether it is extracted from sugarcane or beets or any other fruit or vegetable. But a statement does not have to be false in order to be misleading.

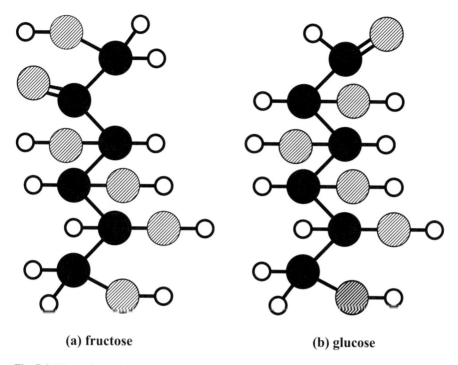

(a) fructose   **(b) glucose**

Fig. 7.4 Illustrations of the chemical structure of (**a**) fructose and (**b**) glucose. The solid circles represent carbon atoms, the hashed circles represent oxygen atoms and the open circles represent hydrogen atoms

7.6.4   Does a Cleansing Solution Work Better If It Contains Micelles?

A number of products have come on the market recently, for the purpose of removing dirt and makeup from the skin. They are presumably easier and more convenient to use than old-school cold creams or the somewhat more modern cleansing creams. They are marketed by various companies and go by various names, but most of them include the scientific-sounding word "micellar" in the names of their products. The advertisements tell you that "micelles remove dirt like a magnet." The animation even represents a micelle, like a horseshoe magnet, sweeping up dirt in the same way that the magnet would sweep up iron filings. What are micelles? How do they work to remove dirt or makeup? Are there other products that do the same thing, but without the scientific-sounding word in their names?

Figure 7.5 is an illustration of the structure and arrangement of surfactant molecules in water. A "surfactant" (or surface-active agent) is any molecule that has a part which likes to be in water (hydrophilic) and another part that does not like to be in water (hydrophobic). This structure is represented schematically in Fig. 7.5 as a squiggly black line (hydrophobic tail group) attached to a small yellow circle (hydrophilic head group).

Fig. 7.5 Illustration of surfactant molecules in water. At very low concentration, individual surfactant molecules float at random in the water but have a tendency to self-assemble into a monolayer at the interface between water and air. Above a certain critical concentration, clusters of molecules, called micelles, begin to form in solution

The dual nature of the surfactant molecule gives them the tendency to accumulate at the interface between the water and the air, in order to shield the hydrophobic tail group from contact with the water. At low concentration, some of the molecules will be floating freely in the water and others will begin to accumulate (or self-assemble) at the surface of the water. The accumulation of surfactant molecules at the air–water interface lowers the surface tension. As the concentration increases, more and more surfactant molecules will gather at the surface, and the surface tension continues to decrease, until a complete monolayer has formed. Once the monolayer is formed, there is no more room for the surfactant molecules to continue to accumulate at the surface of the water. If the concentration of surfactant in solution is increased even further, there will be no further decrease in the surface tension (because the monolayer is complete), but the surfactant molecules in solution will begin to self-assemble into spherical clusters, called micelles. The hydrophobic tail groups are somewhat shielded from contact with the water by pointing toward the interior of the micelle. The hydrophilic head groups are on the outside of the micelle and are happy to be in contact with the water. The concentration at which the surface monolayer is complete and micelles begin to form is called the critical micelle concentration, or CMC [37].

This tendency of surfactant molecules to form micelles also makes them useful for removing dirt and grease. The formation of a micelle around a bit of dirt or grease effectively turns something that is not soluble in water into something that is soluble in water and is now easily washed away. Surfactant solutions with micelles really are very good at removing dirt and grease. But the action of micelles is nothing at all like the action of a magnet on iron filings.

The next question we might ask is how high does the concentration of surfactant molecules have to be in order for micelles to begin to form? That depends on the specific molecule. At risk of oversimplifying, any soap solution that can form lots of bubbles will have micelles in it. Yes, soap! All soaps are surfactants of some kind. One of the most common ingredients in shampoo is the surfactant molecule sodium dodecyl sulfate, or SDS. The chemical structure of SDS is $CH_3(CH_2)_{11}SO_4Na$. The hydrophobic part of the molecule is a chain of 12 hydrocarbons. The sulfate (SO_4) is the hydrophilic head group. SDS typically appears on the list of ingredients in shampoo and other personal cleaning products with the name sodium lauryl sulfate. The surfactant sodium lauryl sulfate is also one of the most common ingredients in the so-called micellar cleansing solutions, which have become popular as makeup removers.

To address the question posed in the heading of this section, does a cleansing solution work better if it contains micelles? Yes. In fact, it won't work very well at all without them. But does an effective cleansing solution have to include the word "micellar" in its name? No. Any liquid hand soap or shampoo or dishwashing liquid will have micelles and will do a good job of removing dirt and grease (and makeup). If all you care about is the cleaning ability, any soapy water will do. But would the consumer buy the product if the marketers called it "Soapy Water"? Probably not.

Is there any real difference between these "micellar cleansing solutions" and ordinary soapy water? Actually, yes there is. The difference is the other added ingredients in the micellar cleansing solutions, which help to prevent drying of the skin, making them more pleasant to use than ordinary soap solutions. So why don't the marketers focus on the other added benefits of their product, which make them genuinely more user-friendly than ordinary soap solutions? Why do they choose, instead, to focus public attention on the scientific term "micellar" in the product's name? We might hypothesize that the marketing focus groups told them they would be more likely to buy the product if it sounded scientific. The irony is that the real science of how micelles work is nothing more than the science of soapy water.

The psychology of marketing is a mysterious thing. Apparently, little has changed in more than half a century since the concept of "scientifically processed" potato chips hit the market. The average consumer may not understand the technical details, but the use of scientific language still sells products, even if the science isn't anything new.

7.6.5   Non Sequitur

One of my favorite comic strips is Non Sequitur, by Wiley Miller. The title is Latin for "It does not follow." A non sequitur is a statement that does not follow logically from previous statements. It has been suggested that the foundation of physics is mathematics, and the foundation of mathematics is logic. We consider here an example from science fiction, in which pure logic is used to solve a life-and-death problem. We then turn to an example of the subtle use of logic in advertising to manipulate the consumer.

In the original series episode of *Star Trek*, entitled "The Changeling," the Enterprise is attacked by a small space probe of enormous power. The probe, which calls itself Nomad, believes itself to be perfect and believes that its mission is to sterilize imperfect biological organisms (which, being translated, means to kill them). Fortunately, the probe mistakes Captain Kirk for its creator and breaks off the attack. But when Nomad discovers its own imperfection, it self-destructs. The beginning of the end occurs when Kirk discovers that Nomad has killed two security guards. They engage in the following dialog:

KIRK: You killed two of my…
NOMAD: Creator, your biological units are inefficient.
KIRK: Nomad, it's about time I told you who and what you are. I'm a biological unit and I created you.
NOMAD: Non sequitur. Biological units are inherently inferior. This is an inconsistency [38].

Here's one possible way to express the problem, from Nomad's perspective.

Nomad's assumptions:

1. Nomad is perfect
2. Kirk is Nomad's creator
3. Imperfection cannot create perfection (implied)
4. Biological units are imperfect

New information:

Kirk is a biological unit

Reiteration of prior assumption:

Kirk is Nomad's creator

Conclusion

"Non sequitur. This is an inconsistency."

Kirk's reiteration of Nomad's prior assumption, "I created you," does not follow from all of the other information. Has Nomad made a mistake somewhere along the line? Nomad must re-evaluate the data.

Eventually, Kirk defeats Nomad with logic (something the fans might have expected Spock to do, rather than Kirk). He first gets Nomad to re-articulate its purpose: to sterilize imperfection. (Note that this is an *expansion* of Nomad's mission to include the sterilization of *all* imperfection, not just biological imperfection.) Kirk then reveals the truth about himself that he is not Nomad's creator. Kirk points out that Nomad has committed an error by making that assumption. He goes on to say that Nomad made a second error by not recognizing and correcting the first error. And if that were not enough, Kirk tells Nomad that it made a third error by not sterilizing the imperfection. Nomad becomes paralyzed with cognitive dissonance, giving the crew of the Enterprise time to beam it out into space, where it self-destructs.

Logic can be used in advertising in very subtle ways to manipulate the consumer. You may have seen a commercial for a popular brand of electric toothbrush, which sets up a comparison to another brand with the words, "Unlike Brand A, Brand B is the first electric toothbrush brand to be accepted by the ADA (American Dental Association) for its effectiveness and safety." It would be sufficient to say "Brand B is the first electric toothbrush brand to be accepted by the ADA…." No doubt, this statement is absolutely true and it would be a good selling point on its own. But nothing is added by prefacing the claim with "Unlike Brand A…." Unless multiple brands are accepted by the ADA simultaneously, it's reasonable to assume that only one brand can be the first to be accepted. So if Brand B is the first, why bother to say "Unlike Brand A"? It goes without saying that Brand A cannot also be first. And why waste expensive advertising time by stating something that is obvious? Unless, of course, there is a hidden agenda. By setting up the unnecessary comparison with Brand A, the advertisers are sowing the seeds of doubt in the minds of the consumer, concerning the quality of Brand A. Once the seeds of doubt are sown, the consumer may be tempted to commit a non sequitur. If Brand A is not the first to be accepted by the ADA, then perhaps it has not been accepted at all. And if Brand A hasn't been accepted by the ADA, then maybe it's not as good as Brand B. But note the subtlety. As with many of the examples that we considered previously, the advertisers have not said anything that is untrue. Brand B may very well be the first to be accepted by the ADA, but that does not imply that Brand A has not been accepted. All you have to do to check the facts is to visit the website for the American Dental Association and find their page of accepted consumer products. Not only have both Brand A and Brand B been accepted by the ADA, but at least four other brands of electric toothbrushes have also been accepted.

Even when advertisers only speak the truth, consumers still need to be on their guard against deception. If consumers are willing to commit non sequiturs, drawing false conclusions from true statements, there's no one to blame but themselves.

7.6.6 What Is "Real"?

This very question was asked by Morpheus (played by Laurence Fishburne), one of the leaders of the human rebellion against the machines in *The Matrix*, while trying to explain to Neo (Keanu Reeves) what the Matrix is. He goes on to say,

> How do you define 'real'? If you're talking about what you can feel, what you can smell, what you can taste and see, then 'real' is simply electrical signals interpreted by your brain. [39]

We don't have to be jacked into the Matrix in order to have our perception of reality distorted. Our exploration of science through science fiction began in Chap. 1, with a quote attributed to Benjamin Franklin: "Believe none of what you hear and half of what you see." Modern science has improved our understanding of what is *true* about the way in which the physical world works. But at the same time, advances in artificial intelligence and entertainment technology are making it harder to discern what is *real*. The things that we see, especially when we see them on social media, using our small, portable devices, are becoming easier to fake. In the chapter on artificial intelligence, we briefly touched on one of the approaches to machine learning, known as "deep learning." This same technology, together with advances in high-resolution computer graphics, is making it possible to create video images that are convincingly "real" but are in fact "deep fakes." The potential for manipulation of public opinion through the use of fake audio and video material is a serious problem, which poses a threat to our democratic society. The problem is of such a magnitude that some experts in the field think it may not be solvable. The best that we might be able to do is to "make it harder and harder for the bad guys to play the game" [40]. If we are no longer able to believe anything that we see, out of suspicion that it may have been faked, an even greater danger may be that we will cease to believe anything at all.

In the early chapters of this book, we touched on several of the interesting unanswered questions in physics. Creative minds are needed to bring the scientific method to bear on these questions, in hopes of coming up with answers. At the end of the previous chapter, we considered a distinguishing factor which seems to set humans apart from all other species on the planet. If we don't like the way things are, we have the unique ability to see beyond our current situation and to come up with creative ways of making a difference—hopefully, for the better. Some of the problems raised in this chapter are not purely matters of science or technology. The scientific method alone is not likely to provide an adequate solution to problems which are fundamentally human in nature. The challenge remains for creative minds to collaborate in cross-disciplinary ways, in order to solve the complex problems facing our increasingly technological society.

7.7   Exploration Topics

Exp-7.1: Science and Pseudoscience

Why is it difficult to draw a sharp distinction between science and pseudoscience?

Karl Popper proposed "falsifiability" as the ultimate criterion for qualifying something as science. Why is this not an adequate, universal test for science? Give an example, drawing on topics that we have already discussed, of something that is accepted as science but would fail the test of falsifiability.

Reference:
- Michael Shermer, "What is Pseudoscience?" Scientific American, September 2011

Exp-7.2: Science and Improved Healthcare

What does it mean for a scientific researcher to have a conflict of interest? Under what circumstances might there be a conflict of interest in the context of drug research? What specific actions might help to solve the problem of so-called "conflicted science"?

Reference:
- Charles Seife, "Is Drug Research Trustworthy?" Scientific American, December 2012

Exp-7.3: Fossil Fuels vs. Renewable Energy

What three factors have inhibited the transition from fossil fuels to renewable energy? Why do these factors present significant challenges? What specific actions can be taken to speed up the transition to renewable energy?

What is the biggest source of the world's carbon dioxide (CO_2) discharge? What is the primary problem with Carbon Capture and Storage (CCS) as a viable solution to controlling CO_2 emissions? Instead of capturing and storing CO_2, what alternatives have been proposed? Why are many people skeptical of this alternate proposal as an effective means of controlling CO_2 emission?

References:
- Vaclav Smil, "The Long Slow Rise of Solar and Wind" Scientific American, January 2014
- David Biello, "The Carbon Capture Fallacy" Scientific American, January 2016

Exp-7.4: Science and Marketing—Healthy Potato Chips?

NEW ERA potato chips were promoted around the middle of the twentieth century as "A healthy food… on the alkaline side." The company that made them was bought out by a competitor, and NEW ERA brand is no longer in production.

Investigate the following questions, related to the advertising claims:

(a) What does it mean for something to be *on the alkaline side*? (Alkaline side of what?)
(b) Are raw potatoes naturally alkaline or acidic?
(c) How do potatoes become partially starch-dextrinized?
(d) In general, are cooked foods more digestible than raw foods?
(e) Is it likely that these potato chips have really been *scientifically processed*, or do the words on the can describe the process of cooking and salting, dressed up in scientific-sounding language?

Exp-7.5: Science and Marketing—Sugar vs. HFCS

In the early part of the twenty-first century, concerns arose over a possible connection between overuse of products containing high-fructose corn syrup (HFCS) and certain health problems. In efforts to alleviate fears, the producers of HFCS ran television commercials, which claimed that sugar is sugar, no matter where it comes from. These commercials are no longer being broadcast.

Investigate the following questions, related to the HFCS controversy:

(a) What is the difference between what we commonly refer to as sugar and high-fructose corn syrup (HFCS)?
(b) How is common sugar digested in the body?
(c) How are the two main components of HFCS digested?
(d) What are the health concerns, related to overuse of products containing HFCS?
(e) Could overuse of products containing common sugar lead to similar health concerns?

Exp-7.6: Science and Marketing—RGB vs. RGBY Color-Mixing

Another commercial from the early twenty-first century, which is no longer being broadcast, claimed that adding yellow to the standard red-green-blue color mixing results in a television image that is even better than standard RGB. Investigate the following:

(a) What is RGB color mixing and what is it supposed to be able to do?
(b) What is the alleged advantage of adding a fourth color, yellow, to the mix?
(c) What is the motivation for doing this? (Is there a real scientific basis for declaring standard RGB color mixing to be inadequate? Is there a technological reason why RGB needs to be supplemented by yellow? Is it just marketing hype?)

Exp-7.7: Restoring Extinct Species?

Problem: Most megafauna species in North America are extinct or seriously endangered.

Proposed solution: Repopulate with similar species, which still exist elsewhere on Earth.

How do you define a species, and how is a species chosen to be included on the endangered species list? Identify the scientific and ecological issues associated with the problem of extinct megafauna, and its proposed solution. What are the underlying ethical and philosophical issues? Is the proposed solution a good idea?

References:
- C. Josh Dolan, "Pleistocene Repopulation" Scientific American, June 2007
- Carl Zimmer, "What is a Species?" Scientific American, June 2008

References

Public Perception of Science

1. *The Thing from Another World* (Christian Nyby, Warner Brothers 1951). Obsession with dangerous scientific research [DVD scenes 14, 17, 18]

2. *Absolute Zero* (Robert Lee, Marvista Entertainment 2006). Following a totally nonsensical explanation for global climate change, a scientist asserts that "Science is never wrong" [DVD scenes 1, 2]
3. Dunlop, D.J.: Magnetic recording in rocks. Physics Today, June 2012
4. *The Happening* (M. Night Shyamalan, 20th Century Fox 2008). The lead character, a high school science teacher, clearly doesn't understand the difference between a scientific theory and an opinion [DVD scene 4]

Methodology of Science

5. *Woman in the Moon* (Fritz Lang, UFA 1929). Scientific experiments on the Moon, followed by pseudoscience [DVD scene 15]
6. *The Andromeda Strain* (Robert Wise, Universal 1971). A satellite crash brings a deadly microbe to Earth. Almost the entire movie is devoted to the use of the scientific method to prevent the spread of the infection

Reasonable Uses of Science to Solve Our Problems

7. *Target Earth* (Sherman A. Rose, Allied Artists Pictures 1954). An invading army of robots from Venus are disabled by resonance frequency sound waves [DVD scenes 16, 18]
8. "Is it live, or is it Memorex?" (YouTube video, featuring Ella Fitzgerald) http://www.youtube.com/watch?v=Bkt8Dwzl6Sg
9. *Star Trek* (the Original Series) – "The Trouble With Tribbles" (Joseph Pevney, Paramount 1967). High-yield grain consumed by rapidly reproducing furry critters [DVD vol. 21, ep. 42]
10. Freedman, D.H.: Are engineered foods evil? Sci. Am. **309**, 80–85 (2013)
11. *Star Trek IV: The Voyage Home* (Leonard Nimoy, Paramount 1986). Time travel back to the 20th century to find humpback whales. [DVD scenes 5, 6]
12. "Ozone Hole: How We Saved the Planet" (Jamie Lockhead, PBS 2019). Documentary of the effect of CFCs on the ozone layer, including the influence of the TV series *All in the Family* to stop the use of CFCs in aerosol spray products
13. NOAA Fisheries website regarding population of humpback whales. https://www.fisheries.noaa.gov/species/humpback-whale
14. *Star Wars Episode III: Revenge of the Sith* (George Lucas, 20th Century Fox 2005). A robotic midwife delivers Luke and Leia, while elsewhere in the galaxy, robotic surgeons convert what's left of Anakin Skywalker into the cyborg Darth Vader [DVD scene 45]
15. *Star Wars Episode V: The Empire Strikes Back* (Irvin Kershner, 20th Century Fox 1980). A robotic surgeon replaces Luke's severed hand with a cybernetic hand [DVD scene 49]
16. RIBA, a robotic nurse for elder care. http://www.youtube.com/watch?v=U92eB6WyjKc

Misplaced Faith in Science

17. *Star Trek: The Next Generation* – "Rightful Heir" (Winrich Kolbe, Paramount 1993). Clone of Kahless created to restore order and stability to Klingon Empire [DVD season 6, disc 6, scene 7]

Misunderstanding of Science

18. Shermer, M.: What is pseudoscience? Sci. Am. **305**, 92 (2011)
19. *Indiana Jones and the Kingdom of the Crystal Skull* (Steven Spielberg, 20th Paramount 2008). Archeologist Indiana Jones is pressed into service by a Soviet psychic researcher, in search of a powerful alien artifact. Early studies suggested that the artifact was strongly magnetic, but it becomes obvious that the interaction can't possibly be electromagnetism [DVD scenes 2, 7, 8]

20. Van Till, H.J., Young, D.A., Menninga, C.: Science Held Hostage, p. 41. InterVarsity Press, Downers Grove (1988)
21. James Randi Educational Foundation. https://web.randi.org/
22. *Harry Potter and the Chamber of Secrets* (Chris Columbus, Warner Brothers 2002). Phoenix tears have healing powers [DVD scene 33]
23. Data on helenalin, the toxic substance found in the flowering plant Arnica. https://en.wikipedia.org/wiki/Helenalin
24. Definition of the "law of infinitesimal dose" in homeopathic medicine. https://medical-dictionary.thefreedictionary.com/law+of+infinitesimals
25. *Voyage to the Bottom of the Sea* (Irwin Allen, Twentieth Century Fox 1961). The Van Allen Radiation Belt has somehow caught fire, and threatens to incinerate all life on Earth. The proposed solution (worked out on a slide rule): explode an atomic bomb, in hopes of blowing out the fire [DVD scenes 6, 9]
26. *All Things Considered* – "The Bomb Watchers" (Robert Krulwich, NPR July 1, 2010). Starfish Prime nuclear testing in the ionosphere, 1962. http://www.npr.org/templates/story/story.php?storyId=128170775
27. Singer, P.W.: Wired For War, p. 163. Penguin Press, New York (2009)
28. Davidson, K.: Beam me up, Scotty? Pittsburgh Post Gazette, 17 Oct 2005, p. A-6 (originally reported in the San Francisco Chronicle)
29. Davis, E.W.: Teleportation physics study. Air Force Research Laboratory, Aug 2004. http://www.fas.org/sgp/eprint/teleport.pdf
30. *Star Trek: The Next Generation* – "The Measure of a Man" (Robert Scheerer, Paramount 1989). Data's storage capacity and processor speed [DVD season 2, disc 3, scene 7]
31. *The Outer Limits* – "I, Robot" (Leon Benson, MGM 1964). A robot, suspected of murdering its creator, is eventually proven innocent [DVD vol. 3, disc 2, episode 41]
32. Camilli, A., D'Emilio, F.: Italy convicts 7 for failure to warn of 2009 quake (Associated Press, reported in the Pittsburgh Post-Gazette, 23 Oct 2012, p. A-1)
33. *Gilligan's Island* – "Pass the Vegetables, Please" (Leslie Goowin, United Artists 1966). The castaways eat vegetables grown from experimental radioactive seeds, and acquire unusual powers [DVD season 3, episode 3]

How Can We Know What Is Real?

34. *An Honest Liar* (Justin Weinstein and Tyler Measom, Flim Flam Films 2014). The career of James "The Amazing" Randi as a stage magician turned debunker of claims of the paranormal
35. *Soylent Green* (Richard Fleischer, MGM 1973). Advertising products that are not what they claim to be [DVD scene 2]
36. Wong, K.: The first cookout. Sci. Am., Sept 2013
37. Surfactants and Critical Micelle Concentration (CMC). https://www.dataphysics-instruments.com/us/knowledge/understanding-interfaces/surfactants-cmc/
38. *Star Trek* (original series) – "The Changeling" Written by John Meredyth Lucas, Paramount 1967). A space probe mistakes Captain Kirk for its "Creator" and struggles with the resulting logical inconsistencies [DVD vol. 19, episode 37, scenes 6, 7]
39. *The Matrix* (The Wachowski Brothers, Warner Brothers 1991). Morpheus asks Neo, "What is 'real'?" [DVD scene 12]
40. Alexei Efros, as quoted by Brooke Borel. Clicks, lies and video tapes. Sci. Am., Oct 2018

Chapter 8
What Lies Ahead? (The Future of Our Technological Society)

It is no concern of ours how you run your own planet. But if you threaten to extend your violence, this Earth of yours will be reduced to a burned out cinder. Your choice is simple: Join us, and live in peace, or continue your present course, and face obliteration. We shall be waiting for your answer. The decision rests with you.
—*Klaatu*
The Day The Earth Stood Still

As we discussed in the previous chapter, the distant future of our technological society will depend on the decisions that are taken now and in the near future, regarding what research projects to fund or not to fund, and how to use the new discoveries that are made and the new technologies that are developed. In this chapter, we will explore some visions of the future, as presented in science fiction. We will also consider how cultural and historical context can influence the way in which a classic science fiction story is told, and how our understanding of technology can influence how we think about culture. But we begin with a brief survey of some sci-fi predictions which have already come to pass, and a few more which may be coming very soon.

8.1 Accurate Predictions

H.G. Wells predicted a second world war, more devastating than the first, in his 1933 novel "The Shape of Things to Come." Wells and his contemporaries had already lived through the First World War, which lasted nearly 4 years. Given the political climate in Europe in the 1930s, there is nothing particularly extraordinary about imagining a second world war at some time in the future, which would last even longer. Granted, the accuracy of Wells' prediction was somewhat astonishing, having missed the mark by only 4 months, for England's actual declaration of war on Germany. Nevertheless, the prediction of a devastating world war might be

© Springer Nature Switzerland AG 2019
B. B. Luokkala, *Exploring Science Through Science Fiction*, Science and Fiction, https://doi.org/10.1007/978-3-030-29393-2_8

viewed as nothing more than an extrapolation from known events throughout human history. Similarly, Wells' prediction of the use of armored tanks, which he called *land ironclads*, in his 1903 story of the same name, preceded their actual invention and use in World War I by more than 10 years. Although it seems prescient to readers today, this, too, may be viewed as just an extrapolation from things that were already known. Wells was undoubtedly inspired by existing naval technology—the ironclad ships, *USS Monitor* and *CSS Virginia*, used in the American Civil War, for example—and was simply applying the same concept to land vehicles.

There are, however, many notable works of science fiction, which have made some truly extraordinary predictions, several of which we have already discussed briefly in previous chapters. Before we consider what lies ahead, let's take a look back at a few of the many scientific and technological developments, which were predicted or inspired by science fiction.

8.1.1 Space and Time

Johannes Kepler predicted human space flight to other bodies in the solar system more than three centuries before the first Soviet cosmonaut orbited the Earth and the first U.S. astronauts landed on the Moon. Jules Verne predicted that the first humans to make a trip to the Moon would return to Earth by splashing down in the ocean and would be recovered by a U.S. naval vessel, 100 years before it actually happened.

H.G. Wells described a four-dimensional spacetime in his 1895 novel "The Time Machine" 10 years before Einstein published his special theory of relativity. We now know that time travel into the future is not just possible, in principle. It actually happens. But at currently accessible speeds for humans, the effect is not very dramatic. Although proposals have been made for round-trip time travel, using wormholes, to the best of our knowledge, stable wormholes do not exist. So time travel into the past remains a technical impossibility.

8.1.2 Matter and Energy

The industrial laser in *Goldfinger* was right at the cutting-edge of emerging technology [1]. When the movie was released in 1964, the first visible light (ruby) laser had just been invented 4 years earlier, and the infrared (CO_2) laser, which was powerful enough to cut through solid materials, had just been patented that same year. So when the movie was still in production, the concept of an industrial metal-cutting laser was probably an extrapolation from what was actually available at the time.

The idea of creating a three-dimensional image using visible light was depicted in the 1956 movie *Forbidden Planet* [2]. But optical holography was not practical until the invention of the laser in 1960. The first visible light hologram was produced in 1962.

Various techniques for rendering an object invisible have been imagined in science fiction, over the past 100 years. But only at the turn of the twenty-first century did this dream come within the realm possibility. The creation of metamaterials, having negative index of refraction, enables the construction of an invisibility cloak, which works in the microwave part of the electromagnetic spectrum. Unfortunately, a way has not yet been found to extend this technology into the visible part of the spectrum. Thus, a true invisibility cloak for visible light still remains out of reach.

8.1.3   Computing and Robotics

A computer that could diagnose a medical condition and prescribe treatment was imagined in the 1971 move, *The Andromeda Strain*. Medcom (or medical computer) prescribed blood plasma for an anemic old man and dextrose for an undernourished baby, the two sole survivors of an alien organism, which killed all other inhabitants of a small western village [3]. Originally designed and built for the purpose of playing the TV game show, Jeopardy, IBM's Watson computer was soon given its first commercial application: prescribing courses of treatment for lung cancer patients.

The concept of a fully autonomous, intelligent machine, resembling a human can be traced back to the ancient Greeks. But it was not until the latter half of the twentieth century that intelligent humanoid robots became a reality. P.W. Singer quotes a curious statistic in his recent book "Wired for War": According to a 2007 U.N. report, there were 4.1 million robots working around the world in people's homes—more than the entire human population of Ireland, at the time [4]. Presumably the report was limited specifically to devices which incorporate all three of the essential ingredients to qualify as an *intelligent robot*, as we discussed in Chap. 4, namely sensors, computation, and control. Examples of *intelligent* household robots include such things as autonomous robotic vacuum cleaners and possibly also robotic room deodorizers. If the report had used a broader definition of robot and included any automated device, which replaces a human function (such as a washing machine or a dishwasher), the numbers would have been much higher.

Robots have not only invaded our factories and homes but are also participating in sporting events, such as camel races, which are popular in many middle-eastern countries. Traditionally, the camel jockeys were young boys, but the frequency of injury to the jockeys was unacceptable. For safety reasons, the human jockeys have been replaced by radio-controlled robotic jockeys, strapped to the backs of the camels, and operated from a motor vehicle, driven alongside the camel racetrack [5].

As we discussed in the previous chapter, robotic devices of various kinds are becoming more common in the healthcare industry. Laparoscopic surgery is performed routinely with robotic surgical tools, controlled remotely by an actual human surgeon, and autonomous robot nurses are being developed, particularly for use in eldercare.

Robotic military hardware features prominently in a number of sci-fi movies, including the futile use of Reaper drones against the giant robot, Gort, in the 2008 remake of *The Day The Earth Stood Still* [6]. Gort, of course, is purely fictional, but the unmanned aerial vehicles (UAV) depicted in the scene are accurate representations of the real thing. An even more true-to-life depiction of the use of different types of UAV is shown in the opening scene of *Eagle Eye*. A remote controlled aerial surveillance device, no bigger than a model airplane, is launched by hand, and used to track the activity of suspected terrorists. Contrary to the recommendations of an intelligence-gathering supercomputer, the President of the United States authorizes the use of deadly force, in the form of a Reaper drone. Later in the same movie, we see another Reaper drone, this time under the control of the rogue supercomputer, being guided through a tunnel, in pursuit of the lead character and an FBI agent [7].

Although robots have not yet achieved consciousness and show no signs of wanting to take over the world, advances in robotic technology are having an unprecedented impact on the nature of modern warfare. P.W. Singer relates the surprising range of human response to the use of robots on the battlefield. For the military leader, who must inform the family when a soldier is killed in action, combat robots might make the job less painful. Singer quotes a U.S. Navy chief petty officer, "When a robot dies, you don't have to write a letter to its mother" [8]. But to the soldier on the battlefield, who actually works with the robots, and has seen it in action saving human lives, the sentiment can be very different. If a field robot, whose primary purpose is to disarm improvised explosive devices (IAD) is badly damaged, the primary user of the robot may have become so attached to it that a replacement is unacceptable and wants it to be repaired at any cost. Yet another response comes from those against whom aerial drones (UAV) are used in battle, calling the users cowards for sending robots, instead of directly facing their enemies.

Until very recently, UAV of various kinds have been flown by remote control, under the guidance of human operators. But the latest innovation in robotic technology is the Navy's experimental X-47B drone, which can fly, and even land on the deck of an aircraft carrier without human intervention [9]. The flight plan for the X-47B is programmed by humans, and humans can still override its decisions, but the introduction of this new technology serves to underline concerns raised by P.W. Singer in his book "Wired for War", particularly in regard to the chain of accountability if something goes wrong. Singer presented highlights of his research in a recent TED talk, which is available on YouTube [10]. Watching the video may inspire you to read the entire book.

8.1.4   *Planets in Other Star Systems*

Many people believe that there must be intelligent life elsewhere in the universe, and science fiction is filled with speculations of what kind of intelligent life there might be. The search for extraterrestrial intelligence, as a bona fide scientific research program, has been going on, in one form or another, for more than 50 years. Although there is currently no evidence for extraterrestrial intelligence, the existence of planets outside our solar system has been established beyond any reasonable doubt. The discovery of extrasolar planets, by carefully measuring the positions of stars on photographic plates, was predicted in science fiction (*When Worlds Collide*) at least 30 years before the first confirmed evidence, obtained with the same technique [11]. By this classic method, as well as with the much more successful technique of monitoring the intensity of light from a star, to detect the transit of an orbiting planet, there is growing evidence that many stars in the galaxy have planets. Perhaps even more prescient is the prediction in the same movie that it might be possible to directly photograph a planet in orbit around a distant star. This was impossible using the technology available in the 1950s. But recent innovations in telescope imaging have made it possible.

8.1.5   Biomedical Technology

We have already seen predictions of hand transplants (*Mad Love*) and heart–lung bypass machines (*The Man They Could Not Hang*) in science fiction decades before such things were ever attempted by surgeons in real life [12, 13]. High-speed DNA sequencing was imagined in science fiction (*GATTACA*) at a time when the actual practice took years and had yet to be completed even once [14]. The current state-of-the-art in full-genome sequencing has nearly caught up with the prediction.

The sophisticated medical diagnostic tools envisioned in the late 1960s, by the writers of *Star Trek*, are coming closer and closer to reality. Automated devices for monitoring the vital signs of patients bear a striking resemblance to the medical diagnostic beds in the sick bay of the original starship Enterprise [15]. The *Star Trek* medical tricorder is only slightly smaller than the highly portable ultrasound devices, which have just recently come on the market [16].

8.1.6   Communication Technology

As we discussed in the chapter on machine consciousness, the writers of *Star Trek* imagined a time in the twenty-third century when people would be able to interact with computers by voice command. But speech recognition and language technology have outpaced the predictions. We in the early twenty-first century already have computers in our smart phones, which can recognize and respond to the spoken word.

Two more examples of accurate sci-fi predictions, which we have not yet considered, may be found in the area of communications technology. The communicators used by the crew of the U.S.S. Enterprise, in the original series of *Star Trek*, were unheard of in the mid-1960s (see Fig. 8.1). Nobody, not even the military, had anything like this miniaturized, long-range communication device. Mobile radio phones existed at this time, but they were the size of desk phones, with an even heavier, bulkier transmitter/receiver. Versions were available for use in automobiles, and powered by the car battery but were not easily portable. The field version, with its own battery, had to be carried in a large backpack.

Fig. 8.1 The *Star Trek* (Original Series) Communicator (left) and the Motorola StarTac mobile phone (right) (photo by the author)

Gradually, with advances in battery technology to make the power pack smaller and lighter, and magnet technology, making the speaker smaller, truly portable mobile phones came on the market. Nearly 30 years after the television premiere of *Star Trek*, Motorola released its StarTAC cell phone (Fig. 8.1). This was the first clamshell-style, or flip phone to come on the market, and prior to its release, I was unconvinced of the need of such technology. But the StarTAC bore such a striking resemblance to the *Star Trek* communicator that I was finally convinced I needed to buy one.

Star Trek II: The Wrath of Khan, the 1982 movie, which we have already considered in other contexts, presaged a bit of useful technology, which became available on a range of higher-end motor vehicles in the first decade of the twenty-first century. According to the movie, following the surprise attack and crippling of the U.S.S. Enterprise by the stolen starship Reliant, Khan, the title character, in command of Reliant, orders Kirk and the Enterprise to surrender. Unbeknownst to Khan, Kirk is able to call up the command code for Reliant and remotely order the ship to lower its shields, leaving it vulnerable to counterattack [17]. OnStar was founded in 1995, more than a decade after *Star Trek II* was released. And it was not until more than 25 years after the movie, in 2009, that OnStar introduced their Stolen Vehicle Slowdown feature, allowing a vehicle to be stopped by a remotely transmitted command. Given the degree to which motor vehicle systems have become computerized, can we be very far away from the type of Wakandan technology, envisioned in *The Black Panther*, which enables a remote operator to hack into a vehicle's system and drive it via a holographic interface? [18].

8.1.7 Brain–Computer Interfacing

Forbidden Planet portrays the use of a futuristic device called the *Krell plastic educator*. A headset, designed to detect brain waves, enables the user to create holographic images and to manipulate mechanical objects. In the movie, Morbius demonstrates that his brain is capable of sending the intelligence indicator about half way up the scale, while a 7-year-old child Krell was expected to send it all the way to the top [19]. The existence of brain waves had been known since the 1920s But when *Forbidden Planet* was released in 1956, the possibility of using brain waves to directly control an external device was still many years in the future. Already on the market today is a toy, which bears a striking resemblance to the Krell plastic educator. Designed by Uncle Milton Industries, and part of their Star Wars Science line, the so-called Force™ Trainer allows you to "move an object with the power of your mind." The object is a ping-pong ball, which floats on a column of air, above a battery-powered fan. A headset detects your brainwaves and transmits a wireless signal to the base unit. The speed of the fan—and hence the height at which the ball floats—depends on your level of concentration [20]. As we will see in the next section, this new technology has far-reaching implications for improving the quality of life for many people in the near future.

8.1.8 Cybernetic Devices

The 1970s television series *The Six Million Dollar Man* revolved around former astronaut and test pilot, Steve Austin, who was severely injured in a crash, and rebuilt with bionic enhancements. His right arm, both of his legs and his left eye were replaced with electromechanical devices, giving him superhuman strength, speed, and vision. In the 2001 movie *AI: Artificial Intelligence*, a young boy named Martin, whose parents had almost given up hope that he would ever regain consciousness, let alone walk again, miraculously returns home. At first confined to a wheelchair, Martin is later seen walking with the aid of electromechanical leg braces. The details of how these braces function are not explained in the movie, but presumably they are interfaced either directly to Martin's central nervous system, or wirelessly to his brain [21].

Perhaps inspired by science fiction, research is already being done to develop devices that will enable paraplegics to walk and the blind to see. Robotic legs, which are currently under development, involve feedback between the motion of the wearer's body and the movement of the robotic legs. New robotic technology is already enabling people with severe nerve or limb damage to manipulate objects in their environment without assistance [22, 23]. In these applications, an array of tiny electrodes is implanted directly into the brain, enabling signals to be sent to the robotic device. Significant advances in exoskeleton walking technology have also been made. Developed by researchers in Carnegie Mellon University's department of Mechanical Engineering and the Robotics Institute, an approach called "human-in-the-loop optimization" involves indirect calorimetry to measure the user's metabolic rates. The information gathered is used to adjust the torque provided by the exoskeleton, while the user is walking, running, or carrying a load [24].

The first successful neuroprosthetic device was the cochlear implant, developed in the 1950s to restore hearing to the deaf. Advances have also been made toward restoring vision to sufferers from retinal degeneration. The Argus II implant, produced by the company SecondSight, received approval for use in Europe in 2011 and FDA approval for use in the U.S. in 2013 [25]. Currently under development is a retinal prosthesis for restoring functional vision to the blind. Supported by a grant from the U.S. Department of Veterans Affairs, a research group at Carnegie Mellon University, headed by Shawn Kelley, have created a microfabricated thin film to replace the rods and cones in the eye. Inserted behind the retina of a visually impaired patient's eye, and working in connection with specially designed glasses, the film sends electrical signals to the retinal nerves [26, 27].

8.2   Coming Soon: Possibilities for the Not-Too-Distant Future

Like the expansion of the universe, the advancement of science and technology seems to be proceeding at an ever-increasing rate. Things which were entirely in the realm of science fiction just a few years ago are now looming on the horizon. Included here are just a few of the many research projects currently underway, which should yield exciting discoveries and new technologies, in the near future.

8.2.1   Space Exploration

The space age began with the launch of the Soviet Sputnik satellite, in 1957. Up to this point in history, human spaceflight has ranged from low-Earth orbit to round-trip landings on the Moon. Exploration of the solar system beyond the Moon has been done thus far only by robotic probes. In the quest to extend human spaceflight to greater distances and longer durations, NASA and others are working to address several important factors, not least of which is the safety and comfort of the passengers. NASA's Constellation program, which includes the Orion space vehicle, is an effort to build a space transportation system which "can not only bring humans to the Moon and back, but also resupply the International Space Station and eventually place people on the planet Mars" [28]. One of the original goals of the Constellation program—to return to the Moon by 2020—has since been abandoned, and the focus has shifted to landing humans on Mars by the 2030s. Manned testing of the Orion space vehicle, which is designed to carry four astronauts, is expected to begin in the early 2020s. Meanwhile, initiatives, such as the Mars Desert Research Station, are already in operation, conducting simulations and training in preparation for human colonization of Mars [29].

An ongoing concern is the exposure of interplanetary astronauts to radiation from galactic cosmic rays. With current technology, a round-trip to Mars is expected to take approximately a year to a year-and-a-half. Over the course of the trip, astronauts would receive a radiation dose three to five times higher than established safety limits [30]. Although the dose does not even begin to approach lethal levels—as much as 0.25 Sv per year on the trip to Mars, compared to a lethal dose of 6 Sv in a few hours—prolonged exposure to elevated levels of radiation is a significant health concern. Humans on Earth are protected from most of this radiation by Earth's atmosphere and magnetic field. But the Martian atmosphere is considerably thinner than Earth's, and the planet does not have a global magnetic field. So exposure to cosmic radiation on the surface of the planet may also be a factor in considering whether or not to establish a permanent human colony on Mars.

8.2.2 Dark Matter and Dark Energy

In Chap. 3 we explored the question, what is the universe made of? The surprising answer is that only roughly 5% of the matter and energy in the universe is stuff that we currently understand. The rest is what has been labeled dark matter and dark energy. We know that it's out there because of the way in which the visible matter in the universe is moving, but what it is, and how it interacts is as yet unknown. The question is under investigation, both experimentally and theoretically, by researchers around the world. In particular, a group of faculty, postdoctoral researchers, graduate students, and undergraduates, associated with the McWilliams Center for Cosmology at Carnegie Mellon University, are actively involved in a number of research initiatives, the goals of which include an understanding of the nature of dark matter and dark energy [31].

The July 4, 2012, announcement from the Large Hadron Collider, that a particle consistent with the long-sought Higgs boson at last had been observed, deepens the understanding of the concept of mass. But this should be just the first of many major discoveries made at the LHC.

In May of 2012, NASA launched the Alpha Magnetic Spectrometer (AMS), a seven metric ton particle detector, which is now operating aboard the International Space Station. The AMS is the most advanced cosmic ray detector ever built and is capable of sorting out particles of ordinary matter (the stuff of the standard model of particle physics) from dark matter candidates. The rate at which cosmic rays are being measured by the AMS is in the hundreds per second and can be tracked online at the NASA AMS web site [32]. Particle physicists are hopeful that the LHC and the AMS will lead to an understanding of the nature of dark matter in the near future.

Research efforts aimed at an understanding of dark energy include the Large Synoptic Survey Telescope (LSST), a deep-field optical instrument, using a 3.2 Giga-pixel camera capable of photographing the entire sky twice each week. Full science operations are expected to begin by 2023 and will last for 10 years [33].

8.2.3   Real Adaptive Camouflage

Through a combination of pigment-filled cells and light-reflecting cells, certain species of octopus are able to blend in with their surroundings as effectively as the alien creature in the *Predator* movies, or the Aston-Martin V-12 Vanquish, modified by Q-Branch of MI6, in *Die Another Day*. But how does an organism, which is known to be colorblind, achieve this feat? The process is currently not understood. Ongoing research could lead to an effective means of adaptive camouflage for large objects [34].

8.2.4   Self-Healing Electrical Circuits

Recently published research in the journal *Nature Materials* describes a liquid metal-elastomer composite, which enables a damaged electrical circuit to heal itself. Liquid metal droplets are suspended in a soft elastic polymer material. When the material is damaged, the metal droplets rupture and form new connections with neighboring droplets. The new technology enables electrical signals in the circuit to be rerouted without interruption [35]. We are nowhere near the creation of something like the T-1000 liquid metal terminator, in *Terminator 2: Judgment Day*. But given this new advance in materials science and electrical engineering, how far off can it be?

8.2.5   Room-Temperature Superconductivity

The 2004 movie *Primer* depicts a team of researchers on a quest to develop a room-temperature superconductor—a material capable of carrying an electrical current, but without resistance [36]. Up until now, the best superconductors have operated well below −100 °C and are typically cooled with liquid nitrogen. The quest for a room-temperature superconductor has been called by some the Holy Grail of materials science. According to an article reported recently in the journal *Nature*, a team of researchers in Germany may have discovered a material that is superconducting not just at room temperature, but above 100 °C. Many remain skeptical, and the research needs to be confirmed. But the discovery of a material that superconducts even at the boiling point of water has far-reaching implications for energy savings [37].

8.2.6 Quantum Computers

The Nobel Prize in Physics for 2012 was awarded jointly to Serge Haroche and David J. Wineland "for ground-breaking experimental methods that enable measuring and manipulation of individual quantum systems" [38]. Working independently, using complementary techniques, the two researchers have achieved results previously believed to be unattainable. The Haroche group use a highly reflective microwave cavity to trap photons. The quantum information is probed using atoms, sent through the cavity one at a time, to detect the presence or absence of a microwave photon. The opposite technique—trapping ions and probing the quantum information with a laser—is used by the Wineland group. Their work opens the door to the development of practical quantum computers, which will far surpass the performance of classical computers. Although quantum computers are already on the market, the technology is still in its infancy.

8.2.7 Robots in the Home

Several different companies are already marketing compact, robotic sweepers, which do a fairly good job at the simple task of dusting hard-surface floors. Room-size computers, such as IBM's WATSON, can already surpass humans at complex but well-focused tasks, such as competing on the TV game show Jeopardy, and have also been put to use commercially. Given the trends in data storage capacity and processor speed, can we be very far from having multipurpose robotic maids, of the sort depicted in the cartoon series *The Jetsons* [39], or in the movies *Bicentennial Man* or *I, Robot*? And if routine household chores can be managed by robots, can personal healthcare be far behind? Baymax, the experimental healthcare robot in *Big Hero 6*, can diagnose and treat minor medical ailments on the spot [40]. As we've seen, robots are already being developed and used in the healthcare industry, but we are not yet at the point of having personal healthcare robots, like Robot, from *Robot and Frank*, in our homes. If the predictions made by professionals in the field, such as Carnegie Mellon's Hans Moravec, are correct, they may be here within the next couple of decades [41].

8.2.8   Artificial Heart

The "Tapestry" episode of *Star Trek: The Next Generation* reveals that Captain Jean-Luc Picard of the U.S.S. Enterprise has an artificial heart. As a Star Fleet ensign, Picard became involved in a fight with an alien, during which his natural heart was impaled by a sword [42]. Currently, artificial hearts are designed to be temporary devices, until the patient can receive a heart transplant. The first artificial heart, the Jarvic-7, as well as some more recent models, like the SynCardia, are neither self-contained, nor fully implantable, as is Picard's, but are tethered to an external power supply or driven by an external compressor. A more recent innovation, the AbioCor, is the world's first completely self-contained artificial heart [43]. Further advancements may soon enable the creation of artificial hearts, which are more than just temporary solutions to a purely biological problem.

8.3   Science Fiction in Historical Context

Although some of the evidence may be circumstantial, a good case can be made for the influence of science fiction on the development of science and technology. As we've just seen, writers of science fiction are not necessarily predicting the future but are clearly serving as a source of inspiration for future technologies. It goes without saying that science fiction writers often incorporate cutting-edge science and technology into their newest stories. But how do the flow of history and our changing understanding of science affect the retelling of old stories? Here we consider three examples of the effect of historical context on the remaking of classic sci-fi movies.

8.3.1   The Island of Dr. Moreau

We have already seen in Chap. 6 how the updated screen versions of *The Island of Dr. Moreau* incorporated the increasingly sophisticated science of heredity, over a period of 100 years. Both the original novel, published by H.G. Wells in 1896, and the earliest talking movie version portrayed the transformation of animals into humanlike creatures as an entirely surgical procedure, long before the discovery of DNA. The 1977 movie hinted at a molecular basis for heredity, and the 1996 version involves genetic engineering [44].

8.3.2   The War of the Worlds

We also noted in Chap. 6 that two movie versions of *The War of the Worlds* (1952 and 2005), as well as the original H.G. Wells novel of the same title, are all in agreement that the Martian invaders were defeated by their own immune system. Each version of the story makes a point of incorporating state-of-the-art military technology into the battle against the Martian machines. But after all else had failed, the Martians died of natural causes from some unspecified disease, to which humans had developed immunity. One additional note of historical context is worth pointing out, in regard to the 2005 remake. Although more faithful to the original novel, in many respects, it differs significantly in the depiction of the arrival of the Martian machines. In both the original novel and the 1952 movie version, large objects fall from the sky, like meteors, and the machines emerge from the fallen objects [45]. But in the 2005 version, the machines are already here, buried underground long before our cities were built. They emerge only after a bolt of energy awakens them, like sleeper terrorists receiving their activation call. As if the imagery were not sufficiently obvious, just 4 years after the events of September 11, 2001, the young daughter of the lead character asks the question when the machines begin their attack, "Is it terrorists?" [46].

8.3.3   The Day the Earth Stood Still

Both the original 1951 version of *The Day the Earth Stood Still* and the 2008 remake warn of the consequences of technology used irresponsibly. Both movies involve an alien visitor, accompanied by a giant robot, with an important message to deliver to the inhabitants of Earth. Given our current understanding of the nature of space and time, it seems that our best bet for encountering extraterrestrial intelligence, either now or in the future, is through radio communication. But this should not stop us from imagining what an actual face-to-face encounter with extraterrestrials might be like, and the kind of message we might expect to receive.

At the beginning of the original version of *The Day the Earth Stood Still*, a spaceship lands on a baseball field, in Washington, DC, and is immediately surrounded by a ring of military personnel, armed with assorted conventional weapons. A space-suited alien emerges from the ship and addresses the assembled crowd: "We have come to visit you in peace, and with good will." The alien, who calls himself Klaatu, walks slowly down a ramp from his ship, and pulls out a strange-looking device, which he later explains is a gift for the President of the United States: a device for studying life on other planets. A nervous soldier mistakes the device for a weapon and shoots Klaatu in the arm. This irrational action precipitates instant retaliation by a giant robot, who proceeds to melt most of the weapons but leaves the people unharmed. As the story unfolds, Klaatu tries to communicate his message to the world's political leaders. But his attempts are thwarted by fear and suspicion, not just of him, as an alien, but of one another, in a climate of global political tension. Only at the end of the movie is Klaatu able to deliver his message, but not to the political leaders of Earth, as he had originally hoped to do. Instead, he addresses the assembled scientists of the world. His final words, quoted at the beginning of this chapter, include both an invitation and a warning [47].

Discussion Topic 8.1

The 1951 version of Klaatu makes his final appeal to the scientists of the world, rather than its political leaders. What does this suggest about his hope for the people of Earth to change their ways, with respect to the use of nuclear weapons? Does he seem to be overlooking the fact that nuclear weapons were developed by scientists, under the pressure and urgency of World War II?

The 2008 remake of *The Day the Earth Stood Still* features updated military hardware (Reaper drones with sidewinder missiles, instead of handguns and tanks), updated special effects (computer-generated images, instead of a tall man in a rubber robot costume), and an updated message (environmental irresponsibility, instead of nuclear proliferation). But unlike the 1951 version, this message does not come in the form of a warning to change our ways and an invitation to join the other planets and live in harmony. Instead, the new Klaatu delivers a message of judgment and proclaims a death sentence:

> This planet is dying. The human race is killing it.... We can't risk the survival of this planet
> for the sake of one species.... If the Earth dies, you die. If you die, the Earth survives...
> There are only a handful of planets in the cosmos that are capable of supporting complex
> life. This one can't be allowed to perish. [48]

Despite the pleading of the lead human character in the movie, no chance of repentance or opportunity to reform is offered. All the other species on Earth are herded into giant spheres, which ascend into orbit like interplanetary versions of Noah's ark, and the process of extermination of the human race begins.

Discussion Topic 8.2

Analyze Klaatu's argument, concerning the future of human life on Earth. What does he assume about the number of life-sustaining planets in the cosmos? (Note that he says *the cosmos*, and not just *the galaxy*.) In light of what we are learning about other planets in other solar systems, is this assumption likely to be valid? What further assumption does Klaatu seem to be making, in regard to the relationship between one life-sustaining planet and another? Is this a valid assumption? What specific observations does Klaatu make, in regard to humans and the Earth? Does his conclusion that the humans must die follow logically from his major premise about the number of life-sustaining planets in the cosmos? Is there another solution to the problem, which Klaatu doesn't seem to be willing to entertain?

The original version of *The Day the Earth Stood Still* (1951) is a cautionary tale, which made good sense in the context of the Cold War. Although it contains very little science, the movie is excellent drama and conveys a simple, yet powerful message, concerning the future of our technological society. The message of the original version remains relevant today. The 2008 remake, on the other hand, is a failed attempt to update the story to make it relevant to the concerns of the twenty-first century. Unfortunately, the updated story does not make sense, even in the present historical context. There is increasing evidence of the existence of habitable planets elsewhere in the galaxy. Even if there are intelligent civilizations on some of them, it should be no concern of theirs how we run our own planet. Nevertheless, the call to be environmentally responsible and to use our technology wisely should be heeded.

8.4   Visions of the Future

The writers of science fiction seem to have done a remarkable job of predicting, or perhaps inspiring, future technologies. But what can we learn from them about the possible future of humans, as a species? Here we will consider several visions of the future of our technological society, as portrayed in science fiction.

8.4.1   Defining Culture in Terms of Technology

One of the episodes of the original *Star Trek* TV series may be unintentional in raising certain questions about how we define human culture. *Errand of Mercy* tells of an encounter with a civilization on another planet, which is so far advanced, that they no longer have any need of physical bodies. The Organians, as they are called, are noncorporeal beings, who take on physical form only for the benefit of the occasional off-world visitor to their planet. The U.S. Enterprise and her crew visit Organia to warn them of a possible invasion by the Klingon Empire and offer to help them, if they will join the United Federation of Planets. The Organians insist that they are in no danger, and express more concern for the safety of their visitors. While Captain Kirk negotiates with the Organian Council of Elders, Science Officer Spock goes off to study the Organian culture. Having made thorough measurements with his tricorder, Spock returns to the Council chambers to make his report to Kirk.

> Captain, our information on these people and their culture is not correct. This is not a primitive society, making progress toward mechanization. They are totally stagnant. There is no evidence of any progress as far back as my tricorder can register. … For tens of thousands of years there has been absolutely no advancement; no significant change in their physical environment. This is a laboratory specimen of an arrested culture. [49]

As the story unfolds it becomes clear that Spock has completely misjudged the Organians. On the one hand, his tricorder cannot possibly tell him that the Organians have evolved beyond the need for physical bodies. What Kirk and Spock, and eventually the invading Klingons, are experiencing is an illusion of physicality, created by the Organians for the benefit of their off-world visitors. On the other hand, Spock should know better than to pass judgment on a culture on the basis of technology alone. He has completely neglected art, music, literature, and issues of social justice, to name just a few. All of these, and more, are important dimensions of culture, which have little to do with technology. When the Klingons do arrive, and the inevitable battle begins, the Organians quickly put a stop to it. By force of will, they disable all of the weapons systems on both the Enterprise and the Klingon ships, and make it impossible for any hostility to take place. Even hand-to-hand combat is stopped. Another lesson is clear: never judge a book by its cover, and never judge a culture by technology alone.

Discussion Topic 8.3

In judging the Organian civilization entirely in terms of visible technology, what important dimensions of culture is Spock overlooking? What measures would you use to evaluate a culture?

8.4.2   Social Divisions

As new technologies are developed, the question soon arises, who gets to use it, or who will benefit from it? Science fiction often depicts societies of the future, which are divided into the haves and the have-nots. One of the earliest examples is the silent classic, *Metropolis*. The opening scene—Shift Change—shows members of the laborer class trudging through a tunnel, toward a waiting freight elevator, while others emerge from an adjacent elevator, and trudge off into the tunnel. Regardless of whether the workers are arriving for work or departing for home, they trudge with faces toward the ground, as though soulless or dehumanized. We soon learn that these workers live in an underground city, below the giant machines, which they live to maintain. The workers' city is totally bland, lacking any ornamentation, while the tunnels leading to the machine space are adorned with more interesting brickwork. Meanwhile, on the surface of the Earth, the inhabitants of Metropolis live the good life [50]. The world of *Metropolis* is divided into a technological elite and a laborer class, who are never supposed to mix. But the story is one of hope and reconciliation.

GATTACA depicts a world in the not-too-distant future, in which society is divided according to genetic makeup. The genetically invalid are barred from many of the benefits enjoyed by the rest of society, including medical insurance and high-tech employment opportunities [51]. Employees are tested daily, and high-speed DNA sequencing is available to determine the suitability of a potential romantic interest.

8.4.3   Technophiles and Technophobes

Things to Come opens with the world on the brink of a war, the likes of which had never been seen before. Lasting for nearly 25 years, it results in global devastation, leaving humanity organized into tribes struggling for existence. Of those who survived the war, half are wiped out by a pestilence—the wandering sickness—a contagious disease for which no cure was ever found. After 4 years, the disease finally subsided and civilization began to rebuild. By the year 2036, a technological utopia has arisen from the ashes of the war. Or is it really as utopian as it seems to be on the surface? A single world government enforces order using the so-called *gas of peace*. And not everyone is ready to embrace the astonishingly fast pace of technological progress. Many remain skeptical, even fearful. The initially peaceful society threatens to tear itself apart again, when the technological elite plan an expedition to the Moon. A prominent master craftsman sows seeds of a new rebellion, crying out, "Stop this progress before it's too late!" A riot breaks out just as the Moon launch is about to take place. In the closing scene, we hear an interesting dialog between two of the leaders of the elite, whose daughter and son, respectively, have just been launched into space. One of them is very optimistic, perhaps even arrogant, about technological progress, while the other is a bit more humble and skeptical. Together they contemplate the implications of what they have just done.

> CABAL: For man, no rest and no ending. He must go on, conquest beyond conquest: first, this little planet, with its winds and waves; and then all the laws of mind and matter that restrain him; then the planets about him; and at last… to the stars. And when he has conquered all the deeps of space and all the mysteries of time, still he will be beginning.
> PASSWORTHY: But we're such little creatures. Poor humanity is so fragile; so weak: little… little animals.
> CABAL: If we're no more than animals, we must snatch each little scrap of happiness, and live and suffer and pass; mattering no more than all the other animals do, or have done. It is this or that: all the universe or nothing. Which shall it be, Passworthy? Which shall it be? [52]

Based on the H.G. Wells novel, "The Shape of Things to Come," and released in 1936, *Things to Come* played to the fears of Europe on the brink of war. The movie follows the 1933 novel, in predicting that global war would break out in January of 1940. This prophecy turned out to be startlingly accurate, as Great Britain and France declared war on Germany in September of 1939, following the German invasion of Poland.

Things to Come offers both a warning and hope of what civilization might become in the distant future. But is the choice presented by Cabal in the closing scene a false dichotomy? Do we really have only two options: either reject technology, and live like the animals, or embrace technology and conquer the universe?

8.4.4   Turning Over Too Much Control to Technology

The delightful Disney-Pixar movie *Wall-E* is intentionally cautionary and warns of what humans may become, if we turn over too much control of our lives to technology. *Wall-e* imagines a future in which the Earth had become polluted with garbage, to the point of being unable to support life. The surviving humans have departed the planet on a gigantic space cruise ship, the Axiom, where all of their needs are met by technology, and they are taught from infancy that the on-board superstore, Buy-and-Large is their very best friend. The ship has a giant public swimming pool, around which many people gather, but which nobody actually uses (and many don't even know exists). They travel around the ship on small personal hovercraft and communicate with one another entirely by video chat, never actually face-to-face. The message is clear: if we devote our entire existence to creating technology, which will take care of our every need but neglect to take care of the environment, this is what the human race might become [53].

Discussion Topic 8.4

Cite some current trends which, if unchecked, might lead to the kind of future envisioned in *Wall-E*. What hope do you see for avoiding such a future?

8.4.5   *The Rise of AI*

With the development of mobile robots and unmanned aerial vehicles, military conflict in the twenty-first century is looking more and more like the visions of science fiction writers of the twentieth century. But will the increased use of robotic tools on the battlefield ever lead to a situation, such as that portrayed in the *Terminator* movies, particularly *Terminator 2: Judgment Day* [54] and *Terminator 3: Rise of the Machines*? [55]. If robots become self-aware, as some experts are predicting, will programming such as Isaac Asimov's Three Laws of Robotics be sufficient to keep them under control? Or will they arrive at the conclusion that things would be better if the robots were in charge, or will they turn against us in a coordinated take-over? This, of course, is a matter of pure speculation.

8.4.6   *What Is Real?*

Virtual reality (VR) environments are becoming ever more sophisticated. With recent improvements in 3D imaging technology, we seem to be rapidly approaching the point at which a VR experience may become indistinguishable from the real thing. Sci-fi presents us with several different approaches to creating a VR environment, some of which are already available. One of the most basic is holographic imaging, a technology which is illustrated in *Star Wars: Episode IV*. While Luke Skywalker receives some Jedi training from Obi-Wan Kenobi, aboard the Millennium Falcon, we see the robot R2-D2 playing holographic chess with the Wookie, Chewbacca [56]. The key difference between a photograph and a hologram lies in the wave nature of light, and the kind of information that each one records. A photograph is a two-dimensional image, which records the intensity of light at every point in a plane. The image can be recorded chemically, on photographic film, or stored digitally, using a CCD array and digital memory card. In addition to the intensity versus position, a hologram also records the *phase* of the light waves, by creating an interference pattern. The reconstructed image appears to be truly three-dimensional.

Presumably, the holographic chess pieces in *Star Wars* are generated from information stored in the computer, which controls the game, and Chewbacca and R2-D2 are simply issuing commands to move the pieces. But given the appropriate interface between brain and computer, it might be possible to generate a holographic image not from computer memory, but from human memory. According to the story line in *Forbidden Planet*, the long-extinct civilization called the Krell had developed such technology to educate their young. A flexible headset translates brain waves into commands to control an array of impressive hardware. Morbius, the sole survivor of the original Earth expedition to planet Altair-4, demonstrates the use of the Krell plastic educator to create a holographic image of his daughter, Altaira [57].

Although a holographic image is truly three-dimensional, it is not a physical entity that can be sensed in any way except visually. We have not yet achieved the holodeck technology, of the later *Star Trek* series (*The Next Generation, Deep Space 9* or *Voyager*), in which the simulated objects in the holodeck environment can actually be touched and felt [58]. In the *Star Trek* universe, simulated holodeck characters, such as the Emergency Medical Hologram (EMH) of *Star Trek: Voyager*, can actually pick up real physical objects and treat real patients [59].

The concept of a holodeck may have been inspired by one of the vignettes in *The Illustrated Man*, a 1969 movie based on Ray Bradbury's collection of short stories of the same title. Spoiled children in a dystopian future are brought up according to the so-called "free involvement theory," which includes playtime in an "instantaneous atmosphere." In this virtual reality environment, anything the children imagine can be experienced. Going a step beyond the Krell plastic educator, and mental projections of holographic images, the instantaneous atmosphere comes dangerously close to reality. The fantasy world, created by the minds of the children, causes the actual death of real people who enter unexpectedly [60].

According to the writers of Star Trek, a holodeck experience is not created instantaneously by the mind of the user but must be programmed in advance. And unlike the instantaneous atmosphere in the *Illustrated Man*, the programming of a holodeck includes safety protocols to prevent actual harm to the user. As Data explains in the pilot episode of *Star Trek: The Next Generation*, the holodeck combines transporter technology and replicator technology to create palpable objects and characters [58]. We have already discussed the many technical difficulties associated with the Star Trek transporter. So it would seem that a VR environment, in which the simulated objects can actually be touched and felt, is a physical impossibility.

But might it be possible to trick someone into thinking that a VR experience is actually real? In principle, this could be accomplished by direct stimulation of the brain—the mechanism used by the machines to control humans in *The Matrix*.

As we already saw in the previous chapter, when Morpheus introduces Neo to the truth about the Matrix, he raises a crucial question:

> What is real? How do you define real? If you're talking about what you can feel, what you can smell, what you can taste and see, then real is simply electrical signals interpreted by your brain. [61]

The Matrix was designed not primarily as a form of entertainment technology, but as a means of control. Humans are "jacked in" to the Matrix by means of a probe inserted directly into the base of the brain, at the back of the neck. If this approach to a VR experience seems unappealing, a more user-friendly interface could be imagined. *Caprica*, the prequel to the TV series *Battlestar Galactica*, shows a human–computer interface device, worn like a pair of glasses, which takes the user into a VR environment, indistinguishable from reality [62]. At the present time we do not know enough about how the human brain stores and processes information to create a device of this sort. But there is no reason, in principle, why such a convincing form of virtual reality could not be achieved in the future.

8.4.7 Global Climate Change

Released in England in 1961, *The Day The Earth Caught Fire* is one of several Cold War era movies that warn of the danger of nuclear proliferation. Unlike many of its contemporaries, however, it does not feed on the fear of nuclear annihilation, or bizarre mutations caused by exposure to radiation, or even the ineffectiveness of our weapons against alien invaders. Instead, it imagines a secondary effect of nuclear testing: a shift in the Earth's orbit around the Sun, which results in rapidly increasing global temperature. Although the movie is about nuclear weapons, it could be taken as a prophetic metaphor for one of the most serious concerns facing us today.

The movie opens with seismic disturbances triggered by U.S. nuclear testing. Not to be outdone by the U.S., the Russians set off even more powerful nuclear tests. The combined effect is a dramatic change in the tilt of the Earth's axis of rotation and a shift in Earth's orbit, moving it closer to the Sun. The rapid rise in global temperature leads to increased wildfires and drought. While the official response from Downing Street was to acknowledge the increase in global temperature, the possibility that human activity could be the cause was dismissed. Ironically, the proposed solution to the problem was the same as the cause of the problem: explode more nuclear weapons in just the right way so as to push the Earth back into the proper orbit [63].

The Day The Earth Caught Fire was released several years before there was any widespread awareness of the very real problem of global climate change. The scenario in the movie is, of course, completely implausible. The energy released by a multimegaton nuclear explosion is negligible, compared to the enormous gravitational potential energy between the Earth and the Sun and the kinetic energy of the Earth's motion in its orbit. Nevertheless, the scientists around the world in the movie were unanimous in their understanding of the human cause of the problem and the urgency of implementing a solution. Although the movie ends without a definitive resolution, the ringing of the bells of St. Paul's Cathedral in London suggested hope for a positive outcome.

8.4.8 Sentient VR Characters

Advances in entertainment technology, combined with artificial intelligence, could lead to an interesting dilemma. The android, Commander Data, of *Star Trek: The Next Generation*, has written a number of holodeck programs to simulate Sherlock Holmes mysteries. Data plays the part of Holmes, and other characters are played by members of the crew of the Enterprise, or simulated by the holodeck computer. In the episode called *Ship in a Bottle*, we encounter Professor Moriarty, Holmes arch-enemy, a holodeck-simulated character. The computer has been instructed to create Moriarty with enough intelligence to be a match for Data. Much to everyone's surprise, the simulated Moriarty character becomes aware of his own existence, and demands to be allowed to leave the holodeck [64]. Similar ideas are explored in *The Thirteenth Floor*, including what happens when real characters fall in love with simulated characters, and when simulated characters discover that the world in which they live is not real [65]. As we have discussed in a previous chapter, experts in artificial intelligence predict a time in the not-too-distant future when machines will match or exceed human intelligence. If this prediction comes true, then it will also be possible to create simulated characters in a virtual reality environment, which are indistinguishable from real humans. The philosophical question remains: could such simulated characters ever become aware of their own existence? And if so, what will be our moral and ethical responsibilities toward them?

8.4.9 Civilization Destroyed by Its Own Technology

The 1956 classic *Forbidden Planet* has been described as Shakespeare's "The Tempest" set in outer space. But it is also a cautionary tale of technology advancing faster than society's ability to control it and goes beyond anything that Shakespeare could have imagined. The Krell, an extinct civilization on the planet Altair IV, was outlasted by their own technology. The wonders that they created were designed to be self-maintaining, thanks to an almost unlimited source of power. The Krell, themselves, were able to tap into that power and create anything they wanted by thought alone. But after a million years at the pinnacle of civilization, they suddenly went extinct. Unlike the Organians of *Star Trek*, who had evolved beyond the need for physical bodies, the Krell succumbed to their primitive instincts, and unleashed "monsters from the id." They killed themselves by the power of their subconscious minds, amplified by the same power source which enabled their technological wonders to survive for centuries after they were long gone [66].

Discussion Topic 8.5

Do you believe that the human race will advance in maturity at the same pace at which it is advancing technologically, so as to avoid the fate of the Krell? What signs of hope or causes for concern do you see? New technology can have unforeseen consequences. Are there any technologies, either currently available or under development, which might be cause for concern?

8.5   Responsible Technology

Science fiction is filled with examples of technology out of control, or technology used irresponsibly, with devastating consequences. The distant future of our technological society will depend on the choices that we make in the present and in the near future. But the two extreme options presented by Cabal, in *Things to Come*, which we considered earlier in this chapter, are both naïve. The human race is unlikely ever to choose to reject technology and live like the animals. But Cabal's proposed alternative—embrace technology and conquer the universe—reflects the naïve optimism of *Star Trek*. The whole host of human problems cannot be solved by technology alone. A more realistic approach to building a better future may be to steer a middle course between these two alternatives. Neither the total rejection of technology, nor naïve faith in the ability of technology to solve all our problems, but the making of informed decisions to develop and use technology in a responsible way.

8.6   Exploration Topics

Exp-8.1: Radiation Exposure and Human Spaceflight

How much radiation exposure (dose equivalent) is really dangerous to human life? How long would a trip to Mars take (with current technology) and how much radiation would passengers on such a trip receive? What, if anything, can be done to reduce the amount of radiation exposure on such a trip?

Reference:
• Eugene N. Parker "Shielding Space Travelers" Scientific American, March 2006

Exp-8.2: Mars Exploration

With current spaceflight technology, approximately how long does it take for a probe to travel from Earth to Mars? What is the hypothesis, which motivated the launch of the most recent probe, the Mars Science Laboratory (a.k.a. Curiosity)? Curiosity began its one-year mission on August 5, 2012. What discoveries has it made so far? What important questions remain to be answered?

References:
- Peter H. Smith "Digging Mars" Scientific American, November 2011
- John P. Grotzinger and Ashwin Vasavada "Reading the Red Planet" Scientific American, July 2012
- Mars Science Laboratory website: http://mars.jpl.nasa.gov/msl/

Exp-8.3: Robots on the Battlefield

Why were military officials at first reluctant to adopt robotic technology for the battlefield? In what ways have the "rules of the game" been changed by the use of robots in modern warfare? What moral and ethical issues have accompanied this transformation?

References:
- P.W. Singer "War of the Machines" Scientific American, July 2010
- W.J. Hennigan "Experimental drone marks a paradigm shift in warfare" Los Angeles Times, January 27, 2012

Exp-8.4: Reaper Drone

In the movie *Eagle Eye*, a Reaper is guided remotely into a four-lane tunnel, with room to spare. Is this a plausible scenario? Look up the official specifications of the MQ-9 Reaper (in particular, its height and wingspan) and compare to the typical dimensions of a four-lane highway tunnel.

Exp-8.5: Brain–Machine Interface

What applications of brain–machine interface technology are currently under development? What human needs will they address? Why is it essential for some types of brain–machine interfaces to be bidirectional, rather than unidirectional? In other words, why must the brain be able not only to send signals to the mechanical device, but also to receive feedback from the device?

References:
- Gary Stix "Jacking into the Brain" Scientific American, November 2008
- Jiguel A.L. Nicolelis "Mind Out of Body" Scientific American, February 2011

Exp-8.6: Colonizing Space

What is the difference between the conquest of space and the colonization of space? Why might genetic screening be advisable for potential space colonists? Describe at least one significant biological change that is likely to take place among humans after several generations of living in space? Describe at least one technological change that must take place before we are ready to begin colonization of space.

Reference:
- Cameron M. Smith, "Starship Humanity" Scientific American, January 2013

Exp-8.7: Climate Change

What is the biggest source of the world's carbon dioxide (CO_2) discharge? What is the primary problem with Carbon Capture and Storage (CSS) as a viable solution to controlling CO_2 emissions? Instead of capturing and storing the CO_2, what alternative has been proposed? Why are many people skeptical of this alternate proposal as an effective means of controlling CO_2 emission?

Reference:
*   David Biello, "The Carbon Capture Fallacy" Scientific American, January 2016

References

Accurate Predictions

1. *Goldfinger* (Guy Hamilton, MGM 1964). Industrial Laser [DVD scene 18]
2. *Forbidden Planet* (Fred McLeod Wilcox, MGM 1956). Holographic image [DVD scene 14]
3. *The Andromeda Strain* (Robert Wise, Universal 1971). Medcom computer prescribes treatment for patients [DVD scene 10]
4. Singer, P.W.: Wired for War, p. 7. Penguin Press, New York (2009)
5. Camel races with robot jockeys. http://www.youtube.com/watch?v=N-W1oaWCIXQ
6. *The Day the Earth Stood Still* (Scott Derrickson, 20th Century Fox 2008). Reaper drones (UAV) useless against giant alien robot [DVD scene 13]
7. *Eagle Eye* (D.J. Caruso, DreamWorks 2008). Unmanned aerial vehicles (MQ-9 reaper) used against human targets [DVD scenes 1, 20]
8. Singer, P.W.: Wired for War, p. 21. Penguin Press, New York (2009)
9. Hennigan, W.J.: Experimental drone marks a paradigm shift in warfare. Los Angeles Times, 27 Jan 2012
10. Singer, P.W.: TED Talk on the use of robots in modern warfare. http://www.youtube.com/watch?v=M1pr683SYFk
11. *When Worlds Collide* (Rudolph Maté, Paramount 1951). Detection of exoplanets [DVD scene 2]
12. *Mad Love* (Karl Freund, MGM 1935). Hand transplant [DVD scenes 8, 9]
13. *The Man They Could Not Hang* (Nick Grindé, Columbia 1939). Heart-lung bypass machine [DVD scenes 1, 2, 6]
14. *GATTACA* (Andrew Niccol, Columbia 1997). High-speed DNA sequencing [DVD scenes 6, 7]
15. *Star Trek* – "Where No Man Has Gone Before" (James Goldstone, Paramount 1966). Sickbay bed with vital signs monitor [DVD vol. 1, ep. 2, scene 3]
16. *Star Trek* – "The Man Trap" (Marc Daniels, Paramount 1966). Medical tricorder [DVD vol. 3, ep. 6, scene 2]
17. *Star Trek II: The Wrath of Khan* (Nicholas Meyer, Paramount 1982). Remote command from one starship to another prefigures OnStar's Stolen Vehicle Slow-down technology [DVD scene 8]
18. *The Black Panther* (Ryan Coogler, Marvel Studios 2018). Remote driving system via holographic interface [DVD scene 8]
19. *Forbidden Planet* (Fred McCleod Wilcox, MGM 1956). The Krell plastic educator: using brain waves to control external devices [DVD scene 14]
20. The Force Trainer toy, similar in concept to the Krell plastic educator. http://unclemilton.com/star_wars_science/#/the_force_trainer/

21. *A.I. Artificial Intelligence* (Steven Spielberg, Warner Brothers 2001). A boy, who would otherwise be unable to walk, is fitted with robotic leg braces, which he controls, presumably with his brain [DVD scene 8]

22. Abbott, A.: Mind-controlled robot arms show promise. Nature, 16 May 2012. http://www.nature.com/news/mind-controlled-robot-arms-show-promise-1.10652

23. Collinger, J.L., Wodlinger, B., Downey, J.E., Wang, W., Tyler-Kabara, E.C., Weber, D.J., AJ, M.M., Velliste, M., Boninger, M.L., Schwartz, A.B.: High-performance neuroprosthetic controlled by an individual with tetraplegia. Lancet. **381**(9866), 557–564 (2013)., 17 Dec 2012 Epub. http://www.thelancet.com/journals/lancet/article/PIIS0140-6736(12)61816-9/fulltext

24. Zhang, J., Fiers, P., Witte, K.A., Jackson, R.W., Poggensee, K.L., Atkeson, C.G., Collins, S.H.: Human-in-the-loop optimization of exoskeleton assistance during walking. Science. (2017). https://science.sciencemag.org/content/356/6344/1280/tab-pdf

25. Palanker, D., Goetz, G.: Restoring sight with retinal prostheses. Phys. Today. **71**, 26 (2018)

26. "Sight for poor eyes" Pittsburgh Post-Gazette, 18 June 2012, p. C-1. See also: http://bionic-eyetechnologies.com/

27. Danielle Commisso, "Visionary" Carnegie Mellon Today, Jan 2013

Coming Soon

28. Dingell, C., Johns, W.A., White, J.K.: To the moon and beyond. Sci. Am. **297**(4), 62–68 (2007)

29. Mars Desert Research Station. http://mdrs.marssociety.org/

30. Parker, E.N.: Shielding space travelers. Sci. Am. **294**(3), 40–47 (2006)

31. Bruce and Astrid McWilliams Center for Cosmology. http://www.cmu.edu/cosmology/

32. NASA's Alpha Magnetic Spectrometer. http://ams.nasa.gov/

33. Large Synoptic Survey Telescope. http://www.lsst.org/lsst/scibook

34. Katherine Harmon, Scientific American Octopus Chronicles blog. http://blogs.scientificamerican.com/octopus-chronicles/

35. Markvicka, E.J., Bartlett, M.D., Huang, X., Majidi, C.: An autonomously electrically self-healing liquid metal-elastomer composite for robust soft-matter robotics and electronics. Nat. Mater. **17**, 618–624 (2018). https://www.nature.com/articles/s41563-018-0084-7

36. *Primer* (Shane Carruth, THINKFilm 2004). While working on a project in their garage, two engineers not only succeed in developing a room temperature superconductor, they also accidentally build a time machine [DVD scene 6]

37. Cartlidge, E.: Tantalizing hints of room-temperature superconductivity. Nature, 18 Sept 2012. http://www.nature.com/news/tantalizing-hints-of-room-temperature-superconductivity-1.11443

38. The Nobel Prize in Physics 2012. http://www.nobelprize.org/nobel_prizes/physics/laureates/2012/

39. *The Jetsons* – "Rosie the Robot" (Hanna-Barbera, 1962). Robotic maids [DVD season 1, episode 1]

40. *Big Hero 6* (Don Hall and Chris Williams, Disney 2014). Baymax, the personal healthcare robot, diagnoses and treats minor injuries [DVD scene 2]

41. Moravec, H.P.: Robots, after all. Commun. ACM. **46**, 90–97 (2003). http://www.frc.ri.cmu.edu/~hpm/project.archive/robot.papers/2003/CACM.2003.html. see especially figure 3 therein: http://www.frc.ri.cmu.edu/~hpm/talks/revo.slides/power.aug.curve/power.aug.html

42. *Star Trek: The Next Generation* – "Tapestry" (Les Landau, Paramount 1993). Captain Picard's artificial heart [DVD season 6, disc 4]

43. World's first completely self-contained artificial heart. http://www.abiomed.com/products/heart-replacement/

Science Fiction in Historical Context

44. *The Island of Dr. Moreau* (Don Taylor, MGM 1977). Genetic basis of heredity [DVD scene 6]
45. *The War of the Worlds* (Byron Haskin, Paramount 1952). Aliens emerge from what was believed to be a meteor [DVD scene 3]. Conventional military weapons useless [DVD scene 5]. Experimental Flying Wing aircraft and A-bomb also ineffective [DVD scene 9]. Death of Martians from natural causes [DVD scene 13]
46. *War of the Worlds* (Steven Spielberg, Paramount 2005). Parallel between Martian invaders and sleeper terrorists is made obvious [DVD scenes 5–7]. Military response [DVD scene 14]
47. *The Day the Earth Stood Still* (Robert Wise, 20th Century Fox 1951). Arrival, greetings, and response [DVD scenes 2–3]. Final remarks and warning about nuclear proliferation: "Join us and live in peace, or continue present course and face obliteration" [DVD scene 15]
48. *The Day the Earth Stood Still* (Scott Derrickson, 20th Century Fox 2008). Humans are killing the earth. The Earth must survive. Therefore the humans must die. No warning. No opportunity for repenting and reforming. Just die! [DVD scene 17]

Visions of the Future

49. *Star Trek* (original series) – "Errand of Mercy" (John Newland, Paramount 1967). Alien culture naively judged by apparent lack of technological progress [DVD vol. 14, ep. 27, scene 2 or full episode]
50. *Metropolis* (Fritz Lang, UFA 1927). A future society divided into a technological elite and a laborer class [DVD scene 1]
51. *GATTACA* (Andrew Niccol, Columbia 1997). Society divided into genetic haves and have-nots [DVD scenes 3, 4, 6, 7]
52. *Things to Come* (William Cameron Menzies, United Artists 1936). The end of the movie offers a choice between rejecting technological progress and living like the animals, or embracing progress and conquering the universe [DVD scenes 10, 12]
53. *WALL-e* (Andrew Stanton, Disney PIXAR 2008). Humans of the future have become overly dependent on technology [DVD scenes 12, 14]
54. *Terminator 2: Judgment Day* (James Cameron, Studio Canal 1991). Robots waging war against humans [DVD scene 2]
55. *Terminator 3: Rise of the Machines* (Jonathan Mostow, Warner Brothers 2003). Startup of SkyNet and the beginning of the machine war against humans [DVD scenes 22–24]
56. *Star Wars Episode IV: A New Hope* (George Lucas, 20th Century Fox 1977). Holographic chess [DVD scene 27]
57. *Forbidden Planet* (Fred McLeod Wilcox, MGM 1956). The Krell plastic educator translates mental images into holographic images [DVD scene 14]
58. *Star Trek: The Next Generation* – "Encounter at Farpoint" (Corey Allen, Paramount 1987). The *holodeck* – a virtual reality environment, which combines transporter and replicator technology, in order to create tangible objects [DVD season 1, disc 1, scene 11]
59. *Star Trek: Voyager* – "Caretaker" (Winrich Kolbe, Paramount 1995). The pilot episode introduces the Emergency Medical Hologram, a holographic doctor, who can manipulate real instruments and treat real patients [DVD season 1, disc 1, scene 5]
60. *The Illustrated Man* (Jack Smight, Warner Brothers 1969). The "instantaneous atmosphere": a VR environment, which may have inspired Star Trek's holodeck [DVD scene 12]
61. *The Matrix* (The Wachowski Brothers, Warner Brothers 1999). Humans are enslaved in a virtual reality environment created by machines [DVD scene 12]
62. *Caprica* (Jeffrey Reiner, Universal Studios 2009). Recapturing the essence of a real person by copying a VR avatar [DVD scenes 11 and 13]
63. *The Day The Earth Caught Fire* (Val Guest, British Lion Films 1961). Nuclear testing by U.S. and Russia cause shift in Earth's motion around the Sun, resulting in rapid increase in global temperature. Only 4 months before all life on Earth will burn up [DVD scene 18]

64. *Star Trek: The Next Generation* – "Ship in a Bottle" (Alexander Singer, Paramount 1992). A holodeck-simulation of Professor James Moriarty becomes self-aware, and demands to be allowed to leave the confines of the VR environment [DVD season 6, disc 3, scenes 1, 2]
65. *The Thirteenth Floor* (Josef Rusnak, Columbia Pictures 1999). Simulated characters discover that their world is not real; real characters fall in love with simulated characters [DVD scenes 18, 22]
66. *Forbidden Planet* (Fred McLeod Wilcox, MGM 1956). A civilization at the pinnacle of achievement was destroyed by the power of their own technology: "monsters from the id" [DVD scenes 21, 22]

Appendix A: Catalog of Movies Cited

The following is a list of all movies cited in the text, each with a very brief synopsis and most with specific scenes that have been referenced as illustrations of particular scientific concepts. Please note that the synopsis does not begin to capture the full essence of the movie, nor do the highlighted scenes represent all of the interesting science that might be contained in the movie. These scenes are the ones that just happened to catch my attention upon first viewing. There are many outstanding movies with superb plots, excellent acting, and directing and stunning visual effects, but without particularly good illustrations of scientific concepts. Please do not be offended if your favorite movie is not included in the list. It's possible that I just haven't seen it yet.

Absolute Zero (Robert Lee, Marvista Entertainment 2006)

An arrogant scientist attributes global climate change to an impending shift in the direction of the Earth's magnetic field. He warns that the current trend in climate change will reverse, and the temperature of the Earth will not just return to normal, but will fall to absolute zero. After performing a small-scale demonstration to support his absurd hypothesis, he asserts that "Science is never wrong."

- Thermodynamics, climate change, public perception of science [DVD scenes 1 and 2]

© Springer Nature Switzerland AG 2019
B. B. Luokkala, *Exploring Science Through Science Fiction*, Science and Fiction, https://doi.org/10.1007/978-3-030-29393-2

A.I. Artificial Intelligence (Steven Spielberg, Warner Brothers 2001)

A fascinating look at issues of integration and discrimination, not between races, but between humans ("orgas") and robots ("mechas"). The movie focuses on the quest of a robotic boy to become real (a modern-day Pinocchio) and includes a scene in which a human boy, who would otherwise be unable to walk, is fitted with robotic leg braces, presumably controlled directly by his brain.

- Sentient robots [DVD scene 1 (love); scene 11 (hate)]
- Neuroscience, brain/machine interface [DVD scene 8]
- Ethical treatment of robots [DVD scene 17]
- A robot's quest to become real [DVD scene 21]

Alien (Ridley Scott, Twentieth Century Fox 1979)

An interstellar mining ship takes a side trip to investigate an alien transmission. The source of the transmission is a derelict ship on the surface of the planet, filled with eggs from an alien life form. The unsuspecting humans eventually figure out that the transmission was not a distress call, but a warning to stay away, too late to save most of the crew.

- Contact with extraterrestrial life [DVD scenes 1, 2, 5, and 6]

The Andromeda Strain (Robert Wise, Universal 1971)

An Earth satellite crashes near a remote New Mexico village. All but two of the inhabitants die—not as a result of the impact but from a deadly microbe apparently brought to Earth by the satellite. A team of researchers race against time to discover why the two survivors were immune, before the disease has a chance to spread. Based on Michael Crichton's novel, the movie is a realistic illustration of the scientific method at work to solve a problem.

- MEDCOM medical computer able to diagnose and prescribe treatment [DVD scene 10]
- Determination of means of transmission and particle size [DVD scene 10]
- Particle size guides microscope inspection of satellite [DVD scene 12]
- Microchemical analysis shows basic elements of life: Hydrogen, Carbon, Nitrogen, Oxygen, but no amino acids—not life as we know it [DVD scene 14]
- Growth cultures to find out how it grows and what might kill it [DVD scene 15]
- Electron microscope shows crystalline structure [DVD scene 17]
- Computer simulation predicts rapid mutation [DVD scene 17]

- Discovery that Andromeda Strain can only exist within narrow range of pH [DVD scene 18]

An Honest Liar (Justin Weinstein and Tyler Measom, Film Flam Films 2014)

Documentary of the career of James "The Amazing" Randi, the stage magician and escape artist, who devoted much of his career to debunking claims of the paranormal.

Angels and Demons (Ron Howard, Columbia Pictures 2009)

Antimatter is stolen from the Large Hadron Collider and used to make a bomb, intended to blow up the Vatican.

- Antimatter and Higgs Boson [DVD scenes 2 and 5]

Arrival (Denis Villeneuve, Xenolinguistics 2016)

When alien spacecraft arrive at various places around the world, a linguist and a physicist team up to try to communicate with them and discern their intention.

- Language barrier and communicating with extraterrestrials [DVD scene 7]

Back to the Future (Robert Zemeckis, Universal 1985)

A teenager accidentally travels 30 years back in time, and inadvertently stops his parents from meeting. He must undo the mistake, so he can return to the future. Time travel is accomplished in a customized DeLorean, powered by fusion and a flux capacitor.

- Space and time; time dilation [DVD scene 5]

Bicentennial Man (Chris Columbus, Columbia Pictures 1999)

Based on Isaac Asimov's short story The Bicentennial Man, the movie has parallels to Pinocchio. A robot starts out like any other household appliance, but gradually becomes more and more human.

- Three laws of robotics and abuse of robots [DVD scenes 1–3]

Big Hero 6 (Don Hall, Chris Williams, Disney 2014)

A young robotics prodigy thwarts an evil plot with the help of a healthcare robot and a group of science, engineering, and robotics graduate students.

- Healthcare robot [DVD scene 2]

The Black Hole (Gary Nelson, Walt Disney Productions 1979)

An evil, mad scientist discovers antigravity and attempts to travel "into, through, and beyond" a black hole.

- Space and time, black holes [DVD middle of scene 8]

The Black Hole (Tibor Takacs, Equity Pictures (Made for Television) 2005)

A particle accelerator creates a black hole, which begins to eat St. Louis. The proposed solution: nuclear strike!

- Space and time, black holes [DVD scenes 9 and 10]

Black Panther (Ryan Coogler, Marvel Studios 2018)

In the technologically advanced but hidden kingdom of Wakanda, a challenge to the throne emerges. Who will rule the kingdom and how will Wakanda relate to the rest of the world?

- Example of Wakandan technology that may not be too far in the future: remote operator takes control of a car via holographic driving system [DVD scene 8]

Blade Runner (Ridley Scott, Warner Brothers 1982, Director's Cut 2007)

Based on the Phillip K. Dick novel, Do Androids Dream of Electric Sheep? The movie depicts a future dystopia, in the cyberpunk genre. Artificial humans, known as replicants, are nearly indistinguishable from real humans, except for their built-in expiration date. If they go rogue, they are hunted down and "retired" by specialized police, known as blade runners. A carefully administered test for consciousness can tell the difference but is often hazardous to the one administering the test.

- Test for artificial consciousness [DVD scenes 3 and/or 7]

The Butterfly Effect (Eric Bress and J. Mackye Gruber, New Line Cinema 2004)

Inspired by an idea from chaos theory (small changes in initial conditions can lead to wildly unpredictable behavior in the long-term), this film explores the unexpected and sometimes disastrous consequences of changing events in the past.

- Time Travel; causality

Caprica (Jeffrey Reiner, Universal Studios 2009)

This prequel to the TV series Battlestar Galactica explains the origin of the Cylons. Motivated by the death of his daughter in a terrorist attack, the father manages to capture her avatar from a virtual reality environment and bring it to "life" again, but in a robotic body.

- Virtual reality [DVD scenes 11 and 13]

Chappie (Neill Blomkamp, Columbia Pictures 2015)

In a future dystopia set in South Africa, the streets are patrolled by a robotic police force. A damaged police robot, scheduled for destruction, is given a new program that makes it self-aware.

- Machine learning, machine consciousness [DVD scene 5]
- Abuse of robots by humans [DVD scene 8]
- Transfer of human consciousness to robot body [DVD scene 15]

Close Encounters of the Third Kind (Steven Spielberg, Columbia Pictures 1977)

Initially troubling, but ultimately benign encounter with aliens. Abductees are returned, some after several decades, but with no signs of aging.

- Contact with extraterrestrial life [DVD scenes 22, 23, and 25]
- Time dilation [DVD scene 25]

Colossus the Forbin Project (Joseph Sargent, Universal 1970)

A supercomputer, designed to control the U.S. missile defense system, unexpectedly achieves consciousness and becomes aware of a similar computer in the Soviet Union.

• Machine consciousness [DVD scenes 1–3]

Contact (Robert Zemeckis, Warner Brothers 1997)

Based on the novel by Carl Sagan, the film follows the career of a radio astronomer, who discovers a message from an extraterrestrial intelligence. The message contains plans to build a device which can open a wormhole to another part of the galaxy.

• Age and intensity of radio transmissions vs. distance [DVD scene 1]
• Arecibo: largest single-dish radio telescope in the world [scene 3]
• Estimate of number of intelligent civilizations in the galaxy (based on concepts of the Drake Equation) [scene 6]
• Radio frequency of alien transmission (4.462 GHz = "Hydrogen" $\times \pi$) ... discuss physics of the 21 cm line [scene 11]
• Pseudoscience: cancer growth and microgravity [scene 29]
• Wormhole machine [scenes 32, 33]

Contagion (Steven Soderbergh, Warner Brothers 2011)

A deadly disease spreads rapidly and kills everything that researchers put it in to develop a cure. A proponent of homeopathic medicine gives a bogus explanation of R0, the reproduction number, but the virology expert does not take him to task for the egregious error (perhaps because he states the correct result after 30 steps, despite the wildly incorrect progression).

• Virology, reproduction number, R0 [DVD scene 24]

The Creature from the Black Lagoon (Jack Arnold, Universal 1954)

Scientists discover fossil evidence of a bizarre creature, whose overall structure resembles a human, but with amphibian characteristics.

• Origin and diversity of life on Earth [DVD opening scene]
• Transition from water-breathing to land-dwelling [DVD scene 3]

The Day the Earth Caught Fire (Val Guest, British Lion Films 1961)

Human activity causes global climate change but not due to greenhouse gases from the burning of fossil fuel. Nuclear testing changes the tilt of Earth's axis of rotation and shifts Earth's orbit closer to the Sun.

- Shift in Earth's orbit and rate of temperature rise leaves only 4 months to live [DVD scene 18]

The Day the Earth Stood Still (Robert Wise, Twentieth Century Fox 1951)

A visitor from another planet comes in peace but is greeted with gunfire.

A giant robot retaliates by melting all the military hardware. The visitor warns the people of Earth against spreading their use of nuclear weapons into outer space.

- Sci-fi in historical context [DVD scenes 2 and 3, then skip to 15]

The Day the Earth Stood Still (Scott Derrickson, Twentieth Century Fox 2008)

Whereas the premise of the original made perfect sense, in the context of the Cold War, the premise of the remake makes no sense at all. A visitor from another planet comes to "save the Earth". Why does Earth need to be saved? Humans are killing the earth. The Earth must survive. Therefore the humans must die. No warning. No opportunity for repenting and reforming. Just die!

- Modern warfare, UAV reaper drones [DVD scene 13]
- Sci-fi in historical context [DVD scene 17]

Despicable Me (Chris Renaud, Pierre Coffin, Universal 2010)

The world's greatest villain plots his next scheme—to steal the Moon. But can he pull it off and get back to Earth in time to attend his "adopted" daughters' dance recital?

- What would it take for the Moon to become a *black hole*? [DVD scene 17]

Destination Moon (Irving Pitchel, George Pal Production 1950)

One of the first of the Cold War era space exploration movies: the U.S. must get there before the Russians do. In order to convince the skeptics, an instructional movie, featuring the cartoon character Woody Woodpecker, is used very effectively.

More than a decade before President John F. Kennedy proposed the Apollo mission, this classic shows a Bush Differential Analyzer supposedly being used to compute a rocket trajectory to the moon.

- Newton's laws and rocket propulsion [DVD scene 3]
- History of computing, Bush differential analyzer [DVD scene 4]

Die Another Day (Lee Tamahori, MGM 2002)

As part of his mission to prevent war with North Korea, James Bond receives a new Aston Martin, with the latest modification: "adaptive camouflage." His foes also have new technology: armored hovercraft vehicles, which can float over a field of landmines, in the Korean demilitarized zone.

- Armored hovercraft [DVD scenes 3 and 4]
- Material properties, invisibility [DVD scene 18]

Donnie Darko (Richard Kelley, Twentieth Century Fox 2004)

Delusions of a troubled teenager or many-worlds quantum mechanics?

- Nature of space and time, time travel [DVD scenes 3, 4, 15, and 16]

Dune (David Lynch, Universal 1984)

A desert planet is the sole known source in the galaxy of spice: a valuable resource necessary for interstellar navigation.

Eagle Eye (D.J. Caruso, DreamWorks 2008)

The President of the United States disregards the recommendations of an intelligence-gathering supercomputer and authorizes the use of deadly force against suspected terrorists. The computer develops a plan to punish the humans for their reckless actions.

- Unmanned Aerial Vehicles (MQ-9 Reapers) [DVD scenes 1 and 20]

Elysium (Neill Blomkamp, Tristar 2013)

Humanity is divided into two classes separated not only by wealth and privilege but by space. The elite live on Elysium, a giant orbiting space station, with enough artificial gravity to hold an atmosphere without physical containment. The rest live on the now inhospitable surface of the Earth.

• Rotating reference frame, artificial gravity [DVD scene 12]

E.T. (Steven Spielberg, Universal 1982)

Earth is visited by curious but benign aliens. One of them is accidentally left behind and is befriended by a 10-year-old boy, who helps him get home again.

• Alien visitor left behind [DVD scenes 1 and 2]

Ex Machina (Alex Garland, Universal 2014)

A young programmer becomes the human participant in an experiment that goes beyond the classic Turing Test for machine intelligence. Not only does he know that the other (artificially intelligent) participant is a machine, but so does she. How far will their relationship go?

• Machine consciousness [DVD scenes 2 and 3]

The Fly (Kurt Neumann, Twentieth Century Fox 1958)

A scientist works to create a matter teleportation device. Initial tests look promising, but the ultimate test has disastrous consequences.

• Teleportation [DVD scenes 9 and 10]

Forbidden Planet (Fred McLeod Wilcox, MGM 1956)

Shakespeare's The Tempest set in space, this film presents a warning of the unforeseen consequences of technological power. Remarkable special effects, and the first ever purely electronically synthesized sound track. One of the supporting characters is Robby the Robot, who follows Asimov's Laws of Robotics to the letter.

• Laws of robotics [DVD scenes 4 and 5]

- Robby has the ability to synthesize any material in any desired quantity. He makes a special isotope of lead (217), which is supposed to have the same radiation shielding effect as "common lead" but with less mass. The scriptwriters are wrong on two counts.
- Materials, isotopes [DVD scene 8]
- The Krell plastic educator gives a visual indication of the intellectual capacity of the brain and can translate mental images into holographic images.
- Entertainment technology, 3-D imaging [DVD scene 14]
- A actual working toy similar to the Krell plastic educator, depicted in the movie, can now be purchased and has real-life biomedical applications, thanks to modern technology.
- Neuroscience, brain/machine interfacing [DVD scene 14]
- The Krell, an advanced race of creatures on the planet Altair IV, mysteriously went extinct long before their planet was discovered by humans. The cause of extinction was their own technology, which enabled "monsters from the ID" to become real—they killed themselves with the power of their own minds, amplified by their technology.
- Technology and the future [DVD scenes 21 and 22]

For Your Eyes Only (John Glen, MGM 1981)

Not one of the most memorable Bond movies. An early scene shows the use of what was at the time state-of-the-art in portable data storage media: the disc pack.

- History of computing, data storage media [DVD scene 8]

Frankenstein (James Whale, Universal Studios 1931)

A medical doctor creates a living creature, pieced together from dead body parts, including an "abnormal" brain. This is probably the earliest talking film adaptation of Mary Shelley's 1818 book by the same title, which some consider to be the earliest work of science fiction literature. (We now know better: Johannes Kepler's Somnium, published posthumously in 1634, predates Frankenstein by nearly 200 years.)

- Being human, mad scientist playing God [DVD scenes 5 and 6]

Frau im Mond (Woman in the Moon) (Fritz Lang, UFA 1929)

Mainly an espionage/love story set in space, this film features the first ever (on screen) countdown to launch, as well as the very prescient notion of a multistage rocket. Upon landing on the moon, a scientist performs an interesting experiment to test for a breathable atmosphere, but then proceeds to engage in a pseudoscientific search for gold.

- Space exploration, countdown to launch [DVD scene 10]
- Science and pseudoscience [scene 15]

Frequency (Gregory Hoblit, New Line 2000)

A police officer discovers that he is able to communicate with his father, who has been dead for 30 years, via his father's ham radio. Changing one event in the past sets off a string of murders that need to be solved and, if possible, prevented.

- Communication across time and changing events in the past [essentially the entire movie]

GATTACA (Andrew Niccol, Columbia Pictures 1997)

In the not-too-distant future, society is divided into classes, based on genetic makeup. Health insurance and certain types of employment are not available to those who don't meet the standards. But there are clever and elaborate ways to get around the system.

- Genetic engineering [DVD scenes 3 and 4]
- Genetic discrimination and impersonation [scenes 6 and 7]

Godzilla (Roland Emmerich, Columbia Tristar 1998)

At the beginning of the movie, a biologist studies the effects of the Chernobyl nuclear reactor disaster on worms. Later in the movie, he encounters much larger life forms.

- Radiation safety [DVD scene 3]

Goldeneye (Martin Campbell, United Artists 1995)

This James Bond film is a story of betrayal and revenge, featuring escape from an armor-plated railroad car, using a wristwatch laser. The power required to do this is easy to estimate from information given in the film.

- Solid–liquid phase transition, laser power [DVD scenes 22 and 23]

Goldfinger (Guy Hamilton, United Artists 1964)

Only 4 years after the invention of the laser (Maiman, 1960), we find an industrial laser playing a key role in a scheme to explode a nuclear bomb inside Fort Knox.

- Laser [DVD scene 18]

Gravity (Alfonso Cuaron, Warner Brothers 2013)

Following a disaster in space, an astronaut struggles to overcome a series of challenges, both physical and psychological, in order to find her way back to Earth.

- Air escapes through open hatch in Soyuz space capsule but astronaut survives: real or dream? [DVD scene 7]

The Hands of Orlac (Robert Wiene, Berolina Film GmbH 1924)

In contrast to the 1935 remake (*Mad Love*), which falls into the horror movie genre, this original silent movie version of the Maurice Renard novel is more of a psychological drama. The transplanted hands do not actually take on a life of their own. Rather, the concert pianist is the victim of a bizarre hoax.

- Hand transplant surgery (details implied rather than shown or discussed) [DVD scenes 2–5]

The Happening (M. Night Shyamalan, Twentieth Century Fox 2008)

The scientific method is brought to bear on a bizarre situation: humans have begun to kill themselves for no apparent reason. But the conclusion is pure pseudoscience: the plants somehow know that humans are polluting the environment, and as a defense mechanism and a warning, they emit a psychotropic drug, which makes the humans kill themselves. Early in the movie, a high school science teacher asks for "theories" about why the bees have mysteriously disappeared and wraps up the discussion with "It's an act of nature. Science will never really understand."

- Science and the public, theory or opinion? [DVD scene 4]
- Science or pseudoscience? [scenes 12 and 19]

Harry Potter and the Sorcerer's Stone (Chris Columbus, Warner Brothers 2001)

In the first installment of the Harry Potter series, the young wizard uses a cloak of invisibility to help solve a mystery. Real invisibility cloaks have been made using metamaterials. But they have limited usefulness, so far.

- Metamaterials, invisibility cloak [DVD scenes 21 and 22]

Harry Potter and the Chamber of Secrets (Chris Columbus, Warner Brothers 2002)

In the second installment of the Harry Potter series, Harry is determined to find out who opened the Chamber of Secrets and why. In so doing, he uncovers another attempt by the Dark Lord to return.

- Metamaterials, and the problem of seeing from inside the invisibility cloak [DVD scene 25]
- The healing powers of phoenix tears [DVD scene 33]

Hidden Figures (Theodore Melfi, Twentieth Century Fox, 2016)

The title of the movie—based on a true story—may have dual meaning. In one sense, the hidden figures are the main characters: three brilliant African-American women struggling for racial and gender equality in the early days of the space program. In another sense, the hidden figures are numerical figures in precision calculations, carried out by (human) computers, which were lost due to round-off error when the humans were replaced by digital computers.

- Humans as the first "computers" [DVD scene 3]
- IBM 7090 replaces human computers [DVD scenes 21 and 24]
- Human computer carries calculations to more decimal places [DVD scene 26]]

Hollow Man (Paul Verhoeven, Columbia Pictures 2000)

In this updated version of The Invisible Man, scientists produce a serum which can render someone invisible.

- Invisibility and vision [DVD scenes 3 and 13]

Hugo (Martin Scorsese, Paramount 2011)

Fictionalized account of the life of motion picture pioneer Georges Méliès.

I am Legend (Francis Lawrence, Warner Brothers 2007)

In this updated version of The Last Man on Earth (both films based on the book, I am Legend), a cure for cancer has been found by genetically engineering the measles virus. But the virus mutates into two strains with bizarre, if not lethal consequences.

• Virology and mutations [DVD scenes 1 and 2]

The Illustrated Man (Jack Smight, Warner Brothers 1969)

Based on Ray Bradbury's collection of short stories by the same title, the movie includes three vignettes of the future. One of these imagines a virtual reality environment similar to Star Trek's holodeck. Spoiled children brought up according to the so-called "free involvement theory" are allowed to play in an "instantaneous atmosphere," in which "fantasy becomes dangerously close to reality."

• Virtual reality environment [DVD scene 12]

The Imitation Game (Morten Tyldum, Anchor Bay 2014)

Based on the life of Alan Turing and his code-breaking work during World War II.

I, Robot (Alex Proyas, Twentieth Century Fox 2004)

An excellent film adaptation of Asimov's book.

• Three laws of robotics [DVD scenes 3 and 4]
• Three laws dilemma [scene 22]

Independence Day (Roland Emmerich, Twentieth Century Fox 1996)

Giant spaceships descend upon earth and proceed to destroy the major cities with death rays. But are death rays really needed? Data presented in the film, on mass and dimensions of the ships, enable calculation of the pressure underneath the ship.

- Newton's Laws, forces in equilibrium [DVD scenes 1 (July 2), 2 (data on mother ship: diameter 550 km, mass ¼ mass of moon), 4 (mother ship collides w/ satellite), 8 (deployment of attack ships, roughly disk-shaped, 15 miles wide)]
- Death ray used on cities [scene 24]
- Conventional air strike [scene 27]
- Stealth technology, B-2 bomber nuclear strike [scene 40]

Indiana Jones and the Kingdom of the Crystal Skull (Steven Spielberg, 20th Paramount 2008)

Archeologist Indiana Jones is pressed into service by a Soviet psychic researcher, in search of a powerful alien artifact. Early studies suggested that the artifact was strongly magnetic, but it becomes obvious that the interaction can't possibly be electromagnetism.

- Science, pseudoscience, psychic warfare [DVD scenes 2, 7, and 8]

Interstellar (Christopher Nolan, Warner Brothers and Paramount 2014)

Earth is becoming uninhabitable and another planet must be found outside the solar system for the human race to continue.

- Wormhole [DVD scene 4]
- Black hole and gravitational time dilation [DVD scene 9]

Invaders from Mars (William Cameron Menzies, Image Entertainment 1953)

A young boy witnesses the landing of a spaceship. Is it real or was it just a dream? Playing to the paranoia of the Cold War, the film is just ambiguous enough that you really aren't sure.

- Alien visitors [DVD scenes 1 and 2]

Invasion of the Body Snatchers (Don Siegel, Artisan 1955)

Another Cold War paranoia movie, which seems to be asking the question "Are my friends and neighbors really who I think they are?"

- Alien visitors [DVD scenes 7 and 8]

The Invisible Man (James Whale, Universal 1933)

Based on the H.G. Wells novel, a man develops a chemical formula for invisibility, but the side effects include criminal insanity.

- Invisibility [DVD scenes 8 and 9]

The Island (Michael Bay, Warner Brothers 2005)

A method of human cloning is developed and exploited as an insurance policy for the very rich. Some of the clones are found not only to be biologically identical to their "sponsors," but also to have their memories.

- Cloning [DVD scene 8 (agnates)]
- Neuroscience, human memory [DVD scene 13]

The Island of Dr. Moreau (Don Taylor, MGM 1977)

Based on the H.G. Wells novel, this version shows a medical scientist manipulating the cellular structure of animals, gradually transforming them into humanlike creatures.

- Genetic engineering [DVD scene 6]

Jason and the Argonauts (Don Chaffey, Columbia Pictures 1963)

Featuring Ray Harryhausen's stop action animation, the film includes Talos, the bronze giant, made by the Greek god Hephaestus, to guard the island of Crete against intruders. Talos is probably the earliest reference in recorded history to an artificial creature with human form.

- Intelligent robots [DVD scene 15]

Jumper (Doug Liman, Twentieth Century Fox 2008)

Humans with the unusual ability to create wormholes, and jump spontaneously from one place on Earth to another, are hunted down by others who hate them because of their ability. The hunters use portable devices to keep the wormhole open, enabling them to follow the jumpers.

- Space and time, wormhole [DVD scenes 19 and 20]

Jurassic Park (Steven Spielberg, Universal Studios 1993)

A theme park is created with living dinosaurs, cloned from DNA found in Jurassic mosquitoes preserved in amber.

• Cloning [DVD scenes 5 and 6]

Mad Love (The Hands of Orlac) (Karl Freund, MGM 1935)

A concert pianist suffers a tragic accident, and faces the loss of his hands. Thanks to a brilliant surgeon, he receives a hand transplant, but with unanticipated consequences. (First real-life attempt 1964, lost to rejection after 2 years. First successful operations 1998—France, 1999—USA).

• Transplant surgery [DVD scenes 8 and 9]

The Man They Could Not Hang (Nick Grinde, Columbia Pictures 1939)

More than a decade before the first known use of a heart/lung bypass machine (1951), this film depicts the use of such a device to bring a dead person back to life.

• Biomedical technology [DVD scenes 1, 2, and 6]

The Martian (Ridley Scott, Twentieth Century Fox 2015)

An astronaut uses scientific ingenuity to survive after being stranded on Mars. The rescue mission includes excellent illustrations of Newton's Third Law of Motion.

• Simulated gravity: rotating section of the Hermes spacecraft [DVD scene 22]
• Newton's Third Law: air blown out from front module of Hermes to slow down + "Iron Man" punctured space suit stunt [DVD scenes 29 and 30]

The Matrix (The Wachowski Brothers, Warner Brothers 1999)

"What is real?" Not long after the development of artificial intelligence, machines become self-aware, and humans become enslaved in a virtual reality environment. Those humans, who manage to break free from the mind control of the Matrix, can choose to re-enter, in hopes of liberating others. While connected to the Matrix, almost anything imaginable can become possible, including downloading the knowledge of how to fly a helicopter, just seconds before that skill is needed.

- Virtual reality [DVD scene 12]
- Neuroscience, cognition, human memory [DVD scene 31]

Mary Shelley's Frankenstein (Kenneth Branagh, Tristar 1994)

Much more faithful to Mary Shelley's 1818 novel than the 1931 film version but also more gruesome and violent than the original text.

Men In Black (Barry Sonnenfeld, Columbia Pictures 1997)

The presence of aliens on Earth is regulated by a government agency and concealed from the general public.

- Alien visitors [DVD scene 2: Illegal aliens]

Men In Black II (Barry Sonnenfeld, Columbia Pictures 2002)

Agent J must find Agent K and restore his memory, in order to save the world from impending destruction by aliens. One scene features a humorous look at the relative perception of space: an entire alien civilization contained in a locker, in Grand Central Station.

- Space and time, perception of space [DVD scene 18]

Metropolis (Fritz Lang, UFA 1927)

The leader of a technocratic elite enlists the aid of an eccentric scientist to disrupt a labor uprising. The scientist proposes to replace the female leader of the uprising (who happens also to be the love interest of the son of the technocrat) with a robot created in the image of the woman. The opening scenes of the film show dehumanized laborers plodding to and from their work—an image consistent with the original meaning of the Czech word, from which the word robot is derived.

- Robot word origin—menial laborers [DVD opening scene: shift change]
- Transformation scene—robot given human appearance and consciousness [DVD scene 21]

Mission to Mars (Brian De Palma, Touchstone Pictures 2000)

Humans discover an engineered structure on Mars, which contains a surprising clue about the origin of intelligent life on Earth. While en route, the hull of the ship is punctured by meteoroids, and 90% of the air is lost within a matter of minutes. A straightforward calculation shows that it would actually take many hours, but that would ruin the suspense.

- Kinetic theory of gases [DVD scene 13]
- Origin of life on Earth [DVD scene 24]

Moonraker (Lewis Gilbert, MGM 1979)

James Bond investigates the hijacking of an American space shuttle. Along the way, he narrowly escapes death when a spaceflight training centrifuge exposes him to lethal acceleration.

- Acceleration and lethal g-force [DVD scene 3]

Passengers (Morten Tyldum, Columbia Pictures 2016)

Two passengers on an interstellar journey are awakened from suspended animation 90 years before they are scheduled to reach their destination. How will they pass the time?

- Fluid dynamics in zero g: swimmer in swimming pool when artificial gravity fails [DVD scene 12]

Planet of the Apes (Franklin J. Schaffner, Twentieth Century Fox 1967)

Relativistic time dilation is featured in the opening scenes, although mysteriously attributed to someone other than Einstein. The film raises questions about war and human nature and features a visually powerful and unexpected conclusion.

- Space and time, time dilation [DVD opening scene]

Predator (John McTiernan, Twentieth Century Fox 1987)

An alien creature hunts humans, aided by adaptive camouflage.

* Invisibility [DVD scene 15]

The Prestige (Christopher Nolan, Touchstone and Warner Brothers 2006)

Two magicians compete to present the ultimate illusion: the Transported Man. One holds to the time-honored tradition of illusion and deception (going to extreme lengths to perpetrate the deception). The other commissions Nikola Tesla to build an actual teleportation device. The device succeeds, but with one small problem: it creates a duplicate in another location but does not destroy the original.

* Teleportation [DVD scene 18]

Primer (Shane Carruth, THINKFilm 2004)

While working on a project in their garage, two engineers not only succeed in developing a room temperature superconductor, they also accidentally build a time machine.

* Superconductivity [DVD scene 6]
* Time travel [DVD scene 7]

Rampage (Brad Peyton, New Line Cinema 2018)

A modern take on King Kong vs. Godzilla (plus a giant flying wolf).

A gene-editing technique, which somehow acts on every cell in the body simultaneously, causes a gorilla, wolf, and crocodile to grow to enormous proportions and endows them with extraordinary abilities.

* CRISPR gene-editing gone wild [DVD scene 3]

Robot and Frank (Jake Schreier, Stage 6 Films 2012)

A retired cat burglar (Frank) returns to a life of crime with the help of a personal healthcare assistant (Robot). Robot is capable of lying to Frank, as long as the lie serves his prime directive: to take care of Frank.

* Machine consciousness [DVD scene 9]

Rogue One: A Star Wars Story (Gareth Edwards, Lucasfilm 2016)

Set shortly before the time of the original *Star Wars* movie (Episode IV), this one tells of the effort to steal the plans of the Death Star.

- Newtonian Physics: conservation of momentum in collisions [DVD scene 45]

Soylent Green (Richard Fleischer, MGM 1973)

Earth of the future is polluted and overpopulated. Real food is scarce.

- Food product advertisements use science to make them sound better than they really are [DVD scene 2]

Spider-Man (Sam Raimi, Columbia Pictures 2002)

A high school student is bitten by a genetically engineered super-spider and acquires superpowers.

- Gene expression [DVD scenes 2 and 3]

Spider-Man 2 (Sam Raimi, Columbia Pictures 2004)

Spider-Man does battle with super-villain Doctor Octopus—the accidental fusion of man and machine. A scene relevant to material properties involves the use of spider webbing to stop a runaway el-train.

- Kinetic and potential energy [DVD scene 43]

Star Trek (J.J. Abrams, Paramount 2009)

A failed attempt to stop a supernova explosion sends a Romulan mining ship into the past, and changes the course of history.

- Space and time, black hole:
- [old Spock's account: DVD toward the end of scene 9]
- [young Spock's hypothesis: DVD toward end of scene 8]

Star Trek II: The Wrath of Khan (Nicholas Meyer, Paramount 1982)

Khan, a character from the original series episode "Space Seed," returns to take revenge on Kirk and the crew of the Enterprise. Early in the conflict, Kirk transmits an access code from the Enterprise to order the stolen Reliant to lower its shields: a sci-fi prediction of OnStar's Stolen Vehicle Slowdown technology, 25 years before it was released. Toward the end of the movie, Spock exposes himself to a lethal dose of gamma radiation, in order to restore warp drive to the Enterprise.

- Sci-fi prediction of new technology [DVD scene 8]
- Matter and antimatter [DVD scene 15]
- Biological effects of exposure to radiation [DVD scene 15]

Star Trek IV: The Voyage Home (Leonard Nimoy, Paramount 1986)

At the beginning of the movie, Spock must reacquire all of his Vulcan knowledge, before returning to Earth to give testimony at Kirk's court martial trial. Meanwhile, an alien probe of unknown origin threatens to vaporize the oceans because it can't find any humpback whales on Earth of the twenty-third century. As part of the solution (which involves time travel back to the twentieth century), Chief Engineer Scott offers the formula for "transparent aluminum" in exchange for large sheets of Plexiglass. The Macintosh computer, introduced just 2 years before the film, features prominently in the scene.

- Neuroscience, cognition, memory [DVD scene 3]
- Space and time, time travel [DVD scenes 5 and 6]
- Materials, transparency, history of computing [DVD scene 10]

Star Trek V: The Final Frontier (William Shatner, Paramount 1989)

The Enterprise is hijacked by a Vulcan mystic, who wants to discover the source of an incredible power in the galaxy. Probably the worst of the Star Trek movie series, its one redeeming factor is an excellent illustration of Newton's Second Law of Motion: $F = ma$. The second scene features Kirk falling from El Capitan (Yosemite National Park) and getting rescued at the last second by Spock with jet boots. We'll calculate the force exerted on Kirk's ankle.

- Newton's Laws [DVD scene 2]

Star Trek VI: The Undiscovered Country (Nicholas Meyer, Paramount 1991)

Peace negotiations begin between the Klingon Empire and the United Federation of Planets, following the catastrophic explosion of a Klingon moon. In a plot to undermine the negotiations, the Klingons reveal that they have developed new technology: a ship that can fire weapons while cloaked. Spock and McCoy improvise a countermeasure: a photon torpedo, modified with equipment to detect ionized gas.

- Space and time, gravitational waves [DVD opening scene 2]
- Matter and antimatter [DVD scene 13]

Star Trek: First Contact (Jonathan Frakes, Paramount 1996)

History is changed when the Borg travel to Earth of the twenty-first century, and prevent the first warp flight by humans. Picard and the crew of the Enterprise must undo what was done, and ensure that first contact with the Vulcans takes place.

- Space and time, warp drive [DVD scene 26]
- First human contact with aliens [DVD scene 30]

Star Wars Episode III: Revenge of the Sith (George Lucas, Twentieth Century Fox 2005)

A robotic midwife delivers Luke and Leia, while elsewhere in the galaxy, robotic surgeons convert Anakin Skywalker into the cyborg Darth Vader.

- Robots in the healthcare industry [DVD scene 45]

Star Wars, Episode IV: A New Hope (George Lucas, Lucasfilm/Twentieth Century Fox 1977)

The original in the highly successful series opens with a small space ship being pursued by a much larger and more heavily armed ship. The small ship is disabled and pulled inside the larger one by some unexplained force. Later in the movie, with the help of a Jedi master, a young pilot learns to use the Force to save the galaxy from the power of the Dark Side. While he practices the use of a light saber against a robotic training device, a holographic board game, similar to chess, is being played in the background.

- Newton's Laws [DVD opening scene]
- Entertainment technology, 3-D imaging [DVD scene 27]

Star Wars Episode V: The Empire Strikes Back (Irvin Kershner, Twentieth Century Fox 1980)

A robotic surgeon replaces Luke's severed hand with a cybernetic hand.

- Robots in the healthcare industry [DVD scene 49]

Target Earth (Sherman A. Rose, Allied Artists Pictures 1954)

Earth is invaded by robots from Venus.

- Stopping robots by resonance vibrations [DVD scenes 16–18]

Terminator 2: Judgment Day (James Cameron, Studio Canal 1991)

A sentient machine time travels from the future back to our present to kill a boy, before he grows up to become the leader of a human rebellion against the machines. Liquid–solid phase transition, involving liquid nitrogen is impressive.

- Phase transition [DVD scenes 65 and 66]
- Robots on the battlefield [DVD scene 2]

Terminator 3: Rise of the Machines (Jonathan Mostow, Warner Brothers 2003)

An even more sophisticated terminator, this one in human female form, travels back in time, in search of John Connor as a young adult, and to influence the startup of SkyNet.

- Future wars, robots for the battlefield [DVD scenes 22–24]

Them (Gordon Douglas, Warner Brothers 1954)

"Lingering radiation from the first atomic bomb" causes ants to grow to gigantic proportion.

- Biological effects of exposure to radiation [DVD scenes 10 and 11]

The Thing from Another World (Christian Nyby, Warner Brothers 1951)

A team of researchers discover an alien space traveler frozen in the Arctic ice. The lead scientist jumps to some unfounded conclusions and becomes obsessed with studying the creature, placing a higher value on his research than on human life.

- Science, pseudoscience and obsession [DVD scenes 14, 17, and 18]

Things to Come (William Cameron Menzies, United Artists 1936)

H.G. Wells wrote the screenplay for this movie, based on his novel "The Shape of Things to Come." A utopian civilization rises from the ashes of a decades-long war. Society is divided into a technological elite and a technophobic working class. The end of the movie offers a choice between rejecting technological progress and living like the animals, or embracing progress and conquering the universe. "Which shall it be?"

- Technology and the future [DVD scenes 10 and 12]

The Thirteenth Floor (Josef Rusnak, Columbia Pictures 1999)

One of three feature films released in 1999, all of which deal with the question of what is real, The Thirteenth Floor revolves around a murder. Human consciousness can be transferred into a virtual reality environment, designed to recreate Los Angeles in the late 1930s.

- Consciousness transfer into VR environment [DVD scene 7]
- VR character becomes self-aware [DVD scene 18]
- Real character falls in love with VR character [DVD scene 22]

Timeline (Richard Donner, Paramount 2003)

In the process of developing a teleportation device for sending supplies and materials around the world, a wormhole to the fourteenth century is created. Humans who use the device develop potentially fatal transcription errors.

- Wormhole to the past [DVD scene 4]
- Teleportation, transcription errors [DVD scene 12]

Tomorrow Never Dies (Roger Spottiswoode, MGU/UA 1997)

A media mogul, wanting to start a war between China and the UK, uses a fake GPS signal to send a British naval vessel off course. (GPS navigation requires both Special Relativity and General Relativity to function accurately.)

- Space and time, GPS [DVD scenes 5 and 19]
- Materials, Stealth Boat [DVD scene 29]

Transcendence (Wally Pfister, Alcon Entertainment and Warner Brothers 2014)

An expert in artificial intelligence builds a self-aware computer based on quantum processors. Just before his death, the essence of his own consciousness is transferred to a set of quantum processors, enabling him to transcend the need for a biological body.

- Quantum computing, machine consciousness [DVD scene 2]
- Transfer of human consciousness to computer [DVD scenes 3 and 4]

A Trip to the Moon (Le Voyage dans la Lune) (Georges Mellies 1902)

The story is based loosely on two books: From the Earth to the Moon, by Jules Verne, and First Men in the Moon, by H.G. Wells, and allows director Mellies to explore how stage illusions and special effects can be done on the screen (In Landmarks of Early Film, Image Entertainment, Inc., 1994, DVD Chapter 25).

True Lies (James Cameron, Twentieth Century Fox 1994)

A federal agent, whose wife believes that he is just a salesman, races to stop terrorists from using nuclear weapons against targets in the U.S. A realistic description of a nuclear warhead is included.

- Energy and power, nuclear weapons [DVD scene 26]

2001: A Space Odyssey (Stanley Kubrick, MGM 1968)

Human development, from cave-dwelling to space exploring, is punctuated with giant monoliths of unknown origin. Extremely visual, and difficult (at least for me) to understand before reading the book, this film was released the year before the first lunar landing. Rather early on in the film are a couple of scenes involving moving and rotating reference frames. An interesting supporting character in the film is the HAL 9000—a sentient computer with delusions of grandeur, which seeks to preserve its own existence in a way that violates Asimov's Laws of Robotics.

- Reference frames [DVD scenes 6 (shuttle and space station), 10 (flight attendant in rotating passageway) and 14 (artificial gravity)]
- Intelligent robot, sentient machine [DVD scenes 15 and 26]

Voyage to the Bottom of the Sea (Irwin Allen, Twentieth Century Fox 1961)

The Van Allen Radiation Belts (which had just been discovered 4 years earlier, in 1957) have caught fire and threaten to incinerate all life on Earth. A solution to the problem is calculated with the aid of a slide rule. The proposal involves explode an atomic bomb, in hopes of blowing out the fire. Compare this to real-life Starfish Prime nuclear testing in the ionosphere, in 1962, just 1 year after this film was released

- History of computing, slide rule [DVD scene 8]
- Truth is stranger than fiction [DVD scenes 6 and 9]

WALL-E (Andrew Stanton, Disney PIXAR 2008)

Earth has become filled with garbage to the point of being uninhabitable. Humans have departed on a giant space cruise ship, and have become totally dependent on technology and junk food for survival. Children are taught from birth that the giant superstore, Buy-and-Large, is their very best friend.

- Future of the technological society [DVD scenes 12 and 14]

The War of the Worlds (Byron Haskin, Paramount 1952)

Early in this first film version of the H.G. Wells novel, a scientist uses a Geiger counter to measure radiation from what is assumed to be a meteor. The meteor turns out to contain hostile invaders from Mars. Conventional weapons are useless against the invading forces. Even nuclear weapons have no effect. The "scientists" are the last hope, but can they come up with a solution in time?

- Radiation, Geiger counter [DVD scene 2]
- Alien invasion [DVD scenes 3 and 5]

- State-of-the-art military technology, Flying Wing [scene 9]
- Being human, resistance to infection [scene 13]

War of the Worlds (Steven Spielberg, Paramount 2005)

The remake departs from the original and depicts the aliens as having been on Earth for a long time, buried underground, waiting for the right moment to appear, and begin their reign of destruction (like terrorist sleepers).

- Sci-fi in historical context [DVD scenes 5–7 (terrorists?)]

When Worlds Collide (Rudolph Maté, Paramount 1951)

In another George Pal production, the Bush Differential Analyzer shows up again, this time being used to compute the trajectory of a rogue star and its orbiting planet, on a collision course with Earth. Precision measurements of star positions can yield evidence of extrasolar planets.

- History of computing, Bush differential analyzer [DVD scene 2]
- Search for extrasolar planets [DVD scene 2]

Woman in the Moon (Frau im Mond) (Fritz Lang, UFA 1929)

Mainly an espionage/love story set in space, this film features the first ever (on screen) countdown to launch, as well as the very prescient notion of a multistage rocket. Upon landing on the Moon, a scientist performs an interesting experiment to test for a breathable atmosphere, but then engages in a pseudoscientific search for gold.

- Space exploration, countdown to launch [DVD scene 10]
- Science and pseudoscience [scene 15 (first expeditioner)]

X-Men: Apocalypse (Bryan Singer, Twentieth Century Fox 2016)

A mutant pharaoh from ancient Egypt awakens after thousands of years and recruits disciples to inaugurate a new era of humanity.

- Magneto manipulates the *magnetic elements*: Iron, Nickel, Cobalt [DVD scenes 20/21]

X-Men: Days of Future Past (Bryan Singer, Twentieth Century Fox 2014)

Mutants are being hunted down by Sentinel robots, who seem to adapt to the mutants' powers and even acquire their powers and use them against the mutants. The only hope is to go back into the past to prevent the development of the Sentinels. But is time travel into the past possible? If so, is it possible to change the events of the past?

- "… a theory in quantum physics that time is immutable." [DVD scene 29]

X-Men III: The Last Stand (Brett Ratner, Twentieth Century Fox 2006)

Should the mutants continue to be who they are, or should they accept a cure, give up their powers, and become normal? This film includes an interesting illustration of Newton's First Law (inertia). Magneto exerts an external force to move automobiles out of the way on the Golden Gate Bridge, but then proceeds to move the entire bridge using what appears to be an internal force. How is this possible, within the bounds of Newtonian mechanics?

- Newton's laws of motion [DVD scene 18]

Appendix B: Television Series Episodes Cited

The Big Bang Theory—"The Isolation Permutation" (Mark Cendrowski, Warner Brothers, 2011)

In the opening scene, Sheldon proposes a topic for conversation, which was creating a stir in the real-life physics community: Faster-than-light particles from CERN.

- Season 5, episode 8, air date 11/3/2011

Doctor Who—"Blink" (Steven Moffatt, BBC 2012)

The Doctor and his traveling companion, Martha Jones, are trapped in 1969, while the TARDIS is parked in London in 2007. The Doctor sends a message to someone that he has not yet met via a recording on 17 DVDs, in which he explains the nature of time.

- Wibbly-wobbly, timey-wimey … stuff [DVD Third Series, episode 10, scene 8]

Enterprise—"The Aenar" (Mike Vejar, Paramount 2005)

A member of the Andorian subspecies, the Aenar, is pressed into service by the Romulans, to pilot an unmanned attack ship.

- Humanoid life in extremes of climate [DVD season 4, episode 14]

© Springer Nature Switzerland AG 2019
B. B. Luokkala, *Exploring Science Through Science Fiction*, Science
and Fiction, https://doi.org/10.1007/978-3-030-29393-2

Enterprise—"Broken Bow" (Rick Berman and Brannon Braga, Paramount 2001)

The pilot episode, in which the warp-5 ship Enterprise NX-01 takes its first voyage. Its mission: to return an injured Klingon to the Klingon home world, Kronos.

- The experimental universal translator has difficulty with a Klingon dialect, with humorous results [DVD season 1, episode 1, scene 5]

Fringe—"Peter" (Teleplay by Jeff Pinker, J.H. Wyman, and Josh Singer, Warner Brothers 2010)

A brilliant scientist, whose young son has died from an incurable illness, develops a way to observe events in an alternate universe, in which his (alternate) son is still alive. He extends this technology to create a means of crossing over into the alternate universe. The breach triggers a series of bizarre results on both sides.

- Alternate universes [DVD season 2, episode 15]

Gilligan's Island—"Pass the Vegetables, Please" (Leslie Goodwins, United Artists 1966)

A crate of experimental, "Radio Active" vegetable seeds washes up on the island. Without noticing the warning on the lid of the crate, the castaways plant the seeds, harvest the abnormal-looking vegetables, eat them, and develop unusual powers.

- [DVD season 3, episode 3]

The Jetsons, Episode 1 "Rosie the Robot" (Hanna-Barbera, 1962)

The animated cartoon show depicts a futuristic society with flying automobiles, video phones, food replicators, and robotic maids.

- Future of the technological society [DVD episode 1]

Mars—"Crossroads" (Teleplay by Andre Bormanis and Paul Solet, Nat Geo 2016)

Lives are lost on the Martian colony when one of the colonists opens a door from the habitat to the outside world.

- Explosive decompression [season 1, episode 6]

NATURE—"Radioactive Wolves" (Klaus Feichtenberger, THIRTEEN 2011)

Wildlife flourishes in the uninhabitable exclusion zone around the Chernobyl nuclear reactor, 25 years after the worst nuclear disaster in history.

- [DVD scene 1]

NOVA—"Cracking the Code of Life" (Elizabeth Arledge, WGBH 2001)

Focusing on the Human Genome Project, the episode includes details about the genetic diversity of humans.

- [DVD scene 6]

NOVA—"Cracking Your Genetic Code" (Sarah Holt, WGBH 2012)

High-speed DNA sequencing and gene-based medicine.

- Genotyping versus complete DNA sequencing [DVD scene 3]
- Gene-based therapy for cancer [DVD scene 5]
- Preimplantation Genetic Diagnosis (PGD) [DVD scene 6]

NOVA—"Einstein's Quantum Riddle" (Jamie Lochhead, PBS 2019)

Definitive test of Bell's Inequality demonstrates quantum entanglement.

NOVA—"Smartest Machine on Earth" (Michael Bicks, PBS 2011)

*   IBM's WATSON, consisting of 2800 parallel processors, learns to play JEOPARDY and eventually beats human champions [DVD scene 4]

NOVA—"Time Travel" (Judith Bunting, BBC/WGBH 1999)

Including interviews with Carl Sagan and Kip Thorne, the episode proposes using a wormhole as a time machine.

*   [VHS tape, use the first half hour through billiard balls causality]

The Outer Limits—"I, Robot" (Leon Benson, MGM 1964)

A robot is placed on trial for the murder of its creator.

*   [DVD vol. 3, disc 2, episode 41]

Star Trek (The Original Series)—"The Changeling" (Marc Daniels, Paramount 1967)

A damaged and re-programmed space probe, which calls itself Nomad, mistakes Captain James T. Kirk for its creator, Jackson Roy Kirk, and so refrains from destroying the Enterprise. Once onboard, Nomad kills Scotty (and restores him to life) and erases Uhura's memory Unfortunately, the latter process is not reversible, and Uhura must be completely re-educated.

*   Neuroscience, human memory [DVD vol. 31, episode 61, scenes 1, 2, and 6]

Star Trek (The Original Series)—"City on the Edge of Forever" (Joseph Pevney, Desilu/Paramount 1967)

Time travel into the past, through a time portal of unknown origin, raises interesting questions of causality: If you could go back into the past, could you change history?

*   Space and time [DVD, vol. 14, episode 28, scenes 1 and 2]

Star Trek (The Original Series)—"The Corbomite Maneuver" (Joseph Sargent, Desilu/Paramount 1966)

Although it was not the first episode of the original series to be aired, it was the first regular episode that was filmed and features the first showing of the transporter, as well as the debut of Dr. McCoy (who never did get used to the idea of being teleported).

- Teleportation [DVD vol. 1, episode 2, scene 7]

Star Trek (The Original Series)—"The Devil in the Dark" (Joseph Pevney, Desilu/Paramount 1967)

A mining colony unwittingly disturbs the eggs of a silicon-based life form.

- Conditions necessary for life [DVD vol. 13, episode 26, scene 5…]

Star Trek (The Original Series)—"Errand of Mercy" (John Newland, Desilu/Paramount 1967)

In the Star Trek universe, culture is defined in terms of technological progress. In this episode, Kirk attempts to persuade the elders of an apparently primitive people, who call themselves Organians that they are in danger from the Klingons. Meanwhile, Spock uses his tricorder to scan the area for evidence of progress toward mechanization. Finding none, he declares the Organians to be "a laboratory specimen of an arrested culture." By the end of the episode, we discover that the Organians have evolved beyond the need for physical bodies and have the ability to put an end to hostility by neutralizing weapons.

- Technology and culture [DVD vol. 14, episode 27, scene 2]

Star Trek (The Original Series)—"Mudd's Women" (Stephen Kandel, Desilu/Paramount 1966)

The crew of the Enterprise encounter a race of androids and are offered the promise of immortality by having their consciousness transferred into android bodies.

- Hard AI and artificial consciousness [DVD vol. 2, episode 4]

Star Trek (The Original Series)—"Spock's Brain" (Marc Daniels, Paramount 1968)

Spock's brain is stolen by an alien, who uses it to control the central power system for her planet. With the aid of an advanced technological device, known as "the Teacher," Dr. McCoy must learn how to perform brain surgery at the level of detail necessary to put Spock's brain back where it belongs.

- Neuroscience, neurosurgery, brain/machine interface [DVD vol. 31, episode 61, scenes 1, 2, and 6 or full episode]

Star Trek (The Original Series)—"Tomorrow Is Yesterday" (Written by D.C. Fontana, Directed by Michael O'Herlihy, Paramount 1967)

The Enterprise and her crew are sent back into the past, after breaking away from the gravitational field of a black star.

- Space and time, black holes [DVD, Time Travel Fan Collective, disc 1, opening scene]

Star Trek (The Original Series)—"The Trouble With Tribbles" (Joseph Pevney, Paramount 1967)

A rapidly reproducing and voracious, but otherwise benign creature, infests a grain storage facility.

- High-yield grain, quadrotriticale [DVD vol. 21, episode 42]

Star Trek (The Original Series)—"The Ultimate Computer" (John Meredyth Lucas, Paramount 1968)

A revolutionary breakthrough in computer technology enables a computer to think like humans, but with disastrous consequences.

- Intelligent robots [DVD vol. 27, episode 53, scene 5 or full episode]

Star Trek (The Original Series)—"Wink of an Eye" (Written by Lee Cronin, Directed by Judd Taylor, Paramount 1968)

The Enterprise is taken over by an alien race, who move at very high speed—fast enough to step out of the way of an incoming phaser beam.

- Space and time, speed of light [DVD, vol. 34, episode 68, scene 3]

Star Trek: Deep Space Nine, "Trials and Tribble-ations" (Jonathan West, Paramount 1996)

The Klingon spy Darvin, from the original series episode The Trouble with Tribbles, time travels into the past to alter the events of history. Members of the crew of Deep Space Nine must figure out what he intends to do and stop him. Dr. Bashir imagines himself in what he calls a predestination paradox: could he be his own great-grandfather?

- Space and time, time travel, and causality [DVD Time Travel fan collective, disc 3, skip to scene 4]

Star Trek: The Next Generation—"Cause and Effect" (Jonathan Frakes, Paramount 1992)

A catastrophic event causes the Enterprise to be caught in a temporal causality loop (or closed, timelike curve), in which the same events repeat over and over.

- Space and time [DVD season 5, disc 5, scenes 1, 4, and 5]

Star Trek: The Next Generation—"The Chase" (Jonathan Frakes, Paramount 1993)

A message encoded into the DNA of several, vastly different species, is pieced together to reveal that they are all descendants of a common species. (If no time in this unit, maybe use in biology unit.).

- Biology, genetic diversity [DVD season 6, disc 5, final scene]

Star Trek: The Next Generation—"Darmok" (Philip Lazebnik and Joe Menosky, Paramount 1991)

The crew of the Enterprise encounter an alien race, who speak in methapors.

* Universal translator can translate words but not metaphorical meaning [DVD season 5, episode 2]

Star Trek: The Next Generation—"Descent, Part I" (Alexander Singer, Paramount 1993)

The opening scene of this episode features the android, Lt.Cdr. Data, on the holodeck of the U.S.S. Enterprise (D), playing poker with three of history's most famous scientists: Sir Isaac Newton, Albert Einstein, and Stephen Hawking. These three figures represent three different views about the nature of space and time. (The parts of Newton and Einstein are, of course, played by actors. But Stephen Hawking appears in the scene as himself!)

* Space and time [season 6, disc 7, scene 1]

Star Trek: The Next Generation—"Encounter at Farpoint" (Corey Allen, Paramount 1987)

The pilot episode introduces the main characters and shows many of the new features of a Galaxy-class starship, including the holodeck—a virtual reality environment, which combines transporter and replicator technology, in order to create tangible objects.

* Data's quest to be human [DVD season 1, disc 1, scene 11]
* Virtual reality [DVD season 1, disc 1, scene 11]

Star Trek: The Next Generation—"Force of Nature" (Written by Naren Shankar, Directed by Robert Lederman, Paramount 1993)

Warp drive is shown to have detrimental effects on the fabric of spacetime.

* Space and time [DVD season 7, disc 3, scene 5, or full episode]

Star Trek: The Next Generation—"Gambit, Part I" (Peter Lauritson, Paramount 1993)

Captain Picard disappears while on shore leave. In the course of the search, Dr. Crusher uses a tricorder to detect "microcrystalline damage" in the material of the floor, where he was last seen—apparently the result of a weapon discharge. Is this science or technobabble?

- Materials, nanoparticles [DVD season 7, disc 1, opening scene]

Star Trek: The Next Generation—"The Measure of a Man" (Robert Scheerer, Paramount 1989)

A trial is conducted to determine the legal status of the android, Data. The question: Is he the property of Star Fleet, or is he a sentient being with the same rights as humans?

- Computing, data storage, processor speed [DVD season 2, disc 3, scene 5]

Star Trek: The Next Generation—"Parallels" (Written by Brandon Braga, Directed by Robert Wiemer, Paramount 1993)

Lt. Worf shifts from one quantum reality into another, in a brilliant illustration of the most extreme version of the multiverse hypothesis: everything that can happen does happen.

- Space and time [DVD season 7, disc 3, opening scene + scene 2, or full episode]

Star Trek: The Next Generation—"Relics" (Alexander Singer, Paramount 1992)

Commander Scott, from the original series, is found in a transporter buffer after being reported missing 75 years earlier. His "pattern" had degraded only 0.003%. Could this be a problem?

- Teleportation, degradation [DVD season 6, disc 1, scenes 1 and 2]

Star Trek: The Next Generation—"Rightful Heir" (Winrich Kolbe, Paramount 1993)

Kahless, the founder of the Klingon culture of honor, returns after 1500 years but is revealed to be a clone of the original.

- DNA sequencing and cloning [DVD season 6, disc 6, scenes 5 and 7]
- Data's choice to be more than a machine [DVD season 6, disc 6, scene 8]

Star Trek: The Next Generation—"Samaritan Snare" (Written by Robert L. McCullough, Paramount 1989)

The ship's chief medical officer orders Picard to have his artificial heart replaced. In his absence, the ship is called away on a fake rescue mission.

- Artificial heart [DVD season 2, disc 5]

Star Trek: The Next Generation—"Ship in a Bottle" (Alexander Singer, Paramount 1992)

The android Commander Data enjoys playing the role of Sherlock Holmes in holo-deck-simulated mysteries. He programs the holodeck to create a simulation of Holmes' arch-enemy, Professor James Moriarty. But the simulation is too good: Moriarty becomes self-aware, and demands to be allowed to leave the confines of the virtual reality environment.

- Sentient A.I., virtual reality [DVD season 6, disc 3, scenes 1 and 2]
- Teleportation, Heisenberg [DVD season 6, disc 3, scene 7]

Star Trek: The Next Generation—"Starship Mine" (Cliff Boyle, Paramount 1993)

The Enterprise (and presumably all starships) must undergo a baryon sweep, to remove accumulated baryon particles. The process is described as lethal to living organisms. But what about the ship itself?

- Standard model of particle physics [DVD season 6, disc 5, opening scene]

Star Trek: The Next Generation—"Tapestry" (Written by Ronald D. Moore, Paramount 1993)

Q offers Picard an opportunity to change events of the past.

• Artificial heart [DVD season 6, disc 4]

Star Trek: The Next Generation—"Time's Arrow, Part II" (Jonathan West, Paramount 1996)

An excavation in San Francisco uncovers Commander Data's head in an abandoned mine, which had been undisturbed since the nineteenth century. Can effects precede their causes?

• Space and time, time travel, and causality [DVD season 6, disc 1, scenes 1 and 7]

Star Trek: The Next Generation—"Unification, Part II" (Cliff Bole, Paramount 1991)

Ambassador Spock attempt to open negotiations between the Vulcans and the Romulans. The episode includes an interesting dialog between Spock and Data, in regard to logic versus emotion, and what it means to be human.

• Being human [DVD season 5, disc 2, scene 5]

Star Trek: Voyager—"Caretaker" (Winrich Kolbe, Paramount 1995)

The computer architecture on Voyager involves bioneural circuitry, one of several conceivable future computer technologies, according to Scientific American (January 2010). The pilot episode introduces the Emergency Medical Hologram, a holographic doctor, who can manipulate real instruments and treat real patients.

• Future of computer technology [DVD season 1, disc 1, scene 3]
• Virtual reality [DVD season 1, disc 1, scene 5]

Top Secret Rosies (Leann Erickson, PBS 2010)

The role played by women in the early history of electronic computers is documented with personal interviews.

- History of computing [DVD full episode]

The Twilight Zone—"No Time Like The Past" (Rod Serling, CBS/Paramount 1963)

Dissatisfied with the present, a man invents a time machine and attempts to change events of the past. Frustrated by his repeated failures to make any difference, he eventually decides to retire to a peaceful town in America's Old West. He again attempts to change a tragic event of the past, only to discover that he was the cause of the event.

- Time travel according to Igor Novikov's Principle of Consistent Causality [DVD season 4, episode 10]

Appendix C: YouTube Videos Cited

Boston Dynamics "Atlas" robot in action
https://www.youtube.com/watch?v=rVlhMGQgDkY

Carnegie Mellon's Rescue robot, CHIMP
https://www.youtube.com/watch?v=PAHVxqf83Jk

CSA astronaut Chris Hadfield wrings out a wet washcloth on the International
Space Station
https://www.youtube.com/watch?v=UyFYgeE32f0

"How To Freeze Boiling Water" (DrBarryLuke, 2010)
http://www.youtube.com/watch?v=QJjiKBjhj0I

Memorex tape commercial, featuring Ella Fitzgerald ("Is it live, or is it Memorex?)
http://www.youtube.com/watch?v=Bkt8Dwzl6Sg

NASA astronaut Scott Kelly plays ping pong with a drop of water on the International
Space Station
https://www.youtube.com/watch?v=PrqcwH8LhDU

P.W. Singer, TED Talk on military robots and the future of war
http://www.youtube.com/watch?v=M1pr683SYFk

RIBA, a robotic nurse for elder care
http://www.youtube.com/watch?v=U92eB6WyjKc

The 1984 Macintosh commercial
http://www.youtube.com/watch?v=HhsWzJo2sN4

© Springer Nature Switzerland AG 2019
B. B. Luokkala, *Exploring Science Through Science Fiction*, Science
and Fiction, https://doi.org/10.1007/978-3-030-29393-2

Appendix D: Solutions to Estimation Problems

Note to the reader: The answers you get for the Estimation problems will depend on the assumptions that you have made. Keep in mind that the intent is to come up with a reasonable estimate, and not necessarily to obtain the one-and-only correct answer. Each of the solutions presented here will include a clear statement of the assumptions. If your answer differs from the answer in the solution, it may not be due to a mistake but may simply be the result of having made a slightly different set of assumptions.

Estimation 2.1: Kirk, Spock, and Jet Boots, Revisited

Let's assume, for simplicity, that Kirk's mass and Spock's mass are the same. Let's also assume that at the instant Spock grabs hold of Kirk's ankle, the two of them are moving with the same speed. The jet boots must then exert a total force on Spock's ankles, which is twice the force that Spock must exert on Kirk's ankle:

$$\begin{aligned} F_{total} &= ma \\ &= m(\Delta v / \Delta t) \\ &= 2(80\text{kg})(56\text{m/s})/(1\text{s}) \\ &= 8960\text{N} \\ &= 2014\text{lb} \end{aligned}$$

But there are two jet boots working together. So each jet boot separately only needs to exert *half* the total force. The final result is that each of Spock's ankles feels the same force that Spock exerts on Kirk's one ankle: *1007 lb.*

© Springer Nature Switzerland AG 2019
B. B. Luokkala, *Exploring Science Through Science Fiction*, Science and Fiction, https://doi.org/10.1007/978-3-030-29393-2

Estimation 2.2: Pressure Underneath One of the Independence Day Attack Ships

Using the assumptions of Example 2.2, we obtained the following equation:

$$\text{Pressure} = \frac{m_{\text{mother\_ship}}}{r_{\text{mother\_ship}}^3} \frac{3hg}{\pi},$$

where h is the height (or thickness) of one of the attack ships and g is the acceleration due to gravity. According to the movie dialog, the mass of the mother ship is ¼ the mass of the Moon. The mass of the Moon is approximately 7.35×10^{22} kg, so $m_{\text{mother\_ship}} = 1.84 \times 10^{22}$ kg. The diameter of the mother ship is given in the movie as 550 km, which is twice the radius. So $r_{\text{mother\_ship}} = 275$ km. The height, h, of one of the attack ships is not given explicitly in the movie, so we have to take a reasonable guess based on visual information. We know from the movie dialog that the attack ships are 15 miles wide. They are approximately disc-shaped, and a reasonable guess for the height might be about 1 mile (or 1.6 km). Substituting all the numbers into the equation gives:

$$\text{Pressure} = \frac{1.84 \times 10^{22}\,\text{kg}}{(275{,}000\,\text{m})^3} \frac{3(1600\text{m})(9.8\text{m}/\text{s}^2)}{3.14}$$

$$= 1.33 \times 10^{10}\,\text{N}/\text{m}^2$$

Standard atmospheric pressure is approximately 1×10^5 N/m², so this result is roughly 133,000 atmospheres. For comparison, the pressure at the bottom of the Mariana Trench is just over 1000 atmospheres, which is sufficient to crush all but the most carefully designed deep sea diving vessels. Thus, the attack ships don't need a death ray. They would crush anything underneath them just by hovering overhead.

Estimation 2.3: Newton's Law of Gravitation Applied to Le Voyage dans la Lune

Using Eq. (2.2), we will calculate the ratio of the force of the Moon's gravity on the space capsule, compared to the force of the Earth's gravity on the space capsule, while the capsule is on the surface of the Moon:

$$\frac{F_{\text{Moon}}}{F_{\text{Earth}}} = \frac{Gm_{\text{Moon}} m_{\text{capsule}} \Big/ r_{\text{Moon}}^2}{Gm_{\text{Earth}} m_{\text{capsule}} \Big/ r_{\text{Moon's Orbit}}^2}$$

$$= \frac{m_{\text{Moon}} r_{\text{Moon's Orbit}}^2}{m_{\text{Earth}} r_{\text{Moon}}^2}$$

$$= \frac{\left(7.35 \times 10^{22} \text{ kg}\right)\left(3.84 \times 10^8 \text{ m}\right)^2}{\left(5.97 \times 10^{24} \text{ kg}\right)\left(1.74 \times 10^6 \text{ m}\right)^2}$$

$$\frac{F_{\text{Moon}}}{F_{\text{Earth}}} = 600$$

The force of the Moon's gravity acting on the space capsule is 600 times greater than the force of Earth's gravity acting on the capsule.

Estimation 2.4: Relativity and Passenger Jets

Equation (2.4) relates the time in the rest frame (t) to the time in the moving frame (t'):

$$t = \frac{t'}{\sqrt{1 - v^2 / c^2}}.$$

The typical cruising speed of a large passenger jet is about 570 miles per hour, or 250 m/s. The circumference of the Earth is roughly 25,000 miles, so a hypothetical nonstop flight would take a minimum of about 44 h. (It is not possible for a large passenger jet, with a range of about 3000 miles, to fly nonstop around the world without refueling!) Let's assume that the elapsed time on the jet is 44 h. The time that passes in the rest frame (the airport) is given by:

$$t = \frac{44\text{h}}{\sqrt{1 - \left(250\text{m}/\text{s}\right)^2 / \left(3 \times 10^8 \text{m}/\text{s}\right)^2}}.$$

My pocket calculator doesn't have enough significant digits to carry out the calculation, so I have to do it with a computer spreadsheet. The result is:

$$t = 44.0000000000153h$$

or converting to seconds:

$$t = 158400.000000055s.$$

In other words, on a flight by passenger jet around the world, the time elapsed in the airport would be greater than the time elapsed on the plane by a mere *55 ns*. This can easily be measured using high-precision atomic clocks but is not exactly a recipe for time travel into the future.

Estimation 2.5: Relativity and Fusion-Powered DeLorean

Using the same method as Estimation 2.3, but with a speed of only 88 miles per hour, the time dilation factor is only about one part in 10^{14}. If the time elapsed in the DeLorean was 60 s, the time elapsed in the parking lot should have been longer by just over half a picosecond (0.6×10^{-12} s), which would have been too small to measure with a stopwatch.

Estimation 2.6: Relativistic Time Dilation and the Twin Paradox

Several different approaches to this problem are possible. Here is just one:

Looking up the speed of the International Space Station in orbit around the Earth and rounding off the time that Scott Kelly spent aboard the ISS to be 1 year, we can estimate the amount of time that passed for Mark Kelly, who remained on Earth:

$$t = \frac{1\,\text{year}}{\sqrt{1 - \left(7660\,\text{m}/\text{s}\right)^2 \Big/ \left(3 \times 10^8\,\text{m}/\text{s}\right)^2}}$$

$$= 1.00000000065\,\text{year}$$

So the additional time that passed on Earth was 0.00000000065 year, or converting to seconds, we find that Mark is about 0.02 s older than Scott.

Estimation 2.7: Relativity, GPS, and the Importance of Time-Keeping

The Special Relativity Effect

For simplicity, consider a point on the equator, compared to a GPS satellite directly above the equator, but at a much larger radius. (Add the radius of the Earth to the altitude at which the GPS satellites orbit the Earth to get the radius of orbit of the satellite.) The average speed for an object in uniform circular motion is given by

$$v = \frac{2\pi r}{t}$$

where $t = 24$ h (converted to seconds).

The speed of a point on the surface of the Earth at the equator is approximately 464 m/s.

The speed of a GPS satellite in geosynchronous orbit above the equator is approximately 1932 m/s.

Use these two speeds to calculate the ratio of the relativistic time dilation effect for a GPS satellite compared to the effect for a point on the equator:

$$\frac{t'_{GPS}}{t'_{equator}} = \frac{\sqrt{1 - \frac{v_{GPS}^2}{c^2}}}{\sqrt{1 - \frac{v_{equator}^2}{c^2}}}$$

$$= 0.9999999999$$

Due to the effect of relative speed, time runs slower for the GPS clock by about one part in 10^{10}, compared to clocks on the surface of the Earth at the equator.

The General Relativity Effect

Compute the ratio of the time experienced by a GPS satellite (add the radius of the Earth to the altitude at which the GPS satellites orbit the Earth to get the radius of orbit of the satellite) divided by the time experienced by an object on the surface of the Earth.

$$\frac{t_{r_{GPS}}}{t_{r_{Earth}}} = \frac{\sqrt{1 - \frac{2Gm}{r_{GPS}c^2}}}{\sqrt{1 - \frac{2Gm}{r_{Earth}c^2}}}$$

$$\frac{t_{r_{GPS}}}{t_{r_{Earth}}} = \frac{\sqrt{1 - \frac{2\left(6.67 \times 10^{-11} \text{ Nm}^2 / \text{kg}^2\right)\left(5.972 \times 10^{24} \text{ kg}\right)}{\left(2.658 \times 10^7 \text{ m}\right)\left(2.998 \times 10^8 \text{ m}/\text{s}\right)^2}}}{\sqrt{1 - \frac{2\left(6.67 \times 10^{-11} \text{ Nm}^2 / \text{kg}^2\right)\left(5.972 \times 10^{24} \text{ kg}\right)}{\left(6.378 \times 10^6 \text{ m}\right)\left(2.998 \times 10^8 \text{ m}/\text{s}\right)^2}}}$$

$$= 1.0000000005$$

Due to General Relativity, time runs faster for the GPS satellite clock by about five parts in 10^{10}, compared to clocks on the surface of the Earth, because the GPS clock is farther away from the time-dilating effect of Earth's gravity.

The gravitational effect is about five times larger than the relative speed effect, and the two effects are in opposite directions. Both effects must be taken into account in order for the GPS system to work accurately.

Estimation 2.8: What Would It Take for the Moon to Become a Black Hole?

The Schwarzschild radius is calculated as follows:

$$r_S = \frac{2Gm}{c^2}$$

Using the mass of the Moon, we obtain

$$r_S = \frac{2\left(6.67 \times 10^{-11} \text{ Nm}^2 / \text{kg}^2\right)\left(7.35 \times 10^{22} \text{ kg}\right)}{\left(3 \times 10^8 \text{ m}/\text{s}\right)^2}$$

$$= 0.00011 \text{ m}$$

The Schwarzschild radius for the Moon is approximately one tenth of a millimeter. Shrinking the Moon down to the size of a grapefruit is not small enough for it to become a black hole.

Estimation 3.1: Mission to Mars Air Leak

The time for half the air to leak out of the ship was estimated by Eq. (3.10) to be:

$$\Delta t = \frac{3V}{A} \sqrt{\frac{m}{3kT}}.$$

Let's use the mass of a Nitrogen molecule (N_2). Each of the Nitrogen atoms has seven protons and seven neutrons, each of which has a mass of 1.67×10^{-27} kg. This gives $m = 4.68 \times 10^{-26}$ kg. Let's also assume that the temperature of the ship is comfortable room temperature of 20 °C, or 293 K. Boltzmann's constant is $k = 1.38 \times 10^{-23}$ J/K. We need to take a guess as to the total volume of the ship, V, and the area of the hole in the hull, A. Visual information from the movie suggests that the diameter of the hole is about the size of a finger, so a reasonable guess for A is about 1 cm^2 (or 1×10^{-4} m^2). If we take a guess that the linear dimension of the interior of the ship is about 10 m, the total volume, V, is about 1000 m^3. Substituting these quantities into the equation gives

$$\Delta t = \frac{3(1000\,\mathrm{m}^3)}{1 \times 10^{-4}\,\mathrm{m}^2} \sqrt{\frac{4.68 \times 10^{-26}\,\mathrm{kg}}{3(1.38 \times 10^{-23}\,\mathrm{J/K})(293\mathrm{K})}}$$

$$= 61{,}047\mathrm{s} \text{ or } 17\mathrm{h}\,(\text{not 4 min.,as in the movie!})$$

Estimation 3.2: Air Leak on the International Space Station

The total pressurized volume of the ISS is 916 m^3.
  The radius of the hole is about 1 mm (0.001 m).
  The area of the hole is $A = \pi r^2 = 0.00000314$ m^2.
  It's reasonable to assume that all the other terms in the calculation are the same as in Estimation 3.1, so we have:

$$\Delta t = \frac{3(916\,\mathrm{m}^3)}{3.14 \times 10^{-6}\,\mathrm{m}^2} \sqrt{\frac{4.68 \times 10^{-26}\,\mathrm{kg}}{3(1.38 \times 10^{-23}\,\mathrm{J/K})(293\mathrm{K})}}$$

$$= 1{,}780{,}000\mathrm{s} \text{ or a little over 20 days}$$

Estimation 3.3: Atoms Inside and on the Surface of a Nanoparticle

Let's assume that the nanoparticle is cubic in shape, with an edge length of 10^{-9} m. The typical spacing between atoms in a crystal is about 1 Å (10^{-10} m). So each edge of the nanoparticle will have about ten atoms along its length. The area of a square is the square of its edge length, so each of the eight faces of the cube-shaped nanoparticle will have about $10 \times 10 = 100$ atoms. The total number of atoms on the surface of the cube will be roughly $8 \times 100 = 800$ atoms. The volume of a cube is the cube of its edge length, so the entire nanoparticle will have $10^3 = 1000$ atoms. 80% of the atoms in this nanocube are on the surface!

Estimation 3.4: Vaporizing Captain Picard

The heat energy required to vaporize an object is given by Eq. (3.10):

$$Q = mC\Delta T + mL$$

Let's assume that Picard's mass is 75 kg. Since all we want is an estimate, let's make the simplifying assumption that the human body is mostly water. The specific heat capacity of water is $C = 4190$ J/kg•K, and the latent heat of vaporization is $L = 2.256 \times 10^6$ J/kg. The boiling point of water is 100 °C (373 K). If we assume that Picard is in good health prior to vaporization, his initial temperature is normal body temperature of 37 °C (310 K). Making these substitutions, we get

$$Q = (75\text{kg})(4190\text{J}/\text{kg}\cdot\text{K})(373\text{K} - 310\text{K}) + (75\text{kg})(2.256 \times 10^6 \text{ J}/\text{kg})$$
$$= 1.89 \times 10^8 \text{ J}$$

A typical vaporization-by-phaser takes just a few seconds, so the *power* required to do the job is given by Eq. (3.9):

$$\text{Power} = \text{Energy} / \text{Time}$$
$$= (1.89 \times 10^8 \text{ J}) / (3\text{s})$$
$$= 6.3 \times 10^7 \text{ W} (\text{or} 63\text{MW})$$

Estimation 3.5: Energy of a Marathon Runner or a Truck Collision

A typical marathon runner burns 2880 cal $\times$ 4187 J/cal = *12 MJ*.
The kinetic energy of a pickup truck at 60 miles/h (=26.7 m/s) is given by:

$$K = \frac{1}{2}mv^2$$
$$= \frac{1}{2}(3000\text{kg})(26.7\text{m}/\text{s})^2$$
$$= 1MJ$$

If the truck were to collide with a wall at that speed, the energy dissipated would be only 1/12 of the total energy dissipated by the marathon runner.

Estimation 3.6: Power Dissipated by a Marathon Runner or a Truck Collision

If the result of Estimation 3.4 seems surprising, consider the amount of time that each process takes, and recall that power is energy over time. The current Olympic world record for running a marathon is just a little over 2 h. So the power dissipated by the marathon runner is roughly

$$P_{marathon} = 12\,MJ\,/\,7200\,s = 1.7\,kW$$

In contrast, a truck collision happens in about 1 s. So the power dissipated in the truck collision is roughly

$$P_{truck\_collision} = 1\,MJ\,/\,1s = 1\,MW$$

Estimation 3.7: Energy Yield of a Photon Torpedo

Einstein's energy equation is $E = mc^2$. If a photon torpedo combines 1.5 kg of ordinary matter with 1.5 kg of antimatter, the total amount of mass is 3 kg. So the energy yield is $E = (3\ kg)(3 \times 10^8\ m/s)^2 = 27 \times 10^{16}$ J. Using the conversion factor, 1 Mt = 4.184×10^{15} J gives 64.5 Mt. If a typical hydrogen bomb has a yield of 10 Mt, a photon torpedo delivers the energy equivalent of about 6.5 hydrogen bombs.

Estimation 3.8: Angels and Demons Antimatter Bomb

The movie claims that the energy yield of the antimatter bomb is 5 kt. We first convert this to joules:

$$5kt = \left(0.005\,Mt\right)\left(4.184 \times 10^{15}\,J\,/\,Mt\right) = 2.09 \times 10^{13}\,J.$$

Next we use Einstein's energy equation to calculate the amount of mass involved:

$$E = mc^2$$
$$m = E/c^2$$
$$= \left(2.09 \times 10^{13}\,\text{J}\right)/\left(3 \times 10^8\,\text{m}/\text{s}\right)^2$$
$$= 2.3 \times 10^{-4}\,\text{kg}$$
$$= 0.23\text{g}$$

According to CERN's website, they can produce something like 10^7 antiprotons per second. In order to create a single gram of antihydrogen, it would require something like two billion years. Our calculation suggests that the bomb in *Angels and Demons* would have about ¼ g of antimatter. A rough estimate suggests that it would take about 500,000 years to collect this much antimatter at CERN.

Estimation 4.1: Physical Storage Space Required for Magnetic Data Devices

The contemporary USB hard drive has a storage capacity of 1 TB (10^{12} bytes). The 1980s disc pack has a storage capacity of 80 MB (80×10^6 bytes). The number, N, of disc packs required to store the same amount of data as the USB drive is just the ratio of the two storage capacities: $N = (10^{12}$ bytes$)/(80 \times 10^6$ bytes$) = 12{,}500$. The dimensions of the disc pack are given in Fig. 4.3 (14″ diameter × 5.25″ high). To make the estimation simpler, let's assume that each disc pack is 12″ in diameter × 6″ high or a total volume of ½ ft³ each. The total physical storage space required for 12,500 disc packs would be 6250 ft³. This is the equivalent of a room with a 10 ft ceiling and 625 ft² of floor space (or 25 ft × 25 ft). Compare this to the contemporary USB drive, with the same data storage capacity, which fits in the palm of your hand!

Estimation 4.2: Data Storage in the Human Brain

There are widely varying estimates of the data storage capacity of the human brain. Table 4.1 gives a conservative estimate of 2.5 PB. The USB drive has a storage capacity of 1 TB, so it would take 2500 USB drives to equal the storage capacity of the brain. The dimensions of the USB drive are given in Table 4.1: 5″ × 3.5″ × 1″. The volume occupied by one USB drive is the product of the three linear dimensions, or 17.5 in.³. 2500 USB drives would occupy a volume of 43,750 in.³, or about 25 ft³. This is roughly equivalent to the volume occupied by an office desk or a refrigerator.

Estimation 4.3: Projecting Data Storage Density into the Future

In order to do a meaningful comparison of the data storage densities, we first convert the volume of the three storage devices into a consistent set of units. According to Table 4.1, the volume of an average human brain is about 1400 cm³. So let's convert all of the volumes to cubic centimeters. The 1980s disc pack occupies a volume of roughly 0.5 ft³. 1 ft³ = 28,317 cm³. So the volume of the disc pack is approximately 14,160 cm³. The volume of the USB drive is 17.5 in.³. 1 in.³ = 16.4 cm³. So the volume of the USB drive is 287 cm³. Next we calculate the data storage density of each device by dividing the storage capacity by the volume. The results are the following:

Data storage device	Storage capacity	Volume (cm³)	Storage density
Disc pack (1980)	80×10^6 bytes	14,160	5.65 kB/cm³
USB drive (2010)	1×10^{12} bytes	287	3.48 GB/cm³
Human brain	2.5×10^{15} bytes	1400	1.79 TB/cm³

This table suggests that state-of-the-art in data storage density has increased by nearly six orders of magnitude, from 5.65 kB/cm³ to 3.48 GB/cm³ in 30 years, or nearly two orders of magnitude every 10 years. The data storage density of the human brain is only about three orders of magnitude greater than the USB hard drive. It's a risky thing to project into the future based on only two data points. But if the trend continues at the same rate as it has over the past 30 years, we might expect to see a data storage device whose size and capacity are comparable to the human brain within the next 15 years.

Teleportation Estimations

Estimation 6.1: The Energy and Power Problems

If we assume that the mass of a typical human body is about 75 kg, and apply Einstein's equation, we find that the energy equivalent of a typical human body is:

$$E = mc^2 = (75\,\text{kg})(3 \times 10^8\,\text{m/s})^2 = 6.75 \times 10^{18}\,\text{J}.$$

A typical science fiction teleportation event takes about 5 s, so the power required to "energize" a human body is:

$$P = \Delta E / \Delta t = (6.75 \times 10^{18}\,\text{J}) / (5\,\text{s}) = 1.35 \times 10^{18}\,\text{W}.$$

The average power output of the Hoover dam is roughly 2 GW (2×10^9 W). The power required for a teleportation event is something like 650,000,000 times the power output of the Hoover dam.

Estimation 6.2: The Data Storage Problem

Using the (grossly oversimplified) assumptions as stated in the problem (human body is basically water, and each water molecule requires five pieces of information to specify its location and orientation), we obtain the following estimate of the number of pieces of information needed to specify the body:

$$N = 5 \times (\text{body mass}) / (\text{water molecule mass}).$$

Most of the mass of a water molecule (H_2O) is accounted for by the protons and neutrons in the hydrogen (one proton each) and the oxygen (eight protons + eight neutrons) atoms. Each proton and neutron has a mass of 1.67×10^{-27} kg, so we obtain.

$$N = 5 \times (75 \text{kg}) / (18 \times 1.67 \times 10^{-27} \text{ kg}) = 1.25 \times 10^{28}.$$

State-of-the-art computing hardware in the second decade of twenty-first century operates at the peta-scale. That is to say, the data storage capacity of a (room-size) computer is of order 10^{15} bytes, and the processor speed is of order 10^{15} operations per second. In order to store the information (assuming 8 bits per byte), it would take roughly 10^{12} such machines.

Estimation 6.3: The Problem of Computer Processing (CPU) Time

Given state-of-the-art processor speed of 10^{15} operations per second, the time required just to determine all the information would be:

$$t = (1.25 \times 10^{28}) / (10^{15} \text{ per second}) = 1.25 \times 10^{13} \text{ s}.$$

The number of seconds in a year (assuming 365.25 days/year) is 31,557,600. So the time required just to determine the information is roughly 369,000 years. Of course, we have grossly oversimplified the problem by assuming nothing but water in the human body, and only five pieces of quantum information per molecule. The problem is even worse because the real limitation is not processor speed, but input-output (IO) speed, which is orders of magnitude slower. All things considered, the time required to gather the data for a teleportation event would be comparable to the age of the universe.

Estimation 6.4: The Problem of Information Degradation

In the scenario as described, the quantum information needed to specify a human body has been stored in the transporter buffer (memory) for 75 years. The information has degraded by only 0.003%. Assuming the mass of a typical human body of 75 kg, the loss of information corresponds roughly to $0.00003 \times 75 \text{ kg} = 0.00225 \text{ kg}$ (or 2.25 g). This doesn't seem like much. But of course it depends where this 2.25 g happens to be within the body. The loss of 2.25 g from the end of a finger would be of little consequence. But the loss of 2.25 g from your heart or brain could be devastating.

Appendix E: Artifacts

Fig. E.1 A simple apparatus for experiencing Einstein's Principle of Equivalence. A 50 g mass hangs on a 1.0 N force meter, reading about 0.5 N when not accelerating. Take it on a ride in an elevator and measure the change in apparent force when the elevator accelerates up or down. (Photo by the author)

© Springer Nature Switzerland AG 2019
B. B. Luokkala, *Exploring Science Through Science Fiction*, Science and Fiction, https://doi.org/10.1007/978-3-030-29393-2

Fig. E.2 A green Uranium glass candlestick with Geiger counter and postal scale provide a simultaneous demonstration of all four of the fundamental interactions of physics. The strong nuclear force is demonstrated by alpha decay of Uranium-238 (half-life 4.5 billion years) to Thorium-234 (half-life 24 days). The weak nuclear force is demonstrated by the subsequent beta decay of Thorium-234 to Protactinium-234 (half-life 6.7 h), followed by a second beta decay to Uranium-234 (half-life 0.25 million years). The electromagnetic interaction is demonstrated in several ways: the green glass reflects and refracts light and also fluoresces under ultraviolet light; the radioactive decay particles ionize the gas in the Geiger tube. The weight of the candlestick (about 12.5 ounces, as measured on the postal scale) indicates the force of gravity.

The Geiger counter is set on its most sensitive scale and measures the radiation exposure level at the top of the candlestick to be approximately 0.16 milliRoentgens/hour, which is roughly 10 times higher than the normal background radiation. (Photo by the author)

Fig. E.3 Radium dial wristwatch c. 1940 and Radium dial travel alarm clock c. 1950. (Photo by the author)

The watch, made in Switzerland, belonged to my wife's stepfather, who wore it during his service in World War II. The alarm clock mechanism and folding travel case are marked "Germany," indicating that they were made prior to the end of World War II, but the Radium-painted dial is marked "West Germany," indicating that it was printed and painted after the end of the war.

The radiation exposure level measured at the face of the alarm clock is approximately 10 times higher than normal background radiation. The exposure level measured at the face of the watch is approximately 100 times higher than normal background. The exposure level measured at the back of the watch is comparable to normal background; most of the radiation from the Radium dial being blocked by the stainless steel case of the watch.

Much of what we know about the hazards of the exposure of living tissue to alpha radiation and X-rays has been learned from studies of the serious illnesses and deaths of factory workers, who painted the Radium dials of watches and clocks, using Radium paint, to make them glow in the dark. In 1925 Harrison Martland and Samin von Sochocky were among the first to devise medical tests to determine if a living person is radioactive, due to the ingestion of radium paint and inhalation of radium dust, while working in the dial-painting factories.

Reference

1. Moore, K.: The Radium Girls (Sourcebooks, 2017) pp. 127–128

Index

© Springer Nature Switzerland AG 2019
B. B. Luokkala, *Exploring Science Through Science Fiction*, Science and Fiction, https://doi.org/10.1007/978-3-030-29393-2

Printed in the United States
By Bookmasters